Proceedings in Life Sciences

Neural Principles in Vision

Edited by F. Zettler and R. Weiler

With Contributions by
D. L. Alkon · H. Autrum · V. Braitenberg
W. Burkhardt · N. M. Case · L. Cervetto
M. G. F. Fuortes · A. Gallego · P. H. Hartline
A. C. Hurley · M. Järvilehto · A. Kaneko
K. Kirschfeld · G. D. Lange · S. B. Laughlin
J. S. McReynolds · J. Moring · K.-I. Naka
G. Niemeyer · J. A. Patterson · J. H. Scholes
H. Shimazaki · U. Smola · N. J. Strausfeld
H.-J. Wagner · R. Wehner · R. Weiler
F. Zettler

With 293 Figures

Springer-Verlag
Berlin Heidelberg New York 1976

Dr. Friedrich Zettler, Zoologisches Institut der Universität
München, Luisenstr. 14, 8000 München 2/FRG

Dipl.-Biol. Reto Weiler, Zoologisches Institut der Universität
München, Luisenstr. 14, 8000 München 2/FRG

QP
474
.N48

ISBN 3-540-07839-8 Springer-Verlag Berlin Heidelberg New York
ISBN 0-387-07839-8 Springer-Verlag New York Heidelberg Berlin

Library of Congress Cataloging in Publication Data. Main entry under title:
Neural principles in vision. (Proceedings in life sciences). "Papers . . . presented
at the symposium . . . held at the Zoological Institute of the University of
Munich during the 15th through the 20th of September, 1975." Includes index.
1. Vision--Congresses. 2. Retina--Congresses. I. Zettler, Friedrich, 1934-
II. Weiler, Reto, 1947- III. Alkon, Daniel L. QP474.N48 591.1'823 76-21804

Offsetprinting: Beltz Offsetdruck, Hemsbach/Bergstr., Bookbinding: Brühlsche
Universitätsdruckerei, Gießen.

PREFACE

Scientific investigation of the retina began with extensive studies of its anatomical structure. The selective staining of neurons achieved by the Golgi method has led to a comprehensive picture of the architecture of the tissue in terms of its individual elements. Cajal, in particular, used this technique to reveal the fundamentals of retinal structure. In the studies that followed, selective staining method continued to be decisive in the analysis of neuroanatomy, and in recent years these techniques have been complemented by electron microscopy.

The complexity of retinal structure that has been revealed demands a functional explanation, and electrophysiology attempts to provide it. But functional analysis, like anatomy, must ultimately be based on the single cell. It is only by using dyes to mark the recording site that one can identify the cells involved. When this succeeds, as it has recently, one can actually fit functional events into the anatomical framework. With these advances, our strategies and tactics toward an understanding of the structure and function of the retina have moved into a new phase.

We feel, however, that an important aspect has been neglected; that is, research on the vertebrate retina and on the retina of insects have proceeded largely in mutual isolation, despite the fact that precise parallels exist in advances at the methodological level. It is for that reason that we organized this symposium, to encourage a dialogue between these separate research groups. Moreover, we believe that this exchange of information may stimulate comparative study of analogous retinal structures. Every retina consists of neural tissue structured to provide a visual representation of the environment. Comparative research is thus an essential component of our evolving analysis of the neural principles underlying this common function.

All the papers in this volume were presented at the symposium Neural Principles in Vision, held at the Zoological Institute of the University of Munich during the 15th through the 20th of September, 1975. We are most grateful to the Volkswagen Foundation for their generosity in financing this international symposium.

Munich, January 1976   F. Zettler / R. Weiler

CONTENTS

3 Molluscs

LIST OF CONTRIBUTORS

ALKON, D.L., Laboratory of Biophysics, National Institute of Health, Bethesda, Maryland 20014 /
USA

AUTRUM, H., Zoologisches Institut der Universität
München, Luisenstr. 14, 8000 München 2 /
W. Germany

BRAITENBERG, V., Max-Planck-Institut für Biologische
Kybernetik, Spemannstr. 38, 7400 Tübingen /
W. Germany

BURKHARDT, W., Max-Planck-Institut für Biologische
Kybernetik, Spemannstr. 38, 7400 Tübingen /
W. Germany

CASE, N.M., Department of Anatomy, Loma Linda University, Loma Linda, California 92354 / USA

CERVETTO, L., Laboratorio di Neurofisiologia del
C.N.R., Via S. Zeno, 51, 56100 Pisa / Italy

FUORTES, M.G.F., Laboratory of Neurophysiology,
National Institute of Neurological and Communi-
cative Disorders and Stroke, National Institute
of Health, Bethesda, Maryland 20014 / USA

GALLEGO, A., Departamento de Fisiologia, Facultad
de Medicina, Universidad Complutense, Ciudad
Universitaria, Madrid-3 / Spain

HARTLINE, P.H., Department of Physiology and Biophysics, University of Illinois, Urbana,
Illinois / USA

HURLEY, A.C., Department of Neurosciences and Marine
Neurobiology Unit, University of California,
San Diego, La Jolla, California 92093 / USA

JÄRVILEHTO, M., Department of Physiology, University
of Oulu, Kajaanintie 52 A, 90220 Oulu 22 /
Finland

KANEKO, A., Department of Physiology, Keio University, School of Medicine, Shinanomachi,
Shinjuku-ku, Tokyo 160 / Japan

KIRSCHFELD, K., Max-Planck-Institut für Biologische
Kybernetik, Spemannstr. 38, 7400 Tübingen /
W. Germany

LANGE, G.D., Department of Neurosciences and Marine
    Neurobiology Unit, University of California,
    San Diego, La Jolla, California 92093 / USA

LAUGHLIN, S.B., Department of Neurobiology, Research
    School of Biological Sciences, P.O.Box 475,
    Canberra, A.C.T. 2601 / Australia

McREYNOLDS, J.S., Department of Physiology, Univer-
    sity of Michigan, Ann Arbor, Michigan 48104 /
    USA

MORING, J., Department of Physiology, University of
    Oulu, Kajaanintie 52 A, 90220 Oulu 22 / Finland

NAKA, K.-I., Department of Physiology and Bio-
    physics, University of Texas Medical Branch,
    Galveston, Texas 77550 / USA

NIEMEYER, G., Neurophysiology Laboratory, Universi-
    täts-Augenklinik, Rämistr. 100, 8091 Zürich /
    Switzerland

PATTERSON, J.A., Hunter College, Dep. Biol. Sci.,
    The City University of New York, 695 Park Avenue,
    Box 1225, New York, N.Y. 10021 / USA

SCHOLES, J.H., The Medical Research Council, Cell
    Biophysics Unit, Kings College London Univer-
    sity, 26-29 Drury Lane, London WC2B 5RL /
    England

SHIMAZAKI, H., Department of Physiology, Keio Uni-
    versity School of Medicine, Shinanomachi,
    Shinjuku-ku, Tokyo 160 / Japan

SMOLA, U., Zoologisches Institut der Universität
    München, Arbeitsgruppe Biokybernetik, Luisen-
    str. 14, 8000 München 2 / W. Germany

STRAUSFELD, N.J., European Molecular Biology Labora-
    tory, Postfach 102209, 6900 Heidelberg /
    W. Germany

WAGNER, H.-J., Abteilung für Klinische Morphologie,
    Universität Ulm, 7900 Ulm / W. Germany

WEHNER, R., Department of Zoology, University of
    Zürich, Künstlergasse 16, 8006 Zürich /
    Switzerland

WEILER, R., Zoologisches Institut der Universität
    München, Luisenstr. 14, 8000 München 2 /
    W. Germany

ZETTLER, F., Zoologisches Institut der Universität
    München, Luisenstr. 14, 8000 München 2 /
    W. Germany

Opening Remarks

H. Autrum

What vision is, we know primarily from ourselves, that is, we know
that we see with our eyes. When animals have anatomically similar
organs, we assume that they serve similar purposes. This comparative
method, when applied consistently, leads us very far: it led Richard
Hesse, for example, to his discovery of the individual light-sensitive
cells in the epidermis of Annelids.

It is astonishing, however, that historically a correct conception of
the anatomical structure of the eye became possible only when people
began to understand how the eye functions. Neither the Greek nor Ara-
bian anatomists had correct ideas concerning the construction of the
eye, and, its function they understood even less. (A drawing by
Andreas Vesalius in 1543 still puts the lens of the eye in the center
of the vitreous body (Fig. 1)).

What was the reason for these difficulties? The physical laws of image
representation by lenses were unknown. Immediately after they were
discovered they were applied to the human eye by Kepler, first in 1604
and then in more detail in 1611 (Dioptrice seu demonstratio eorum quae
visui et visibilibus propter conspicilla non item prida inventa acci-
dunt; Augustae Vindelicorum, typ. D. Franci, 1611). Before Kepler the
lens was considered to be the organ which, in some mystical way, converted light into visibilia. Goethe was the last great scientist who
believed that no physical process is involved when the visible world
is transformed into what is actually seen (1827: "Wär nicht das Auge
sonnenhaft, die Sonne könnt es nie erblicken"; Zahme Xenien, 3).

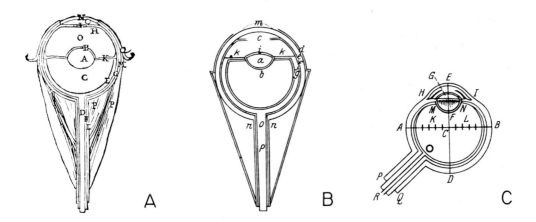

Fig. 1A-C. Diagrams of the human eye. (A) By Andreas Vesalius, De Corporis Humani
Fabrica (1543) (B) By F. Plater, De Corporis Humani Structura et Usu (1583)
(C) By Christophorus Scheiner. In: Oculus Hoc Est: Fundamentum Opticum (1619)

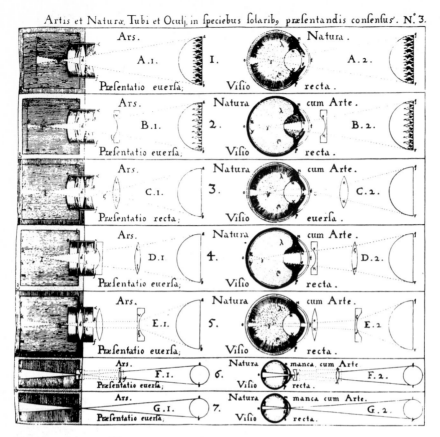

Fig. 2. Diagram illustrating the working mechanisms of the dioptrical apparatus of the eye. From Christophorus Scheiner, Rosa Ursina (1626-1630)

Kepler showed that the lens forms an image of the environment of the retina. Modern and physiological optics thus begin with biophysics. Kepler was the first biophysicist, and after him came an uninterrupted line of biophysicists, including Pater Christoph Scheiner (1619: Oculus Hoc Est: Fundamentum Opticum; Oenoponti (Innsbruck), ap. D. Agricolam), who first gave experimental evidence for the inverted image on the retina (Fig. 2); Athanasius Kircher (Fig. 3); Heinrich Müller (1820-1864), Professor at Würzburg, who, collaborating with Kölliker, gave the first correct description of the layers of the retina and the arrangement of rods and cones; Max Johann Sigismund Schultze (1825-1874), Professor at Halle and Bonn, the founder of the double function of the retina and of the duplexity theory; Ramón y Cajal, who was the first to recognize the neurons as individual structures in vertebrates and insects, and to describe their wiring.

The biophysical analysis of the eye culminated in the work of Hermann von Helmholtz, who broadened the Three-Component Theory of Thomas Young (1802). Helmholtz was a student of Johannes Müller who began the comparative examination of the physiology of vision. In 1826 he published his book "Zur vergleichenden Physiologie des Gesichtssinnes des Menschen und der Thiere" (C. Cnobloch, Leipzig). He formulated the first theory of vision in insects. His student, Helmholtz, received

Fig. 3. Diagram of the
formation of the retinal
image. From Kircher, Ars
magna Lucis et Umbrae
(1646)

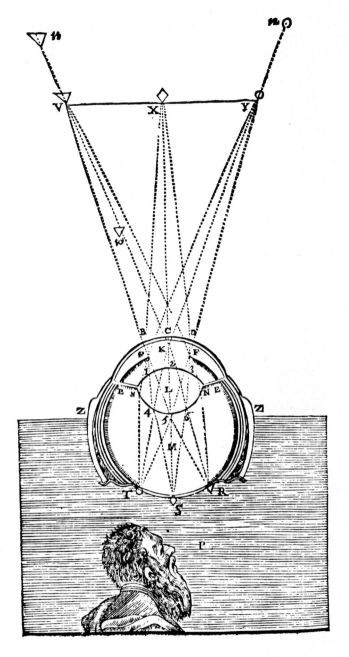

his Ph.D. in 1843 with a dissertation on the physiology of the invertebrate nervous system (De fabrica systematis nervosi evertebratorum,
Inaug. Diss., Berolini, 1843).

To Kepler and Helmholtz the eye was a physical device. Of course,
Helmholtz did not have the anatomical and physiological knowledge of
our days. He is supposed to have said that - from the standpoint of
the physicist - the eye is a very badly constructed instrument.

While that certainly holds for the stimulus-conducting part of the eye, Helmholtz did not consider that behind this part lie the neurons of the retina and higher centers, i.e. a complicated computer, which converts the incomplete physical image into biologically, though not necessarily physically, relevant perceptions. With this computer we recognize friends and colleagues; with it we have discovered the computer itself. Moreover Helmholtz could have confused this computer by the simple means of alcohol. Today we know that alcohol has very little influence on the receptor cells, but it does influence the neuronal wiring quite strongly (Bernhard, C.G., Knave, B., Persson, H.E., Acta physiol. scand. 88, 373, 1973). But Helmholtz was a pronounced teetotaler.

In 1865/66 Frithiof Holmgren discovered the electro-retinogram, and immediately saw the significance of this method: "No doubt a great many questions concerning physiological optics may be treated according to this method, and among them are such as can hardly be solved in any other way now known to us. Among them I count, for instance, the problem of the time course of excitation in the retina" (Upsala Läkareförenings Förhandlingar 6, 419-455, 1870/71). Since Holmgren's paper was written in Swedish, Sir James Dewar and McKendrik (who later became Professor of Physiology at Glasgow) discovered the ERG independently of Holmgren in 1873. As early as 1874 or 1875, they also recorded the ERG of the compound eye for the first time.

I do not wish to follow the history of the physiology of vision any further. This symposium is to show how far we have advanced during these last hundred years: Quite far when we add up the details, but not very far if we consider how much remains to be discovered beyond our horizon. The horizon begins already at the dioptric apparatus and photoreceptor cells. Some daredevils have even ventured forward to the second-order neuron; some dare to go up in a balloon and try to peek beyond the wiring diagram of the first optic ganglion of the fly; among us there might even be some astronauts who are trying to gain an overall view.

In conclusion let me thank all those who made this meeting possible: the Stiftung Volkswagenwerk generously furnished the financial means for this symposium; Dr. Zettler and Mr. Weiler organized it. Probably all the papers we will discuss here have been supported by various grants. Two examples come immediately to mind: the German contributions were supported by the Deutsche Forschungsgemeinschaft; and our Munich laboratories and lecture halls were built in 1932 for Professor von Frisch by the Rockefeller Foundation.

In spite of the fact that the Bavarian language is more difficult than English for me, let me open this symposium in Bavarian: Pack mas (i.e.: grasp your nettle).

1 VERTEBRATES

1.1 Patterns of Golgi-Impregnated Neurons in a Predator-Type Fish Retina

H.-J. WAGNER

1.1.1 Introduction

Attempts to disentangle the neural network of the vertebrate retina
have dealt primarily with the first synaptic layer where the terminals of receptor cells are linked to the dendrites of bipolar cells,
often mediated by the horizontal cell processes. This basic arrangement has been found in all vertebrate groups (see Stell, 1972) and
has recently been analyzed in great detail in the rabbit (Sjöstrand,
1974), in the goldfish (Stell, 1975; Stell and Lightfoot, 1975) and
in the rudd (Scholes, 1975). In every case, a pattern of highly selective connectivity emerged in the outer plexiform layer (OPL) between visual and bipolar cells (Scholes, 1975) or between receptors
and horizontal cells (Stell, 1967; Stell and Lightfoot, 1975).

Although far more difficult to assess, the second synaptic layer of
the retina has also been shown to be organized in a precise scheme
of functional links, including the contacts of bipolar terminals and
ganglion cell dendrites with amacrine cell processes acting as interneurons (see Stell, 1972). Because of the far greater variety of cell
types entering the inner plexiform layer (IPL) as compared to the
OPL a more detailed analysis of this area has succeeded only in identifying a simply structured retina as in the case of <u>Necturus</u>
(Dowling and Werblin, 1969) and remains to be done in more complex-
ly organized types.

In certain lower vertebrates, particularly teleost species, there is
a special kind of pattern. Governed by a symmetrical, mosaic-like arrangement of the visual cells it shows repercussions in the organization of the deeper retinal layers. As has been proposed by Engström
(1963), Anctil (1969) and Wagner (1972) on the basis of comparative
works the visual cell mosaic may be regarded as a prerequisite for a
highly developed motion perception. Indeed, most of these species have
been shown to possess a remarkable visual capability enabling them to
live as predators feeding on quickly moving prey. Furthermore, current
studies with the microspectrophotometer (Ali, pers. comm.) seem to
stress the functional importance of the mosaic arrangements of receptor cells with regard to their spectral sensitivity in different
photic environments.

In the species studied so far with Golgi techniques, no such symmetrical arrangement of visual cells was present. Therefore, in a comparative analysis of different species we are attempting to follow the
subsequent neural connections that are determined by different types
of visual cell mosaics. In addition to the morphological studies,
physiological experiments on the spectral sensitivities of the different rods and cone types will be carried out.

In this chapter some aspects of the neural organization of the retina
of a cichlid fish Nannacara anomala will be described as an example
for a mosaic type retina. After a brief review of the types of recep-

tors and neuron types as revealed by conventional double Golgi impregnation some recent findings with the Golgi EM technique concerning the connections between visual and horizontal cells will be presented.

1.1.2 Observations

1.1.2.1 Cell Types: Light Microscopy (Fig. 1)

1.1.2.1.1 The Outer Plexiform Layer (OPL)

The OPL consists basically of a network of processes of receptor pedicles, horizontal and bipolar cells the majority of synaptic contacts being localized in the cavities of the bell-shaped visual cell terminals.

Morphologically, three types of receptor cells can be distinguished. Rods (a) are characterized by the long myoid and the rod like terminal swelling which only occasionally bears a basal process as found in other lower vertebrates by Cajal (1894). The cones are easily divided into double (b) and single (c) cones on the basis of the different length of the axonal processes and the greater diameter of the basal processes in single cones. There seems to be no structural difference between the two components of double cones. Several contacts have been observed, linking several double cones as well as double cones to single cones. Rod basal processes may contact other rods or cones; no cone basal process was seen in contact with a rod.

Two kinds of horizontal cells were identified in tangential and transversal sections. The sclerad type (d) resembles closely the brush-like cell of Cajal (1894) sending most of its processes towards the endfeet of visual cells. The dendritic field covered by this type is only a third or a half in diameter when compared to that of the second internal horizontal cell type (e) which is located more vitreally than

---

Fig. 1a-z. Montage of camera lucida drawings of Golgi-impregnated neural elements ▷ in the retina of Nannacara anomala X 550.

a-c) visual cells; outer segments are not depicted because they were hidden in pigment epithelium (a) rods (b) double cones (c) central single cone.

d-e) horizontal cells; (d) sclerad horizontal cells (e) internal horizontal cells.

f-h) bipolar cells; (f) large bipolar cells (g,h) small bipolar cells.

i-r) amacrine cells; (i) cable-like cell (j) irregular stratified cell (k) radiating stratified cell (l) simple asymmetrical cell (m) two-layered asymmetrical cells (n) two-layered cell (o) simple radial cell (r) branched radial cell (p) diffuse cell (not shown) (q) candelabrum-like cell.

s-x) ganglion cells; (s) varicose stratified cells (t) smooth stratified cells (u) multilayered cell (v) diffuse cell (w) radial cells (x) two-layered cells.

y) cells with displaced nuclei; (y_1) horizontal processes in the outer plexiform layer; (y_2) horizontal processes in the inner plexiform layer (displaced ganglion cell?).

z) glial cells (not shown).

Triangles: cells found in Nannacara only.

Open triangles: cells possibly mentioned by Dogiel as "mitralisförmige Zellen"

the first one. More details about the morphology and connectivity of these cells will be described later.

Bipolar cell types have to be classified on the shapes of their dendritic arborizations as well as on the position and form of the axonal terminal swellings. In the OPL, it is easy to identify large bipolar

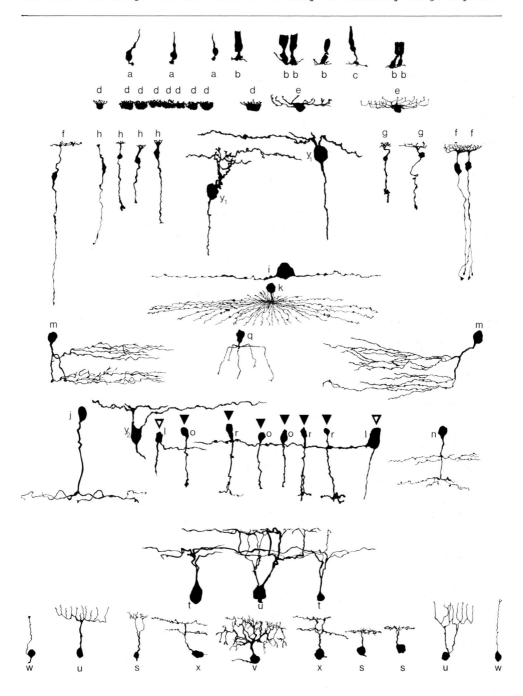

cells (Cajal, 1894) with dendritic field dimensions of 40-50 μm (Fig. 1f) thus ranging between the values given by Parthe (1972) for the large and small rod bipolar cells. Two other types of bipolar dendrites do not reach these diameters. The cell type g covers an area with a diameter up to 30 μm whereas the last type (h) has only very short branches which do not exeed 5 μm in length. No displaced bipolar cells have been found in Nannacara. As for the connectivity of these cell types, only preliminary observations have been made. As in other vertebrates, large bipolars receive the majority of their input from rods (Cajal, 1894; Boycott and Dowling, 1969; Kolb, 1970; Parthe, 1972; Scholes, 1975). The small h-type bipolars seem to contact cones only similar to the "selective cone bipolars" of Scholes (1975). Cells of the g-type have been seen linked to cones but not to rods; therefore it is not clear whether they correspond to Scholes' "mixed bipolar" or to Cajal's small cone bipolar cell.

In addition to the neural elements mentioned, there is a cell type with nuclei at various levels of the inner nuclear layer (INL). It exhibits horizontal processes in the OPL with several varicosities and a length of more than 100 μm (Fig. 1,y_2). It has an axon-like process which crosses the IPL and seems to continue among the ganglion cell axons. Determinations of its function must await further analysis.

Patterns in the Outer Plexiform Layer. Tangential sections at levels sclerad or vitread of the OPL reveal the mosaic-like morphology of the Nannacara retina (Fig. 2). The equal double cones are arranged in a very regular square mosaic, each square unit containing one central single cone (Kuenzer and Wagner, 1969). No additional single cone located in the "corners" of the squares is found in Nannacara even during early stages of the eye development (Wagner, 1974). At the level of the sclerad horizontal cells, the visual cell mosaic determines the arrangement of the cell somata. The distance between the nuclei of these cells is constant and equals the distance between the central single cones. Serial sections showed that the soma of each sclerad horizontal cell is located just underneath one central single cone. This spatial arrangement can be found in numerous species in which a square pattern including a central single cone is present (Wagner, 1975). Until now, internal horizontal cells and bipolar cells, could not be shown to fit into this pattern.

1.1.2.1.2 The Inner Plexiform Layer

In analogy to the first synaptic layer, the endfeet of bipolars in the IPL meet the dendrites of ganglion cells, often by the intermediate of amacrine cells. However, unlike the OPL, synaptic contacts are distributed in a certain number of horizontal layers; the exact number of layers differs among different species. Cajal (1894) indicated five layers in the carp, Scholes (1975) found six in the rudd and in Nannacara as many as seven layers are present (Fig. 3A), some of which may be further subdivided (Wagner, 1973). The layers are formed to a great extent by stratified processes of amacrine and ganglion cells. The same pattern of ramification characteristic of a certain cell type may be found independently on different layers of the IPL.

The terminal swellings of bipolars lead to certain levels of the IPL. Terminals of large bipolars always end in the most vitread layer, just above the ganglion cell nuclei. Small "cone" bipolars, however, seldom reach that far; instead, their axons terminate in one or several layers more sclerad in the IPL, sometimes emitting short basal processes.

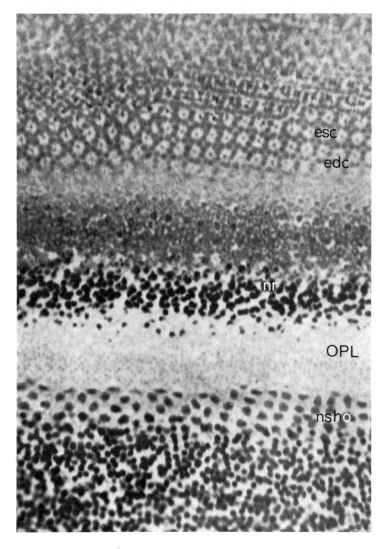

Fig. 2. Tangential section at the level of the outer plexiform layer (OPL) of
N. anomala showing the patterns of receptor cells (square mosaic of cones) and
horizontal cells. edc: ellipsoids of double cone, esc: ellipsoid of central single
cone, nr: nuclei of receptor cells, nsho: nuclei of sclerad horizontal cells.
5 μm paraffin section, Azan; x 450

Amacrine cells are the group of neurons with the widest variety of
cell types in the retina of Nannacara. Among ten different types
that may be distinguished on the basis of their patterns of ramification, there are five which are either uni- or bistratified, two
asymmetrical, one diffuse and two which are more radially orientated.
The cable-like cell (i) is almost linear, never branched and always
located at the most sclerad (1st) level of the IPL. Processes of the
unistratified cells (j and k) may be found from the second to the
seventh layer. The horizontal arborizations of the j-type are irregular, branched and almost without varicosities whereas in cell k the

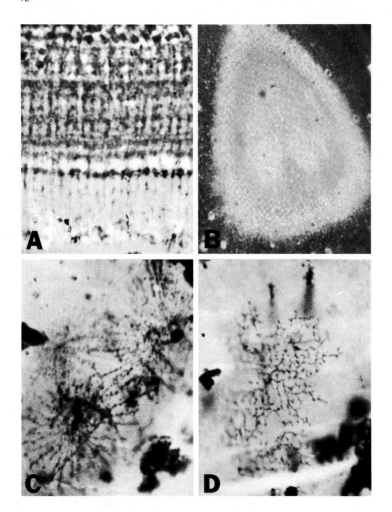

Fig. 3A-D. Patterns in the inner plexiform layer (IPL) of the Nannacara retina
A) In transverse sections there is a distinct horizontal stratification (7 layers)
which is overlapped by radially orientated light and dense bars. 5 μm paraffin
section; Azan; x 1,575. B) Tangential section of the sclerad part of the IPL.
The horizontal layers of the transverse section appear as concentric rings whereas the light and dark stripes appear as light "holes" and a network of dense lines
crossing at right angles. 5 μm paraffin section; Azan; x 400. C) Tangential section
of the IPL showing the processes of a radiating stratified amacrine cell (type k).
Golgi impregnation. x 500. D) Tangential section of the IPL showing the dendrites
of a s-type stratified varicose ganglion cell. Note the net-like array of the arborizations. Golgi impregnation. x 500

straight, unbranched, and varicose processes radiate from the single
main stem covering an area more than 100 μm in diameter. Fine, irregular and varicose branches characterize the bistratified cells m
and n. The m-type amacrine is unusual because at both levels the processes start only in one direction spreading over a fairly narrow
angle of $5°-10°$. Another asymmetrical type is the 1-cell which resembles closely the "mitralisförmige Zelle" of Dogiel (1888). Opposed to

these layered and horizontally organized types is the diffuse amacrine cell (p) the varicose processes of which give off numerous branches covering the total width of the IPL. The last category of amacrine cells exhibits a dominantly radial orientation of processes. They may start after a short horizontal course (q) or run down right from the soma (o,r). In these last types, the process may be unbranched (o) or branched (r), in each case terminating in the seventh layer in close vicinity to ganglion cell somata.

Ganglion cell types, although not as diversified as amacrine cells, exhibit a striking similarity in the dendrite-arborization patterns to those of amacrine cells. However, no asymmetrical types, present in cyprinids (Cajal, 1894), are present. Both types of unistratified cells differ in that the s-type cell has fine varicose processes in contrast to the smooth and stouter branches of the t-type ganglion cells. The dendritic pattern of the bistratified (x) cell is similar to the s-cell. Radially orientated ganglion cells (w) are not em- branched and bear an apical swelling. The u-type cell shows a morpho- logy similar to the q-type amacrine and the diffuse pattern of the v-cell appears as the mirror image of the p-type amacrine. Some dis- placed ganglion cells (y_1) with nuclei in the IPL have stratified dendrites similar to the t-type cell.

Patterns in the Inner Plexiform Layer. Tangential sections of the IPL reveal a regular array of more or less dense structures. "Holes" with a diameter of 2.5 μm spaced at constant distance from one another are separated by stripes of dense fiber material which results in a grid-like structure (Fig. 3B). Closer examination of transverse sec- tions (Fig. 3A) shows that the horizontally orientated layers of stra- tified processes are overlapped by a radially orientated pattern of light and dark stripes corresponding to the grid pattern in tangential sections. Although all structures underlying this network have not yet been identified, there is evidence, that at least the dense parts lin- ing the "holes" are formed by the arborizations of j-type amacrine and, above all, s-type ganglion cells. The dendrites of these cells resemble a net with processes dividing at about 90° when viewed tan- gentially (Fig. 3D).

Comparative Aspects. Inferring any functional significance from these more taxonomical observations appears problematic without further ul- trastructural or physiological evidence at hand. Available results from other investigators (Meller and Eschner, 1965; Raviola and Raviola, 1967; Witkovsky and Dowling, 1969; Boycott and Dowling, 1969; Dowling and Werblin, 1969; Dubin, 1969, 1970; Witkovsky and Stell, 1971) sug- gest, however, that the high degree of integration accomplished in the IPL is dependent on the degree of specificity of synaptic special- izations achieved by, e.g., ribbon, dyad, serial or reciprocal synapses. Yet there is little doubt that the degree of complexity is also deter- mined by the number of cell types participating in the process of transmitting information in the IPL. Therefore, the different types of amacrine and ganglion cells as found in Nannacara shall be short- ly compared to the cell types in some other lower vertebrates.

Except for the more primitive structured retina of Necturus, the hori- zontally layered organization of the IPL is accompanied by the presence of stratified amacrine and ganglion cells. Most frequently found are irregular stratified (j) amacrine and varicose stratified (s) ganglion cells (Cyprinids, frog: Cajal, 1894; Perca, Salvelinus: Wagner, un- publ.); but cable-like (i), radiating (k) amacrine cells as well as smooth ganglion cells also occur in most of these species. The same holds for both diffuse amacrine (p) and ganglion (v) cells. Asymmet- rical amacrine cells seem to be limited to teleosts. They have been

described by Parthe (1967) and can also be observed in the perch whereas their occurence in cyprinids appears doubtful. On the other hand,
asymmetrical ganglion cells are observed in cyprinids as well as per-
ches and trouts. The most characteristic elements for the IPL of
Nannacara seem to be the radial cells. Unbranched, O-type amacrine
cells are lacking in any of the other species; only in the perch is
a branched radial cell (r) present. As for ganglion cells of the
w-type, they also occur in the perch; so far, they have been regarded
as a cell characteristic of the primate foveal region (Polyak, 1941).

In conclusion, it is necessary to emphazise that with increasing progress of physiological or ultrastructural techniques, new types of
neurons may be found further differentiating the morphologically defined types listed above. This is all the more likely in light of the
studies of Kolb (1970) and Scholes (1975) about bipolar connectivity
and of Dowling and Werblin (1969) who reported that ganglion cells in
Necturus, though morphologically not discernible, belonged to two different types when judged by their synaptic input. It seems, however,
safe to assume that in a retina organized like that of Necturus, only
a reduced set of neurons is present when compared to other lower vertebrates. For most of the more visually orientated fishes, the "basic"
set of cells in the IPL includes stratified as well as diffused amacrine and ganglion cells. The greater variety of amacrine cells, apparently accompanied by a larger number of layers in the IPL (like
Nannacara or the perch), is very likely to bring about an increasing
amount of amacrine-mediated information processing when transmitted
from bipolar to ganglion cells. This means that the degree of peripheral, i.e. retinal integration, should be much higher than in species
with only a few amacrine cell types or few sublayers in the IPL.

Furthermore, the occurence of radially orientated neurons in cichlids
and percids points to the possibility of a "point to point" transmission and to a spot-like effect of amacrines on single ganglion cells.
Bearing in mind the life habits of these animals, it seems feasible
to assume that these radial cells are the structural basis for a high
visual acuity essential for predators.

1.1.2.2 Connectivity of Horizontal Cells: Light and Electron Microscopy

1.1.2.2.1 Rod and Cone Horizontal Cells

Although structure and function of horizontal cells in some species
of lower vertebrates has already been successfully investigated
(Naka and Rushton, 1966; Stell, 1967; Svaetichin, 1967; Kaneko, 1970;
Fuortes and Simon, 1974; Tomita, 1975; Stell, 1975; Stell and Lightfoot, 1975), the puzzling diversity of this group of cells among different classes of vertebrates or even within the teleosts will continue to make them interesting enough for further studies. There is
general agreement that, in fishes, one kind of horizontal cell contacts rods only giving rise to the L-type S-potential. In most cases,
as in the goldfish (Stell, 1967) or in several South-American perci-
forms (Parthe, 1972) including Nannacara (Wagner, 1973), these cells
occupy the most vitread row of horizontal cells.

As for the remaining types of horizontal cells, three more rows are
often distinguished in routine semithin sections of certain species.
For the goldfish, Stell and Lightfoot (1975) have demonstrated them
to contact cones of different spectral sensitivities in a very specific pattern suggesting that each cell type generates a different kind
of C-type S-potential. There are, however, other species which do not
possess three different cone horizontal cells. In Nannacara, only one

horizontal cell type seems to be present besides the rod horizontal
cell being situated at the sclerad border of the INL.

1.1.2.2.2 Shapes of Cone Horizontal Cells in Nannacara

The dimensions of this cell type vary in different regions of the
retina; in the central part, it is high and narrow in appearance
whereas near the ora serrata it looks flat (Fig. 4A,B). These differences are readily explained by variations in the density of cone
populations causing the single cone distance to be 30% greater in
the periphery compared to the area. The average diameter of horizontal cells in the periphery is 25.4 μm (height 6.5 μm) as opposed to
16.7 μm (9 μm) in the area.

In transverse thick sections, the processes arising from the soma may
be divided into three groups: there is a central tuft which is clearly separated from a more peripheral group, both running sclerad towards the direction of the receptor pedicles. In addition, there are
some short processes running horizontally in the OPL and ending with
a short terminal swelling (Fig. 4C). Longer processes observed in the
OPL could not be traced back to sclerad horizontal cells; it is most
likely, that they originate from y_2-type cells (cf. above). In contrast to other vetrebrate horizontal cells (Polyak, 1941; Kolb, 1970;
Kaneko, 1970; Stell, 1975) no long horizontal "axon" in the cone horizontal cells of Nannacara could be demonstrated in the great number
of cells examined.

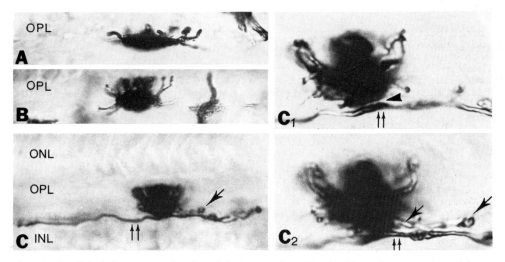

Fig. 4A-C. Golgi-impregnated sclerad hoirzontal cells in 80 μm transverse sections.
Photomicrographs, x 1,550. OPL: outer plexiform layer, ONL outer nuclear layer,
INL: inner nuclear layer. A) At 120 μm from the ora serrata the shape of the horizontal cell is flat, with a diameter of 22.6 μm and a height of 6.5 μm. B) In the
temporal area sclerad horizontal cells have a more compressed aspect with a diameter
of 16.1 μm and a height of 9.0 μm. C) Sclerad horizontal cell with short horizontal
process (arrow). Note a large horizontal process at the level of the inner margin of
the soma (double arrows). C_1,C_2) Higher magnification (x 2,150) of the same cell shows
that this process (double arrows) does not originate from the impregnated cell (arrow
head in C_1). The short process starting from the lateral part of the soma and has a
terminal swelling (arrows, C_2; see also Fig. 3C). Note the arrangement of sclerad
processes which may be divided into a central tuft and a lateral ring

1.1.2.2.3 Fine Structure of Cone Horizontal Cells

Closer examination of the terminals of the radial dendrites of sclerad
horizontal cells in semithin sections reveals that a great number of
them exhibit dichotomous branching whereas others do not. All, however, seem to end inside different cone terminals (Fig. 5A-C). Assessment at the electron microscopic level reveals that the different
aspects of the terminals reflect different relationships to the ribbon
synapses inside a cone pedicle. A tangential section of a cone pedicle
(Fig. 6A) shows that only one of the horizontal cell processes flanking the synaptic ridge is filled with Golgi stain. This may be taken
as an indication that the unstained process originates from a different cell than the stained one. On the other hand, there may also be
two stained processes at each side of the synaptic ribbon (Fig. 6B)
indicating that both arise from the same cell, since there are no
other impregnated cells nearby. Similar observations have been made
by Kolb (1970) in the cat and Stell and Lightfoot (1975) in the goldfish, both using Golgi-techniques. Sjöstrand (1974), however, confirms
never having found two horizontal cell processes from the same cell
facing one ribbon synapse in his reconstruction of serial sections
of the rabbit cone pedicles.

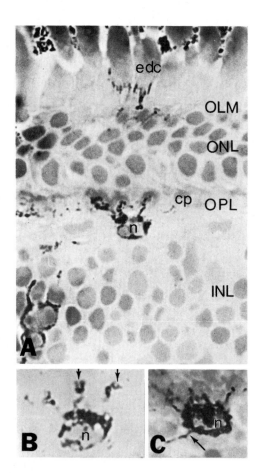

Fig. 5A-C. Golgi-impregnated sclerad
horizontal cells in semithin transverse sections. OLM: outer limiting
membrane ONL: outer nuclear layer
OPL: outer plexiform layer INL: inner
nuclear layer edc: ellipsoids of
double cones cp: cone pedicle
n: nucleus of sclerad horizontal
cell. x 1,500. Note that the nuclei
of impregnated cells are unstained.
A) Processes of the impregnated cell
penetrate into 3 different overlying
cone pedicles. B) 2 of the processes
show dichotomous terminal branching
(cf. Fig. 4B). C) A horizontal process originates from the inner edge
of the cell (arrow). Cf. Fig. 2C

Fig. 6A and B. EM micro-
graphs of ribbon synapses
in cone pedicles cp;
sr synaptic ribbon.
x 34,000. A) Tangential
section. Only one of the
horizontal cell proces-
ses flanking the synap-
tic ridge is impregnated
by Golgi stain. This in-
dicates that they ori-
ginate from 2 different
horizontal cells.
B) Transverse section.
Both processes flanking
the synaptic ridge are
stained. This indicates
that they originate from
the same cell.
Cf. Fig. 3B

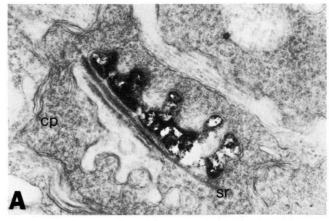

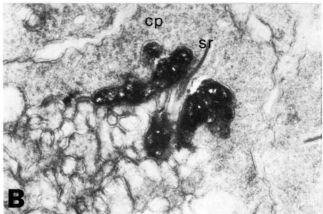

In order to study the spatial distribution of the processes of sclerad
horizontal cells, eight different cells were identified in 80 μm plastic sections and then reorientated for semi-serial sectioning in the
tangential plane. Alternating semi- and ultrathin sections resulted
in two complimentary sets of series (Fig. 7D and E). At the level of
the OPL, just sclerad of the perikaryon (Fig. 7E) one distinguishes
one group of processes forming an inner ring which corresponds to the
central "tuft" as seen in transversal sections. In addition, there is
a second, more peripheral, concentric ring of processes the diameter
of which is twice that of the inner ring. Finally, in the very center
of both rings, there is another single process originating from the
same impregnated cell. An average of 17.8 dendrites per cell has been
observed forming both circles and central spot.

1.1.2.2.4 Connectivity Pattern of Cone Horizontal Cells

Single Cells. In cross sections of cone pedicles the relationship of
this arrangement to the cone mosaic can be analyzed (Fig. 8). Every
cone pedicle forming one unit of square mosaic, i.e. four double cones
and one central single cone is contacted by dendrites of the impregnated cell, thus accounting for the central spot and the inner ring.

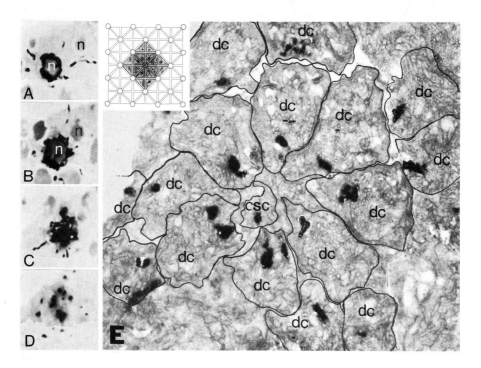

Fig. 7A-E. Tangential sections of a sclerad horizontal cell. A-D) Serial semithin sections at ca. 1 μm interval. n: nuclei of sclerad horizontal cells. X 1,200. E) EM micrographs of the same cell at the level of Fig. 5D. dc: pedicles of double cones. csc: pedicle of central single cone. x 7,600. Pedicles containing an impregnated horizontal cell process are labeled and outlined with ink. The plane of the section passes through the stem of the horizontal cell processes at a level where the branching for the ribbon synapses has not yet taken place. Note that 16 double cones and 1 central single cone are contacted by this cell. The 8 double cone pedicles which enclose the central single cone correspond to one unit of square mosaic pattern. The remaining pedicles facing by two each pair of the double cones in the center form a larger outer square. The mosaic pattern of cone pedicle as well as the distribution of ribbon synapse contacted as reconstructed from semi-serial sections are shown in the inset-scheme which has the same orientation as the micrograph

Fig. 8. Tangential section of a group of photoreceptor terminals contacted by one ▷ Golgi impregnated sclerad horizontal cell. Pedicles containing stained processes: outlined by ink. Processes in the central square: arrow heads pointing towards the center; processes in the outer square: arrow heads pointing towards the periphery. Stained processes in the central single cone: small arrow heads. Two rod spherules rs outlined by dotted lines are not contacted by this horizontal cell. Note that apart from the regular 2 x 8 double cones two additional cones are contacted. In the left and upper part of the montage, the relation of impregnated processes to ribbon synapses may be observed (cf. Fig. 4). Either one or two profiles facing a synaptic ribbon sr may be stained; incidentally, one single process may contact two different ribbon synapses. EM micrograph; x 11,200. The inset shows the schematical pattern of cone pedicles and ribbon synapses as reconstructed from serial sections of the whole cell

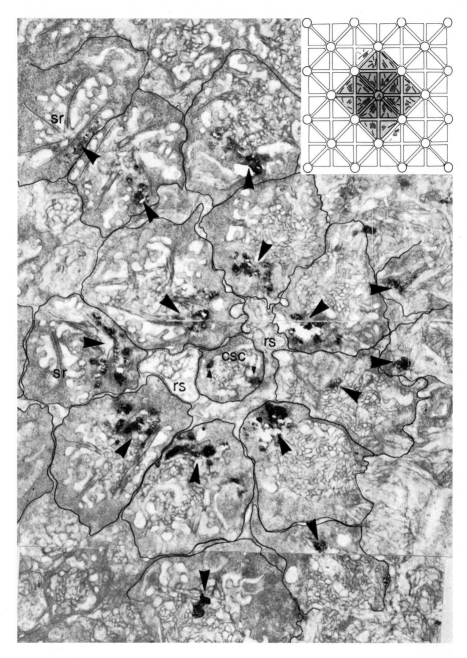

In the schematical representation, where individual double cone pedicles are depicted as triangles and central single cone pedicles as circles, the input of one horizontal cell to this square unit is more easily recognized. In addition, at least eight more individual double cones are linked to this horizontal cell. They are localized along the sides of the central square, each cone pedicle apparently belonging to a separate set of double cones. These processes have been observed before as constituting the peripheral ring (Fig. 4C). Finally, one or

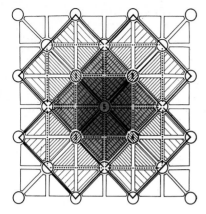

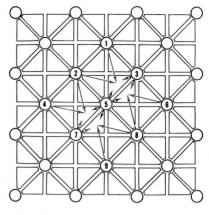

Fig. 9.   Fig. 10.

Fig. 9. Schematical drawing of cone pedicles demonstrating the overlap pattern resulting when 5 neighboring horizontal cells would be stained. The sketch is based
on observations of impregnations of 2 adjacent cells (like cells 2 and 3) and 2
overlapping cells (like cells 1 and 2). As follows from serial sections the perikaryon of each horizontal cell (1-5) is located just underneath each single cone;
it has not been depicted. The central square of double cones contacted by each cell
is surrounded by <u>solid lines</u>. The group of cones contacted by cell 5 (<u>center</u>) is
slightly shaded. Every cone contacted by this cell is also contacted by two or
three of the other cells. This results in a complete and multiple overlapping of
the area "covered" by cell 5 by each of the neighboring 4 cells

Fig. 10. Schematic representation of the number of sclerad horizontal cell processes
entering each of the central square of double cones overlying cell 5. Every double
cone pedicle is contacted by processes of 4 different sclerad horizontal cells

two cone pedicles outside this basic pattern of two concentric circles
and center may also be contacted (as seen in Fig. 8) but this is not
generally the case. Rod spherules, by contrast, are never invaded by
processes of this type of horizontal cell (Fig. 8).

As a further step, the relation of impregnated dendrites to ribbon
synapses was examined. Preliminary studies seem to indicate that in
each double cone there is a fairly constant arrangement of synaptic
ribbons (see Fig. 8, top left). One ribbon faces the flat side of the
terminal whereas three more radiate from the middle of the first one
towards the periphery. Ideally, there are four synaptic ribbons in
double cones, whereas in single cones, not more than two ribbons are
found. Quantitative assessment of the number of synaptic ribbons in
double cones results in a figure of 4.2 (n = 150) and 1.7 for single
cones (n = 50). In the total cone pedicles contacted by the dendrites
of the cone horizontal cell studied so far, there was an average of
28.4 processes per cell facing a synaptic ribbon. With a mean of 17.8
cone terminals invaginated by each cell this amounts to 1.6 dendrites
of a given horizontal cell being synaptically linked to 4.2 ribbons
per cone pedicle. Corrected for double cones, the figure is 1.7 per
each pedicle. This corresponds well to the observation that in the
cone terminals of the inner circle there are mostly two stained processes addressing generally the same synaptic ribbon, whereas in the
outer ring there is more often only one single process per pedicle.

Neighboring Cells: Pattern of Overlapping. In two cases, neighboring
cells were stained, thus permitting the study of the pattern of overlapping. The results are schematically represented in Fig. 9, where
cone pedicles contacted by five directly adjoining sclerad horizontal

cells are shown. The areas "covered" by the dendrites of the four peripheral cells (1-4) overlap each other so that the outer ring of processes of one cell is superimposed to a part of the inner ring (square
unit) of the next cell. This leads to a complete and multiple overlapping of the dendritic field of the central cell (5).

On this basis, the number of processes that originate from different
horizontal cells and converge towards one single pedicle of double cone
was analyzed and the results depicted in Fig. 10 for the square unit
overlying cell 5. It is obvious that each cone pedicle is contacted
by dendrites from four different sclerad horizontal cells. Two of these
processes are part of the inner circles of two different cells whereas the two remaining belong to the outer rings of dendrites of two
other cells.

1.1.2.2.5 Discussion

Color Vision with One Type of Cone Horizontal Cell. Unfortunately, we
are only just beginning to study the spectral sensitivity of the cone
types present in the Nannacara retina, but even before first results
are available, the following findings and observations would suggest
that Nannacara possesses a bichromatic vision based on cones which
respond to red and green but not to blue light.

In a list of photopigments found in various teleosts (Ali and Wagner,
1975), it appears that the majority of cichlid fishes have paired pigments which are based to a great extent on the aldehyde of vitamin A_1
(rhodopsin). There are a few specimens with a pure rhodopsin-type retina (Muntz, 1973) and these were all sampled in their natural habitat.
On the other hand, all specimens obtained from pet shops exhibit mixed
pigments with various amounts of porphyropsins (Schwanzara, 1967;
Bridges, 1972). As Nannacara was kept under artificial conditions, it
seems fair to assume that this fish, too, has both types of photopigments.

Although morphological criteria are poor clues for the determination
of the spectral identities of the cone types it seems to be a common
feature that, at least in the rudd (Scholes, 1975) and the goldfish
(Stell and Lightfoot, 1975), red and green cones are of similar length
and shape whereas the blue cones are considerably shorter and more displaced towards the outer limiting membrane. This observation is supported by theoretical evaluations relating the vertical separation of
the cone population to the axial dispersion of red, green and blue
lights (Eberle, 1967; Scholes, 1975). In the case of Nannacara, however, there is only one type of single cone present which is nearly
as long as the double cones. Although it is not possible to determine
the chromatic properties of these cones, the lack of a second type of
single cone makes it fairly probable that there are no blue cones in
this retina.

Finally, ecological and behavioral considerations corroborate the
assumption of Nannacara being a dichromatic fish. Though the colora-
tions of the male and female fish are very variable according to their
activities, there is never a shade of blue on them. Furthermore, when
studying some behavioral responses of young Nannacara, Kuenzer (1968)
found them quite capable of "telling" red from yellow and green but
much less capable of discriminating blue. This agrees with observations of Lythgoe (1975) that, in a slightly murky environment, radiation of short wavelengths is of far less importance than longer wavelengths to the visual perception of fishes.

The assumption that Nannacara is a bichromatic fish is only based on circumstantial evidence; however, it seems to be a feasible one if one takes into account the findings concerning the cone horizontal cells. Summarizing their observations in the trichromatic goldfish, Stell and Lightfoot (1975) show that the sclerad (H1) horizontal cell is linked to all three chromatic cone types whereas the next (H2) type contacts blue and green cones and the last (H3) type invaginates only blue cones. In the absence of blue cones, however, the second and third type become obsolete and a single type of cone horizontal cell with synapses to red and green cones would be sufficient to compute opponent colors and generating biphasic potentials. In this context, it is interesting to note, that in the trout which normally has three types of both cones and cone horizontal cells, a loss of at least one cone type and probably two rows of horizontal cells can be observed during the development of albino specimens in micrographs published by Ali (1964).

The absence of other types of cone horizontal cells in Nannacara other than the sclerad one may be demonstrated by a simple quantitative evaluation of the number of horizontal processes invaginating each double cone pedicle. In 150 double cones, an average number of 7.1 dendrites was found being in contact to an average of 4.2 synaptic ribbons (cf. above). On the other hand, because of the overlapping pattern, there are four different horizontal cells converging towards every cone terminal, each cell contributing 1.7 ribbon-facing processes on the average. This results in a theoretical value of 6.8 horizontal processes per double cone pedicle which is in fairly good agreement with the figure of 7.1 which was determined in independent measurements. As for the central single cones a similar correspondence is found between the actual number of dendrites and those calculated after counting the impregnated processes. This indicates that there seems to be no room left for more processes which might come from different horizontal cell types.

The Pattern of Overlapping. The functional significance of the pattern of overlapping observed in the dendritic fields of the cone horizontal cells in Nannacara remains open to speculation as long as there are no electrophysiological measurements. For the time being, two suggestions may be made: (1) This pattern might serve for blending various opponent color responses which would result in the coding of hue difference. (2) As electrical coupling has been reported to exist between neighboring horizontal cells (Yamada and Ishikawa, 1965; Stell, 1972), different velocity or mode of signal propagation may exist between the indirect lateral electrotonic transmission and the direct chemical transmission to several cells due to the overlapping which might be of importance for the coding of lateral inhibition.

1.1.3 Conclusion

This paper attempted to present some neural aspects of the retina of an apparently dichromatic teleost. The peculiarity of the species chosen consists in characteristic mosaic-like pattern in the OPL and IPL.

As for the mosaic of cones and cone horizontal cells it was shown not to be a pure coincidence resulting perhaps from packing problems as proposed by Locket (1975) for deep-sea fish retinae. The fact that the same pattern is found in the horizontal cell dendrites and in the connectivity of both cell types resulting in a very regular overlapping

array of dendritic fields strongly suggest that is has a definite functional significance. At the same time, this mosaic permits a quantitative approach to the "wiring pattern" more easily than nonmosaic retinae. Therefore, after examining some physiological questions such as
the color sensitivity of the cone types, other mosaic-type retinae with
different or more cone types will be studied. On the other hand, the
similarity of the patterns of arborizations of amacrine and ganglion
cells leading to a geometrical organization of the IPL is far from
being understood. However, comparative data suggest that the degree
of complexity of the operations performed in this synaptic layer is
clearly related to the number of cell types sending processes to this
layer. As has been suggested for the OPL, a highly regular arrangement
would further increase the capacity of the IPL.

After the recent studies in the visual system of insects (Part II of
this volume), it would come as no surprise if repercussions of this
mosaic-like arrangement of neurons in the retina would also be found
in the higher visual centers of the vertebrate CNS.

Acknowledgements. Most of the experimental part was carried out at
the Inst. f. Anatomie of the Universität of Regensburg and at the
Dépt. de Biologie de l'Université de Montréal. I thank Drs. E. Lindner,
M.A. Ali and Ch. Pilgrim for helpful discussions.

1.1.4 References

Ali, M.A.: Retina of the albino splake (Salvelinus fontinalis x X. namaycush).
    Can. J. Zool. 42, 1158-1160 (1964)
Ali, M.A., Wagner, H.-J.: Visual pigments: phylogeny and ecology. In: Vision in
    Fishes. Ali, M.A. (ed.), New York: Plenum 1975, pp. 481-516
Anctil, M.: Structure de la rétine chez quelques téléostéens marins du plateau con-
    tinental. J. Fish. Res. Bd. Can. 26, 597-628 (1969)
Boycott, B.B., Dowling, J.E.: Organization of the primate retina: light microscopy.
    Phil. Trans. R. Soc. London 255B, 109-184 (1969)
Bridges, C.D.B.: The rhodopsin-porphyropsin visual system. In: Handbook of Sensory
    Physiology. Photochemistry of vision. Dartnall, H.J.D. (ed.), Berlin-Heidelberg-
    New York: Springer 1972, Vol. VII/I, pp. 417-480
Cajal, S.R. y.: Die Retina der Wirbelthiere. Wiesbaden: Bergmann, 1894
Dogiel, A.S.: Über das Verhalten der nervösen Elemente in der Retina der Ganoiden,
    Reptilien, Vögel and Säugethiere. Anat. Anz. 3, 133-143 (1888)
Dowling, J.E., Werblin, F.S.: Organization of the retina of the mudpuppy, Necturus
    maculosus. I. Synaptic structure. J. Neurophysiol. 32, 315-338 (1969)
Dubin, M.W.: The inner plexiform layer of the retina: A quantitative and compara-
    tive electron microscopic analysis in several vertebrates. Ph. D. Dissertation,
    The Johns Hopkins University, 1969
Dubin, M.W.: The inner plexiform layer of the vertebrate retina: A quantitative
    and comparative electron microscopic analysis. J. Comp. Neurol. 140, 479-506
    (1970)
Eberle, H.: Cone length and chromatic abberation in Lebistes reticulatus. Z. Vergl.
    Physiol. 57, 172-173 (1967)
Engström, K.: Cone types and cone arrangements in teleost retinae. Acta Zool.
    (Stockholm) 44, 179-243 (1963)
Fuortes, M.G.F., Schwartz, E.A., Simon, E.J.: Colour dependence of cone responses
    in the turtle retina. J. Physiol. 234, 199-216 (1973)
Fuortes, M.G.F., Simon, E.J.: Interactions leading to horizontal cell responses
    in the turtle retina. J. Physiol. 240, 177-198 (1974)

Kaneko, A.: Physiological and morphological identification of horizontal, bipolar and amacrine cells in goldfish retina. J. Physiol. 207, 623-633 (1970)

Kaneko, A.: Receptive field organisation of bipolar and amacrine cells in the gold-fish retina. J. Physiol. 235, 133-153 (1973)

Kolb, H.: Organization of the outer plexiform layer of the primate retina: electron microscopy of Golgi-impregnated cells. Phil. Trans. R. Soc. Lond. 258B, 261-283 (1970)

Kuenzer, P.: Die Auslösung der Nachfolgereaktion bei erfahrungslosen Jungfischen von Nannacara anomala (Cichlidae). Z. Tierpsychol. 25, 257-314 (1968)

Kuenzer, P., Wagner, H.-J.: Bau und Anordnung der Sehzellen und Horizontalen in der Retina von Nannacara anomala. Z. Morph. Tiere 65, 209-224 (1969)

Locket, N.A.: Some problems of deep-sea fish eyes. In: Vision in Fishes. Ali, M.A. (ed.). New York: Plenum 1975, pp. 645-655

Lythgoe, J.N.: Problems of seeing colours under water. In: Vision in fishes. Ali, M.A. (ed.). New York: Plenum 1975, pp. 619-634

Meller, K., Eschner, J.: Vergleichende Untersuchungen über die Feinstruktur der Bipolarzellschicht der Vertebratenretina. Z. Zellforsch. 68, 550-567 (1965)

Muntz, W.R.A.: Yellow filters and the absorption of light by the visual pigments of some Amazonian fishes. Vis. Res. 13, 2235-2254 (1973)

Naka, K.-I., Rushton, W.A.H.: S-potentials from colour units in the retina of fish (Cyprinidae). J. Physiol. 185, 536-555 (1966)

Parthe, V.: Cêlulas horizontales y amacrinas de la retina. Acta Cient. Venez. Supl. 3, 240-249 (1967)

Parthe, V.: Horizontal, bipolar and oligopolar cells in the teleost retina. Intern. Symp. Visual Processes in Vertebrates, Santiago 1970. Vis. Res. 12, 395-406 (1972)

Polyak, S.L.: The retina. Chicago: Univ. of Chicago, 1941

Raviola, G., Raviola, E.: Light and electron microscopic observations on the inner plexiform layer of the rabbit retina. Am. J. Anat. 120, 403-426 (1967)

Scholes, J.-H.: Colour receptors, and their synaptic connexions, in the retina of a cyprinid fish. Proc. R. Soc. Lond. 270B, 61-118 (1975)

Schwanzara, S.A.: The visual pigments of freshwaterfishes. Vis. Res. 7, 121-148 (1967)

Sjöstrand, F.S.: A search for the circuitry of directional selectivity and neural adaptation through three-dimensional analysis of the outer plexiform layer of the rabbit retina. J. Ultrastruct. Res. 49, 60-156 (1974)

Stell, W.K.: The structure and relationships of horizontal cells and photoreceptor-bipolar synaptic complexes in goldfish retina. Am. J. Anat. 121, 401-424 (1967)

Stell, W.K.: The morphological organization of the vertebrate retina. In: Handbook of Sensory Physiology. Fuortes, M.G.F. (ed.). Berlin: Springer 1972, Vol. VII/I.B

Stell, W.K.: Horizontal cell axons and axon terminals in goldfish retina. J. Comp. Neurol. 159, 503-520 (1975)

Stell, W.K., Lightfoot, D.O.: Color-specific interconnections of cones and horizontal cells in the retina of the goldfish. J. Comp. Neurol. 159, 473-502 (1975)

Svaetichin, G.: Cêlulas horizontales y amacrinas de la retina: propiedades y meca-nismos de control sobre las bipolares y ganglionares. Acta Cient. Venezolana, Suppl. 3, 254-276 (1967)

Tomita, T.: Microelectrode study of the physiology of neurons in the fish retina. In: Vision in fishes. Ali, M.A. (ed.). New York: Plenum 1975, pp. 69-79

Wagner, H.-J. Vergleichende Untersuchungen über das Muster der Sehzellen und Hori-zontalen in der Teleostier-Retina (Pisces). Z. Morph. Tiere 72, 77-130 (1972)

Wagner, H.-J.: Die nervösen Netzhautelemente von Nannacara anomala (Cichlidae, Teleostei). I. Darstellung durch Silberimprägnation. Z. Zellforsch. 137, 63-86 (1973)

Wagner, H.J.: Die Entwicklung der Netzhaut von Nannacara anomala (Cichlidae, Tele-ostei) mit besonderer Berücksichtigung regionaler Differenzierungsunterschiede. Z. Morph. Tiere 79, 113-131 (1974)

Wagner, H.-J.: Comparative analysis of the patterns of receptor and horizontal cells in teleost fishes. In: Vision in Fishes. Ali, M.A. (ed.). New York: Plenum 1975, pp. 517-524

Witkovsky, P., Dowling, J.E.: Synaptic relationships in the plexiform layers of carp retina. Z. Zellforsch. 100, 60-82 (1969)

Witkovsky, P., Stell, W.K.: Gross morphology and synaptic relationship of bipolar cells in the retina of the smooth dogfish, Mustelus canis. Anat. Rec. 169, 456-457 (1971)

Yamada, E., Ishikawa, T.: The fine structure of the horizontal cells in some vertebrate retinae. Cold Spr. Harb. Symp. Quant. Biol. 30, 383-392 (1965).

1.2 Comparative Study of the Horizontal Cells in the Vertebrate Retina: Mammals and Birds

A. GALLEGO

1.2.1 Introduction

H. Müller (1851) first described horizontal cells in the teleost retina; Cajal (1893) gave them the commonly used name of horizontal cells. Between the time of Müller's initial description and Cajal's study, these cells were given several names by different authors: "tangentiallen fulcum Zellen" (W. Müller, 1874), "cellules basales" (Ranvier, 1882); "sternenförmige Zellen" (Dogiel, 1884); "konzentrische Stützzellen" (Schiefferdecker, 1886); "cellule superficiali" (Tartuferi, 1887).

Since Cajal's study (1893) it has been generally admitted that horizontal cells in tetrapoda retina were axon-bearing neurons of two types: "external" and "internal" horizontal cells. Cajal also showed that the dendrites and the axon endings of both cell types contact the "feet" (synaptic bodies) of the visual cells; he believed that the external horizontal cell was related to rods and the internal to cones. Such a structural arrangement, quite different to Cajal's ideas on the relationship between neurons, was for him a "paradox" (Cajal, 1933a) which he interpreted considering these cells as "energetic centers" of the retina. In the avian and reptiles retina Cajal referred to the internal type as the "brush-shaped" horizontal cell and he named "stellate" horizontal cell to the external type

In his description of the teleostei retina, Cajal distinguished three layers formed by cells which he called "external", "median" and "internal" horizontal elements. According to Cajal's description the external horizontal cells relate to the "membrana fenestrata" of Krause (1884) and to the "intermediary concentric cells" of Schiefferdecker (1886). The second row of horizontal cells or Cajal's median horizontal elements are arranged in an almost continuous layer below the external cells which they resemble in their morphological features; like the external horizontal cells they have a large "horizontal expansion" (an axon) whose destination Cajal was not able to determine. The median horizontal cells were previously described by Schwalbe (1874), Reich (1875), Retzius (1881), Schiefferdecker (1886) and Krause (1884). According to Krause they behave in a way similar to the external type forming beneath them the "membrana perforata". The internal horizontal elements are very long, thick structures, situated horizontally below the median layer; they are spindle-shaped structures which Schiefferdecker designated by the name of "kernlose konzentrische Zellen" (concentric cells without nucleus). According to Krause they would form his "stratum lacunosum".

Complete impregnation of horizontal cells in tetrapoda including soma, axon and axon endings was achieved by Cajal (1893) in the chicken retina and a few years later by Marenghi (1901) in the calf retina (Fig. 1); Cajal's description has been confirmed (Gallego et al., 1975a) but the existence of the several collateral branches of the axon described by Marenghi is doubtful. Up to a few years ago we

Fig. 1A,B. (A) Horizontal
cell of chicken retina
(Cajal, 1893, Plate IV,
Fig. 4). (B) Horizontal
cell of calf retina
(Marenghi, 1901, Plate V,
Fig. 6)

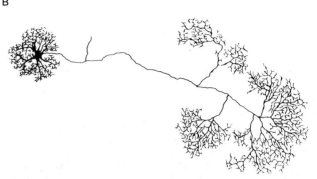

lacked information about the axon length and structure of the axon
endings in other retinas.

In his studies Cajal used mainly the Golgi method, first fixing the
retina by his "enroullement" technique; although the tissue was sectioned in several planes this method produced limited data on the distribution of the various cell types in the central and peripheral retina and on the connections between different retinal neurons. During
the last decade positive advances in the knowledge of retinal structure were made with the aid of the following new techniques:

1. Electron microscopy (Missotten, 1965) and especially Stell's
   Golgi-EM technique (1965) showed that the dendrites and axon endings of the horizontal cells penetrate the synaptic bodies of
   cones and rods forming the lateral elements of the triad.

2. Silver impregnation (Gallego, 1953) and Golgi staining of the entire retina and its study in whole flat mounted preparations led
   to the discovery in low mammal retinas of a new type of horizontal
   cell without axon (Gallego, 1964).

3. The intracellular recording combined with procion yellow injection
   (Kaneko, 1970) established the basis for a precise correlation between electrophysiological and structural data.

Using these techniques a comparative study of the vertebrate retinas
leads to the conclusion that two basic and clearly different types
of horizontal cells can be considered: "short axon horizontal cells"
and "amacrine or axonless horizontal cells".

Cajal's (1899) classification of neurons in "long axon", "short axon"
and "axonless" neurons is still valid: he gave the name "long axon
neurons" to the nervous cells described by Golgi as "motor" or type I
neurons, characterized by an axon whose presynaptic fibers were located either in a different neuronal pool than the one where the soma
lies or outside the nervous system; the Golgi "sensitive" or type II
neurons, named by Cajal "short axon neurons" have an axon whose presynaptic fibers are distributed in the same neuronal pool where the
soma is located; these cells have been also named "Schaltzellen" or

28

"schaltneurons" (v. Monakow, quoted by Cajal, 1902, 1903), "inter-
nuncial neurons" (Lorente de Nó, 1933) and finally interneurons.
Cajal also described a third type of axonless neurons and among them
the "amacrine cells" of the inner plexiform layer in the retina.

The axon-bearing horizontal cells fulfill the definition of short
axon neurons as the axonless do that of the "amacrine" nervous cells
(a, not; makros, long; is, inos fiber). But not only the structural
data differentiate both types of cells: their connections with the
visual cells and the membrane contacts between themselves are different, as are probably also their functional properties.

1.2.1.1 Short Axon Horizontal Cells (Fig. 2)

These were first described by Cajal (1893) and are characteristic of
the tetrapoda retina; their dendrites invaginate into the cone triads
and their axon endings penetrate the synaptic complex of rods, both
in mammals (Kolb, 1970, 1974; Gallego, 1971, 1975a,b; Gallego and
Sobrino, 1975) and in nocturnal birds (Gallego et al., 1975a; Gallego,
1975b). In diurnal birds (Gallego, 1975c) reptiles (Gallego and Pérez
Arroyo, 1975) and amphibian (Pérez Arroyo and Gallego, 1976) retinas
the preliminary findings show that the dendrites of this type of cell
contact some cone synaptic bodies and the axon endings contact either
the rods or a different type of cone.

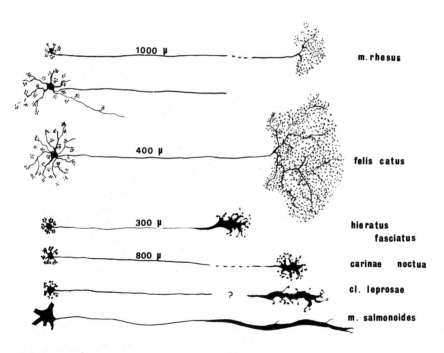

Fig. 2. Schematic drawings of the short axon horizontal cell in flat-mounted Golgi
preparations of primates (Macaca rhesus, central and peripheral retina); low mammals
(felis catus); birds (diurnal, Hieratus fasciatus, and nocturnal, Carinae noctua)
and reptiles (Clemmys leprosae). Horizontal cell of a teleost (Micropterus salmo-
ides)

All the tetrapoda retina studied up to now show a single type of short axon horizontal cell and not two types - external and internal - as stated by Cajal. The dendrites of this cell type overlap to a great extent; such overlapping has not been observed in the axon endings. Gap-junctions between the membranes of neighbor short axon horizontal cell bodies have not been detected.

Different animal classes and even species show morphological variations in this type of cell which refer to the size of their dendritic arborization, to the number of cones connected to it, to the axon length, to the structure of the axon endings and to the number of rods connected to them.

1.2.1.2 Axonless Horizontal Cells (Fig. 3)

In the tetrapoda retina this cell type was first described in low mammals by Gallego (1964, 1965) and by Gallego and Pérez Arroyo (1975) in the turtle. The axonless horizontal cell has not been found up to now in primates, birds or amphibian retina. All kind of fishes have this type of axonless horizontal cell: in the teleostei retina it is represented by the rod horizontal cells and in the selachii retina by the two layers of horizontal cells connected to rods (Gallego, 1975b).

This cell type forms a plexus which extends all over the retina; it has no axon and neighbor cells show large gap-junctions between them which suggest electrical coupling. It is also a characteristic of the

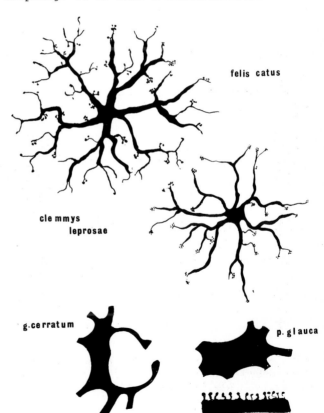

felis catus

clemmys leprosae

g.cerratum

p.glauca

Fig. 3. Schematic drawings of the axonless horizontal cell in flat-mounted Golgi preparation of mammal (felis catus), reptile (clemmys leprosae) and two selachii (gynglimostoma cerratum and prionace glauca) retinas

axonless horizontal cell that their cell processes invaginate into
the synaptic complexes of a <u>single type</u> of photoreceptors, either
cones as in cat and rabbit retina (Kolb, 1974; Gallego, 1975a,b,c) or
rods as in teleostei and selachii retina (Stell, 1967; Gallego, 1975b).

1.2.1.3 A Third Type of Horizontal Cell (Fig. 2, Micropterus salmoides)

Stell's discovery (1975) in teleostei retina that the internal horizontal cells are in fact the axon endings of the cone horizontal cells
led us to consider a third type of horizontal cell. This cell type displays some of the features of each cell type previously described. It
has an axon as the short axon cell; on the other side gap-junctions
are seen between neighbor cells as in the axonless horizontal cell type.
Furthermore this cell type deserves to be considered as a third type
of horizontal cell because its spindle-shaped ending does not reach
the synaptic bodies of any photoreceptors but lies at the vitreal side
of the inner granular layer. The connections of these axon endings remain unknown.

1.2.2 Horizontal Cells of the Retina of Mammals and Birds

In comparison with the large horizontal cells of fish retina, which
are arranged into two to four well-defined plexus layers (Krause's
"membrane perforata" and "membrane fenestrata"), the horizontal cells
of amphibians, reptiles, birds and mammals, are small and arranged in
one to two layers that are not always well defined as seen in perpendicular sections of the retina. During the last century the horizontal
cells of terrestrial vertebrates were studied by Kölliker (cow, 1867),
Rivolta (horse, 1871), Golgi and Manfredi (horse, 1872), Dogiel (man,
1884), Ranvier (cat, 1882). To this list should be added the studies
of Krause (1884). Schiefferdecker (1886), Tartuferi (1887) and Cajal
(1893) who studied these cells in a variety of vertebrates.

Cajal's generalization concerning the remarkable structural similarity
of the retina among vertebrates: "The only anatomical variations which
occur are those in the relative thickness of the layers and in the form
and thickness of the rods and cones ..." (Cajal, 1893), was perhaps pre-
mature. The idea was first challenged by Polyak (1941) who described
several new varieties of bipolar and ganglion cells in the monkey retina; he also described a single type of horizontal cell in the monkey
retina, as had been mentioned previously be Dogiel (1884) in human retina. A new contribution to the knowledge of horizontal cells was made
by Gallego (1964) who initially described, in lower mammals, a new type
of horizontal cell that lacks an axon ("amacrine" horizontal cell);
this cell type forms a plexus that spreads throughout the retina. Since
this time other investigators have studied the horizontal cells of the
retina in several mammalian species and their connections with the photoreceptors.

1.2.2.1 Mammals

There now is general agreement that only one type of axon bearing horizontal cell is found in the <u>primate</u> retina, a short axon horizontal
cell (Kolb, 1970; Gallego, 1971; Boycott and Kolb, 1973; Gallego,
1975a,b) and not the two types described by Cajal in vertebrates and
by Boycott and Dowling (1969) in the primate retina.

Two clearly different types of horizontal cell have been described in <u>lower mammals</u>: the horizontal cell with an axon studied by Cajal (1893) and the "amacrine" horizontal cell described by Gallego (1964). The presence of the latter type has been confirmed by several authors (Leicester and Stone, 1967; Honrubia and Elliot, 1969; Fisher and Boycott, 1974; Boycott, 1975) in the retina of both cat and rabbit.

With the use of electron microscopy, and in particular with Stell's technique (1965), it has been possible to establish the connections between the horizontal cells and the photoreceptors in the mammalian retina. The dendrites of the short axon horizontal cell form the lateral elements of the cone triad and the axon terminals penetrate the synaptic bodies of rods in the primate (Kolb, 1970; Gallego, 1971; Gallego and Sobrino, 1975), cat and rabbit (Gallego, 1971; Kolb, 1974; Gallego, 1975a,b,c) retinae. The "amacrine" or axonless horizontal cell establishes connections only with the synaptic bodies of cones in cat and rabbit retinas (Kolb, 1974; Gallego, 1975a,b,c).

1.2.2.2 Birds

Since Cajal's studies there have been no further significant data concerning the horizontal cells of the avian retinas. Consistent with his general description of the retinal structure of vertebrates, Cajal (1893, 1904, 1933b) described two types of horizontal cell in the avian retina, the "stellate" and the "brush-shaped" horizontal cells. His distinction was based on (1) the different location of the cell type in relation to the outer plexiform layer, and (2) on certain morphological features. However it has been shown recently (Gallego et al., 1975a) that in both diurnal and nocturnal bird retinas, there is only one type of axonal horizontal cell, the brush-shaped horizontal cell. Its axon terminals are structures which were erroneously described by Cajal as the stellate type of horizontal cell.

The relationship of the horizontal cells with the photoreceptors in the avian retina, are more difficult to analyze because of the different types of photoreceptors and the position of their synaptic bodies in relation to the outer plexiform layer. The presence of rods and single and double cones in the bird retina was noted by Schultze (1866, 1867) who observed red, yellow and colorless droplets in the cones. Photoreceptors, especially the counting of colored oil droplets in order to determine the proportion of different cones and their topographical distribution, have been the subject of several studies (see Crescitelli, 1972; Muntz, 1972). A comprehensive review on the visual cells in several bird species was made by Rochon-Duvigneaud (1943) using light microscopic techniques. Nevertheless later studies (Meyer and Cooper, 1966; Morris and Shorey, 1967; Morris, 1970; Meyer and May, 1973) on the bird visual cells using phase contrast and electron microscopic techniques have contributed immensely to the current knowledge on this subject.

Morris (1970) described different types of visual cells in the chicken retina: (1) rods with no oil droplets, (2) double cones and (3) three types of single cones. Meyer and Cooper (1966) and Morris and Shorey (1967) disagree about the color of the oil droplet found in cones of the chicken retina. The former authors reported a red droplet in single cones, a golden-yellow one in the "chief" member of the double cones and a greenish droplet in the "accessory" member. However, Morris and Shorey (1967) maintain that single cones have either a red or a yellow droplet. Furthermore they state that the chief member of the double cones has a green droplet and that the accessory member does not contain any droplet.

Cajal (1893) described the outer plexiform layer of the "Gallinaceous" retina as having three distinct strata or "concentric plexuses": (1) the external plexus was formed by the basal rod filaments and the dendritic expansions of certain bipolar cells; (2) the intermediate plexus was formed by the vitread end, or synaptic bodies, of the "straight" cones and the ascending dendrites of other bipolar cells; (3) the internal plexus was formed by the basal filaments of the "oblique" cones and the dendritic expansions of bipolar cells. Further information on the structure of the outer plexiform layer in other bird species is not mentioned by Cajal. According to Morris and Shorey (1967) the synaptic bodies of chicken photoreceptors are located at two different levels in the outer plexiform layer. The rod and double cone synaptic bodies are in the external or sclerad level, whereas the synaptic bodies of the single cones are in the internal or vitread level. Recently we have confirmed the three-layered stratification of the outer plexiform layer in the diurnal bird retina (Gallego et al., 1975b) as described by Cajal (1893). The external (sclerad) row is formed by the synaptic bodies of rods, chief and accessory members of double cones; the intermediate row is formed by the scattered synaptic bodies of single cones and the internal row consists of the synaptic bodies of the "oblique" cones. In contrast to the stratification of the outer plexiform layer found in diurnal birds, the photoreceptor synaptic bodies in nocturnal birds lie in a single row in the outer plexiform layer.

The present paper contains the results obtained in the study of several species of mammals and birds. It will be followed by the study of the reptiles and amphibians retina (Pérez Arroyo and Gallego, 1976); the ciclostomes retina (Sobrino and Gallego, 1976) and the teleostei and selachii retina (Gallego and Pérez Arroyo, 1976).

1.2.3 Materials and Methods

1.2.3.1 Animals

Our studies on mammals have been performed in the dog, cat, rabbit, guinea pig, rat and monkey retinas. The studies on the avian retinas have been performed in three diurnal species Gallus domesticus, Hieratus fasciatus and Milvus milvus and in two nocturnal ones, Carinae noctua and Asio flameus. In both cases the retinas were obtained immediately after killing the animals.

1.2.3.2 Light Microscopy

The following histological procedures have been used: Cajal's silver nitrate technique (1903); Balbuena's method (1922) and the Gallego (1953) technique of silver impregnation of the entire retina. Some retinas were prepared by the Erlich-Dogiel technique of supravital staining with methylene blue as described by Cajal and De Castro (1933).

A few retinas were observed in toto by the Nomarsky optics after enucleation of the eye. Enzymatic digestion of cat and rabbit retinas with a proteolitic enzyme obtained from streptomyces and further sucrose gradient centrifugation allowed the analysis of isolated cells by Nomarsky optics.

The classical Golgi method and its Lasansky (1971) and Colonnier (1964) modifications applied to the entire retina were widely used.

1.2.3.3 Electron Microscopy

Cells impregnated by the Golgi method were selected, reembedded and oriented for ultrathin sectioning and further examination in the electron microscope (West, 1972).

Conventional electron microscopic studies were carried out on pieces of retina fixed with 3% glutaraldehyde in 0.1M phosphate buffer at pH 7.4 - 7.5 for 4 - 24 h. They were then washed with 10% saccharose in 0.1M phosphate buffer and postfixed with 2% osmium tetroxide in 0.1M phosphate buffer for 2 h at 4°C. The pieces were then stained en bloc with 2% uranyl acetate in cold distilled water, dehydrated in graded mixtures of methanol and water, infiltrated with propylene oxide and embedded in Epon. The ultrathin sections were stained with aqueous uranyl acetate followed by lead citrate (Reynolds solution) and examined in a Philips 200 electron microscope.

All the retinas were observed in both sectioning planes: vertical and transversal, that is parallel or perpendicular to the centripetal pathway.

1.2.4 Results

1.2.4.1 Mammalian Retina

In the retinae of the mammals studied we have found two clearly different types of horizontal cells: horizontal cell with axon and the axonless horizontal cell.

1.2.4.1.1 Horizontal Cell with Axon

This type of cell, which corresponds to the classical horizontal cell described by Cajal, shows common features in all the mammals studied and some differences, especially related to the extent of the area covered by the dendrites, to the length of the axon and to the extent of the axon terminal arborizations.

Primates. In the primates retina, the soma and the arrangement of the dendritic processes, show morphological differences between the cells located in the proximity of the fovea and those situated in zones of the peripheral retina. The soma in the zones near to the fovea (Figs. 5,6,7) is small, 8 to 10 μ in diameter, and it gives out 5 to 8 dendrites directed vertically towards the synaptic bodies of the photoreceptors. The area covered by these dendritic processes is practically circular with a diameter of 25 to 30 μ.

Each dendritic process ends in a small cluster of knobs; these knobs are usually 0.3 to 0.5 μ in diameter and each cluster, which covers an area of 5 to 7 μ of diameter, consists of seven to ten knobs. As shown by Kolb (1970), the terminal knobs of dendrites form the lateral components of the triads in the cone synaptic bodies. In our preparations, the synaptic body of the cones situated between 4 and 6 mm from the center of the fovea is 6 to 8 μ in diameter. Each cluster of terminal knobs connects with the synaptic body of a single cone which is also joined by the dendrites of the individual and flat cone bipolars. In flat whole mounts of Golgi-impregnated retinas, the confluence of the dendritic terminals of two neighbor horizontal cells (Figs. 6,7) as well as of the dendrites of bipolar and horizontal cells (Figs. 8,9) can be demonstrated.

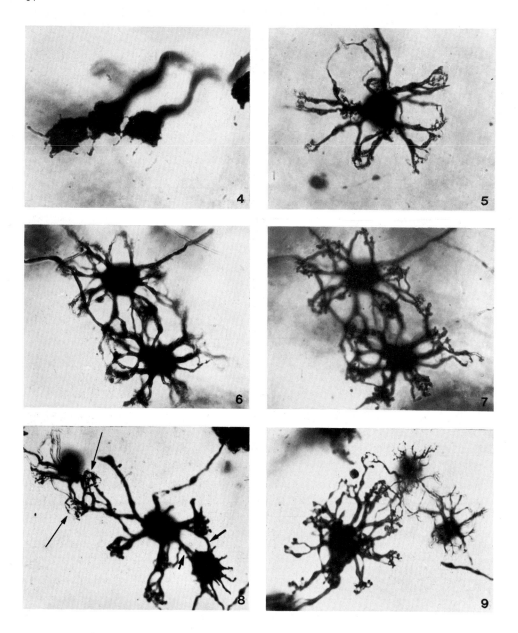

Figs. 4-9. Light micrographs of the monkey retina (M. rhesus) prepared by the Golgi-Colonnier method in flat whole mounts:(4) Synaptic bodies of cones which show basal filaments. (5) Short axon horizontal cell. (6,7) Two horizontal cells at two different focus levels show the dendritic overlapping.(8) Short axon horizontal cell whose dendrites form clusters of knobs joining with two dendrites of a bipolar cell (thin arrow). Two dendrites of the horizontal cell (thick arrows are directed towards the synaptic body of a cone whose basal filaments are clearly seen.(9) Confluence of the dendritic endings of a bipolar and those of a short axon horizontal cell

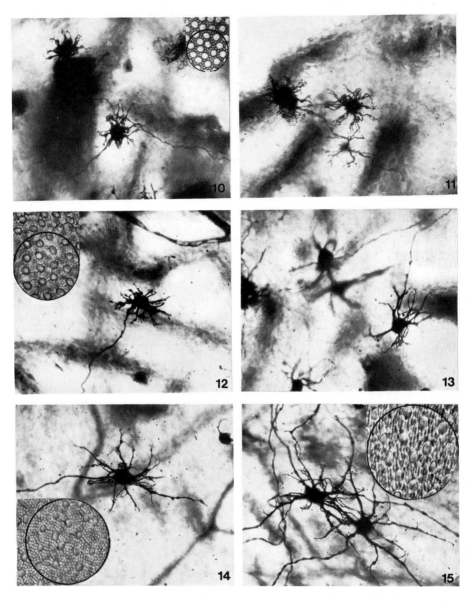

Figs. 10-15. Light micrographs of the monkey retina in flat-mounted Golgi-Colonnier preparation. Short axon horizontal cells located at different distances from the fovea are shown: 2,5 mm (10); 4,5 mm (11); 6,5 mm (12); 8,5 mm (13); 10,5 mm (14); 12,4 mm (15). The insets show the retinal area covered by the dendrites of each cell and the cones included in such area. The circumference shown in the inset of (1) is 25 μ in diameter

The morphology of the soma and of the dendrites of the horizontal cells varies according to their distance from the fovea. In the areas near to the "ora serrata" the horizontal cells (Fig. 15) look quite different to those situated in the proximity of the fovea: the cell body is slightly bigger in size, and the dendrites thicker and longer, spread out nearly horizontally over oval shaped areas with diameters of the order of 60 to 100 µ. It is important to point out that the horizontal cells situated more than 8 - 10 mm from the fovea usually show one or two dendrites, much longer than the rest, about 100 and 150 µ long. Along the shorter dendrites and in the base of the long ones, thin fibers which end in small clusters, formed by thinner and scarcer knobs than those which form the clusters of horizontal cells in the proximity of the fovea, show out. Small collateral branches arising from the longer dendrites and ending in a small knob or sometimes in a small cluster of knobs, can be frequently observed.

The study of retinae in toto when horizontal cells situated at different distances from the fovea are stained, clearly shows the gradual transition of the morphology of the cells situated in the proximity of the fovea to that of the ones located in the more periphereal zones (Figs. 10-15). The area covered by the dendrites gets bigger with the distance from the fovea; so cells located 4.5 mm from the latter have dendrites which cover a circular area of about 40 µ of diameter; those situated 6.5 mm from the fovea cover a circular area of about 50 µ of diameter. From a distance of about 8.5 mm the area covered by the dendrites is about 60 µ if the long dendrite, which begins to show out, is not included; in the periphery, 10.2 and 12.5 mm from the center of the fovea, the areas covered by the dendrites are circular, about 70 µ of diameter or more (Figs. 14,15), if we only include in the measurement the short dendrites. In all the zones of the retina, the areas covered by the dendrites of the horizontal cells overlap; this overlapping is quite ample in the peripheral cells (Fig. 15).

Each dendritic cluster connects with the synaptic body of a single cone. As has been mentioned, very often two dendrites of neighbor cells form one cluster of knobs that connects one cone synaptic body; such arrangement explains the commonly found situation in Golgi-EM analysis where the dendritic terminals of one horizontal cell form only one of the two lateral components of the triads (Kolb, 1970).

The best way to set the number of cones connected with one horizontal cell is by counting the number of clusters that its dendrites have. The reliability of this procedure depends on the success of the staining. In our best preparations the data obtained agree with Boycott and Kolb's (1973). Each parafoveal horizontal cell contacts six to nine cones and the most peripheral one up to 30 - 40 cones.

A more precise study is needed to determine the number of cones in the area covered by the horizontal cell dendrites in each retinal zone; the techniques used, due to retraction of the tissue after fixation, staining, dehydration and mounting of the specimen, can give contradictory results.

Only seldom have we been able to impregnate the axon of the horizontal cells completely, up to its axonic terminals; up to now this impregnation has only been successful in the horizontal cells located in the zone between 4.5 and 6.5 mm from the fovea and in the temporal retina. In these cases (four cells) the axon was between 1100 and 1400 µ long, slightly longer than the axons described by Ogden (1974) in the retina of the monkey Aotes. In preparations in which we have obtained a complete impregnation of a great quantity of horizontal cells it was seen that the axons spread in all directions and do not follow any regular way of distribution (Gallego, 1975b, Fig. 59).

The axon endings are similar to those of the horizontal cells in the retina of the remaining mammals. The axon follows a nearly straight or flexuous course and thickens and ramifies at its end (Fig. 16);

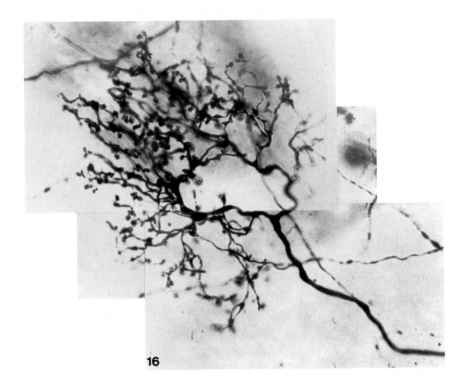

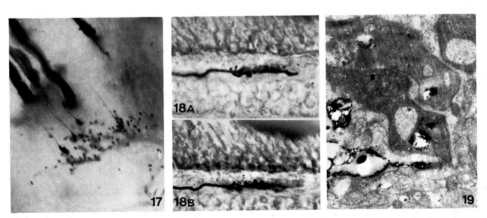

Figs. 16-19. Light micrographs of the monkey retina (Gallego, 1975b). (16) Axon terminal of a short axon horizontal cells shown by a montage of micrographs. Flat mount of a Golgi preparation. (17) Axon terminal knobs of the horizontal cells appear to make contact with the synaptic bodies of several rods. Golgi preparation in whole flat mount. (18) Axon terminal of a horizontal cell at two different focus levels which has been reembedded for EM analysis. Perpendicular section of Golgi-impregnated retina. (19) Electron micrograph of an ultrathin section of the axon terminal shown in (18). The rod synaptic bodies show the silver chromate precipitate

from these terminal branches set out thin fibers which generally expand terminally to a knoblike structure thicker than the small knobs that terminate the dendritic clusters. In our preparation the axon terminal considered in toto covers retinal areas of different size

whose minor and major axis are about 50 - 90 x 100 - 180 μ. When counting the knobs of an axon terminal that covers an area of 5000 μ² we
have found 70 to 100 knobs/1000 μ². This means that the horizontal
cells in which we completely stain soma, axon and terminals a relation through them was established between six to nine cones and
350 - 500 rods.

The connections of the horizontal cells with axon in the primate retina have been identified by the study of the Kolmer "Kristaloide" em-
ploying the electron microscope (Gallego, 1971) and with the aid of
the combined Golgi-EM method (Kolb, 1970; Boycott and Kolb, 1973;
Gallego and Sobrino, 1975): the terminal clusters of the dendrites
form the lateral components of the cone triads and the knobs of the
axon terminals penetrate into the synaptic complexes of the rods
(Fig. 19).

Nonprimates. In the retinae with no fovea of the inferior mammals,
we have systematically found in Golgi preparations the horizontal
cell with axon whose structure has been well known in light microscopy since Cajal's studies (1893). However, we must point out that
in the cat, between the horizontal cells of the "area centralis" and
its proximity and the more peripheral ones, there also exist slight
morphological modifications due to the dispersion of the dendritic
tree. In these retinae, the horizontal type which is characteristic
of the parafoveal zones in the primate retina does not exist: the cat
horizontal cells closely resemble those in the intermediate or peripheral zones of the latter.

In the cat retina the soma of the horizontal cell with axon is about
10 μ in diameter and bears a great number of dendrites (Figs. 20-22)
which are thinner and longer than the ones of the parafoveal horizontal cells and of the intermediate regions in the primate retina. They
are also more arborized, forming thin terminal branches which show a
small cluster of knobs. The number of knobs in each cluster is lower
than in the primate retina, which might mean that in the synaptic bodies of the cones in the inferior mammals, at least in the cat, the
number of triads per cone is lower than in the primates.

In the horizontal cells studied we have found in their dendrites,
which are distributed in a circular area of 100 to 150 μ in diameter,
an average of 35 to 45 clusters per cell, which leads us to believe
that they synapse with the same number of cone synaptic bodies.

When a lower mammal horizontal cell is completely stained (soma, axon
and axon terminals) (Fig. 20), we can observe the strongest distinctive characteristics with the corresponding cells in the primate retina which are as follows: (1) the lesser length of its axon (which is
usually not more than 400 μ long; (2) the larger size of the axon terminal arborization which covers retinal areas of the order of
150 x 250 μ; (3) the number of knobs per area of equal size is larger
than in the primate retina.

In electron microscopy studies we described for the first time in the
cat and rabbit retinae (Sobrino and Gallego, 1970; Gallego, 1971),
the distinctive characteristics between the axonless and the horizontal cell with axon. In perpendicular sections the horizontal cell
with axon shows a large soma with thin dendrites directed towards the
feet of the photoreceptors whereas the axonless horizontal cell shows
thick cellular processes, whose size makes it difficult to identify
the limits of the soma; moreover the cell body is located at a slightly higher level than that of the horizontal cell with axon. In parallel sections to the retinal surface the morphological difference be-

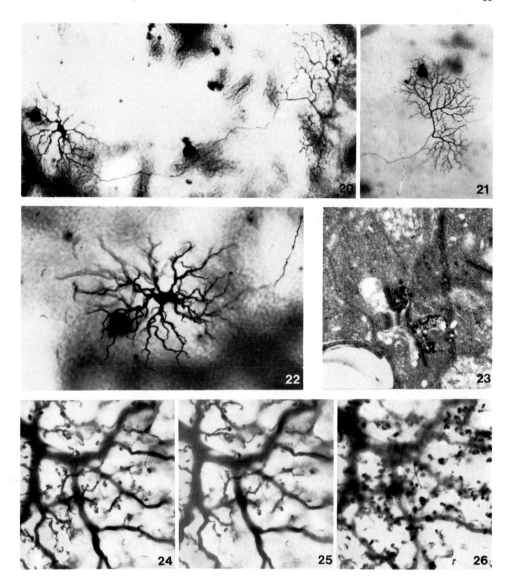

Figs. 20-26. Light microscopy of the cat retina in whole flat mounts of Golgi-Colonnier preparations (Gallego, 1975a modified). Fig. 23, West technique. (20) Short axon horizontal cell completely impregnated.(21) The axon terminal of a short axon horizontal cell.(22) The cell body of a short axon horizontal cell.(23) Electron micrograph of a Golgi impregnated axon terminal which was re-embedded and sectioned for EM analysis. The rod synaptic body shows the silver chromate precipitate.(24-26) Detail of a horizontal cell axon terminal at three different focus levels showing the knobs to be counted

tween both types of cell is very clear in regard to the disposition and thickness of their dendrites or cellular expansions. In the rabbit retina the cell with axon shows a bigger soma than the axonless cell, with thin dendrites directed towards the feet of the photoreceptors.

It shows a round nucleus without indentations or irregularities in
the membrane. The cytoplasm, which is clearly defined, shows rosette-
shaped ribosomes, a scarcely rough endoplasmic reticulum and a small
number of mitochondria as well as Golgi vesicular complexes. The cytoplasm shows scattered tubuli which are more evident in the dendrites
which also contain some poliribosomes.

The connections between the horizontal cells with axon and the photoreceptors in the inferior mammals, are identical with those described
in the primates. The dendrites of the horizontal cells with axon in
cat and rabbit penetrate into the synaptic bodies of the cones, as has
been shown with electron microscopy in young animals (Gallego, 1971),
and confirmed by Kolb (1974) using the Golgi-EM method. The thin knobs
of the axon terminal penetrate into the synaptic body of the rod
spherulae (Fig. 23) as seen by Kolb (1974).

We have studied three completely stained cells (soma, axon and axon
endings), and in these 35 to 45 cones synapse with the horizontal cell
dendrites, which covers a circular area 125 to 150 μ in diameter. The
great number of knobs in the axonic terminal makes counting difficult;
however, in well-impregnated cells (Figs. 24-26) it can be carried
out with accuracy. In these cells the number of knobs/100 μ^2 was
eight to eleven. Bearing in mind that each synaptic knob corresponds
to one rod, and each dendrite to the synaptic body of one cone, it
may be inferred that about 25 to 35 cones are related with 3000 to
4000 rods through the horizontal cell with axon.

1.2.4.1.2 Axonless Horizontal Cells (Amacrine Cells of the Outer Plexiform Layer)

This cell type has been described for the first time in the retina of
lower mammals using both the light (Gallego, 1964, 1965) and the electron microscope (Sobrino and Gallego, 1970; Gallego, 1971) as a different type of the Cajal classical horizontal cells, that is to say as a
different type of horizontal cell without axon.

Nonprimates. The technique of silver impregnation of the entire retina,
used by us for the first time (Gallego, 1953), reveals the existence
in lower mammals of a layer of horizontal cells which covers the entire retina. Early studies were carried out in cat and rabbit retina,
where it was demonstrated that these cells extend from the peripheral
zone to the papilla (Figs. 27-30) forming a well-defined layer, even
in the "area centralis" of the cat's retina. The neurofibrillar methods, especially Balbuena's and Cajal's reduced silver, impregnate this
cell type easily. Only rarely are they stained by the Golgi method,
which may explain why they were not noticed by Cajal in his study of
the vertebrate retina.

Due perhaps to the fact that our first description was published in
French, it is not familiar to English-speaking authors. In our study
carried out in 1964, these cells were thus described: "The nuclei of
these cells are located in the outer row of the inner granular layer.
They are poor in chromatine, and have a lobulated mulberry-shaped
nucleolus. The silver methods show that they contain a great number
of neurofibrils. They have three or four thick protoplasmic processes
divided into branches that become gradually thinner, and whose terminal
endings are sometimes difficult to follow in light microscopy. These
thin branches end in the plexus (neuropile) known as the outer plexiform layer. The protoplasmic processes of neighboring cells intermingle
and establish abundant contacts among themselves, but intracellular
communication does not exist between them, at least inside the limits

Figs. 27-30. Light micrograph of the cat retina prepared by Gallego's technique (Gallego, 1975a). The plexus formed by the axonless horizontal cells is shown at different distances from the area centralis.(27) Peripheral retina.(28-30) Intermediate zones between the peripheral retina and the area centralis, temporal side

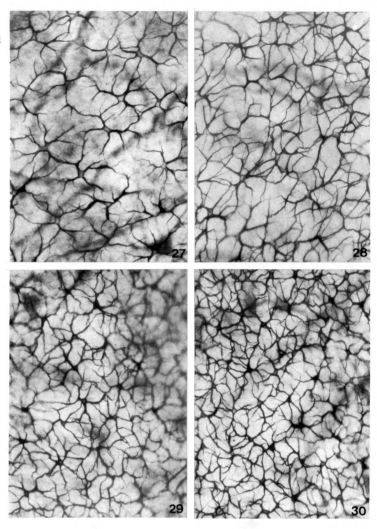

allowed by the light microscopy. The outer dendritic processes of the bipolar cells and of the Müller cells cross through the free spaces left between them. These cells do not have an axis cylinder; we can certify this after the meticulous study in preparations in which the thinnest axis-cylinders of other cells were stained."

Before reaching the conclusion that this cellular type was different from the horizontals with axon, we studied their distribution in the cat retina (Orellana and Gallego, 1959), verifying that their number in the cat peripheral retina was $100 - 140/mm^2$; $250 - 300/mm^2$ in the yuxtapapillar zone; and $500/mm^2$ in the area centralis.

By using the Golgi method and in whole-flat mounted preparations these cells can be observed isolated (Figs. 31-34). We have verified that they are more frequently stained when using the Colonnier modification, then giving partial images of the plexus which has been clearly shown with the neurofibrillar methods. The somata of these cells have dia-

42

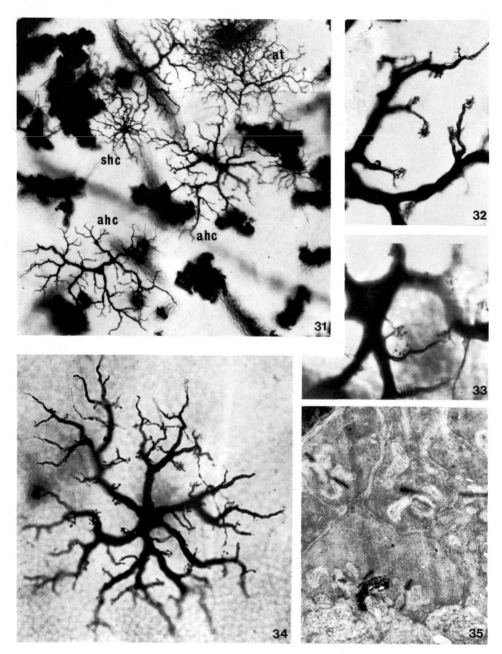

Figs. 31-35. Light micrographs of the cat retina in flat mounted Golgi-Colonnier
preparations. (35) Golgi-EM electron micrograph. (31) In this micrograph two axonless horizontal cells ahc one short axon horizontal cell shc and a horizontal cell
axon terminal at are shown. (32,33) The clusters of knobs of an axonless horizontal
cell are shown in detail. (34) Axonless horizontal cell. (35) Ultrathin section
through cone synaptic body showing the silver chromate precipitate in one of the
lateral elements of a triad. Golgi-EM study of an axonless horizontal cell

meters of about 10 - 14 μ and from which several cell processes, three
to six in number extend, which branch dichotomously; originally these
processes are as thick as the cell body, making it difficult to discern the morphological limits of the latter. From these cell processes
extend, from time to time, thin fibers, directed towards the feet of
the photoreceptors, which may be shown in well-impregnated neurofibrillar preparations. However, the terminal thin knobs are only impregnated
by means of the Golgi method. The number of terminal knobs in each
fibrilla is one to four.

The retinal area covered by the cell processes of the axonless horizontal cells varies according to the different retinal location. In
the peripheral zone the area is 150 to 200 μ in diameter, but in the
area centralis not larger than 50 μ. In any case the area covered is
practically circular, causing considerable overlapping between the
areas covered by contiguous cells.

There are morphological differences between these cells depending on
the animal studied; the cell body is bigger in the rabbit with fewer
but thicker dendrites whose ramifications form a thicker plexus than
that of the cat. In the dog these cells closely resemble those of the
cat, whereas in the guinea pig they are more similar to the corresponding cells in the rabbit; in the bull, the axonless horizontal cells
have a bigger soma and shorter dendrites.

Electron microscopy examination of these axonless cells in the cat and
rabbit retinae (Sobrino and Gallego, 1970) shows that they can be identified, in perpendicular sections, by the size of the cell body which
is bigger than that of the bipolar cells, and by the size of their
cell processes which extend radially from the perikaryon. The size of
these cell processes broader and thicker than those of the horizontal
cells with axon, makes it difficult to determine the limits of the
cell body. They appear as clear cells, located in the outer limit of
the inner granular layer. Horizontal sectioning reveals their typical
stellate shape and the very thick cell processes which extend from the
soma and form a very evident plexus.

The nucleus is irregular in shape, with frequent indentations of fine
granulated chromatine, and a well organized nucleolus. The cytoplasm
contains a few mitochondria, primarily located in the vicinity of the
nucleus. In its proximity two Golgi complexes of the vesicular type
can be seen. A characteristic of these cells is the absence of a rough
endoplasmic reticulum and free poliribosomes. The scarceness of cytoplasmic structures gives to those cells their clear aspect when observed in electron microscopy.

The perikaryon and cell processes are full of tubuli and thin filaments that cannot be followed for long distances, thus suggesting an
undulated course. Between these filaments and tubuli, small granules
(glycogen?), dense core vesicles and multivesicular bodies can be discerned. The plexus formed by the axonless horizontal cell bodies and
their processes shows numerous membrane-to-membrane contacts of the
gap junction type between the processes themselves, between the processes and the cell body and between nearby cell bodies.

This type of axonless horizontal cell contacts the cone synaptic bodies and according to Kolb (1974), the terminal branches of the cell
processes form the lateral components of the cone triads. Our studies
with electron microscopy lead us to the conclusion that on the contrary, the thin endings of their cell processes would penetrate into
the rod synaptic complexes (Gallego, 1971). However, Golgi-EM studies
using the West technique convinced us of the correctness of Kolb's

description; the axonless horizontal cell dendrites penetrate into the cone synaptic complexes and form the lateral elements of the triads.

Primates. With neurofibrillar methods it has not been possible to stain the axonless horizontal cells in the primate retina. However, by using the Golgi method (Colonnier modification) we have been able to show the existence at the inner granular layer, mainly at its sclerad level, of axonless cells which could correspond to the horizontal axonless cells found in the retina of lower mammals (Gallego, 1975b, Figs. 60-62).

The soma of these cells is 7 - 9 μ in diameter and gives out three to five cell processes, which branch several times to cover retinal areas of about 100 - 150 μ in diameter. In some instances thin collaterals terminated by a thick knob have been seen arising from the cell processes. Perpendicular sectioning reveals this cell type preferentially located at the level of the outer plexiform layer, but it can also be seen at different levels in the inner granular layer. As a first approach they resembled microglial cells but their cell processes, spreading always horizontally and extending thin branches with terminal knobs, have never been seen in our studies on microglia. To demonstrate that these cells are horizontal cells it is necessary to show their relation with the photoreceptors. For this purpose we have reembedded Golgi impregnated cells for further analysis in electron microscopy. The results obtained up to now are not conclusive.

Avian Retina. In both diurnal and nocturnal birds we have found only the type of horizontal cell with axon, namely the "brush-shaped" horizontal cell; the structure named by Cajal "stellate" horizontal cell is in fact only the enlarged axon terminal of the referred cell. There are some differences between the horizontal cell with axon in the nocturnal retina, and the corresponding cell in the diurnal retina that refer to some morphological features of both the soma and the axon terminals as well as to the axon length (Gallego et al., 1975a).

Diurnal Birds. In perpendicular sections of Golgi-stained retinae two clear zones, below the photoreceptor synaptic bodies, can be distinguished when no horizontal cells appear impregnated. When impregnation of the horizontal cells does occur the internal, or vitread zone is shown to be formed by the horizontal cell bodies of the brush-shaped type described by Cajal, whereas the external or sclerad zone is occupied mainly by the structures described by Cajal as "stellate" horizontal cells (Fig. 41).

In chicken and eagle retinae the horizontal cells (Fig. 42) show short dendrites directed towards the synaptic bodies of the photoreceptors, the soma is 8 - 10 μ in diameter. The dendrites terminate in small clusters of knobs (Figs. 43-47). The axon terminal appears located more externally than the body of the horizontal cells and is a flattened structure (Figs. 44-48), which shows very thin isolated terminal fibers directed towards the photoreceptor synaptic bodies.

In some cases the horizontal cells show dendrites not directed towards the photoreceptors but proceeding laterally and even towards the inner granular layer where the synaptic bodies of single cones, both straight and oblique, are located.

In flat whole mounts of Golgi-impregnated retinas a complete impregnation of the horizontal cells, including soma, axon and axon terminals, can be obtained. The diameter of the retinal area covered by the dendrites is approximately circular and about 30 μ in diameter; very

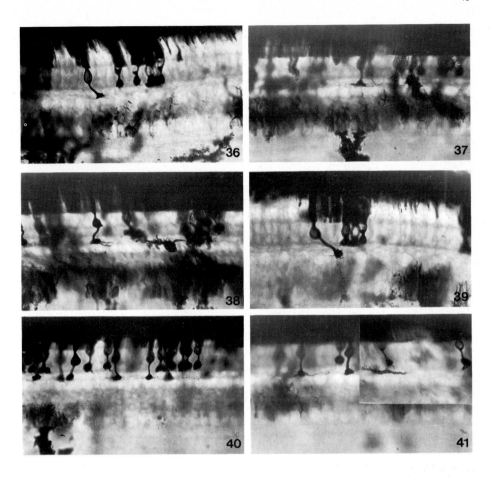

Figs. 36-41. Light micrographs of perpendicular sections of the eagle retina prepared by the Golgi method.(36,39) Two oblique cones.(37,38,41) Straight cones.
(40) Rod and double cone pedicles lie at the external row and straight cone pedicles
at the intermediate row.(38) Short axon horizontal cell ("brush-shaped" cell).
(41) Inset: axon terminal of a brush-shaped horizontal cell

often elongated horizontal cell bodies with dendrites arising from the
opposite poles can be seen (Fig. 45); regardless of the cell body
shape (elongated or round) the characteristic of this cell type is
the termination of its dendritic tree, six to eight dendrites per
cell, in clusters of four to six small knobs, similar to those found
in the horizontal cell dendrites of the mammal retina. The area covered by each cluster of knobs is 4 - 6 μ in diameter, that is approximately the same size of the area of the cone synaptic bodies, both
single and double cones.

The axon is short; in complete impregnated cells, its length ranges
between 100 - 300 μ. In all probability the shortness of the axon makes
it easier to stain these cells completely in the avian retina, what
was achieved for the first time by Cajal (Fig. 1). The axon terminals
characteristically appear as wide enlargements of the axon. When the
impregnation falls short of these axon terminals and they are examined under the light microscope as isolated elements, their size

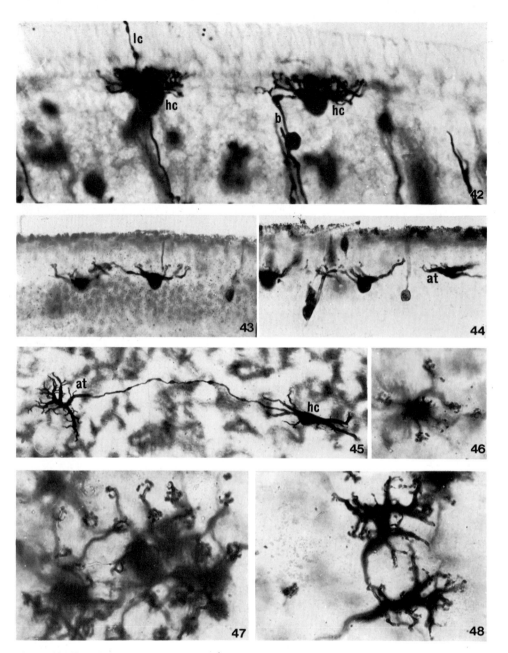

Figs. 42-48. Light micrographs of the eagle retina prepared by the Golgi-Colonnier method.(42-44) Perpendicular sections.(45-48) Whole flat mounts (Gallego, 1975b). (42) Two brush-shaped horizontal cells hc. A Landolt club lc of a superimposed bipolar cell is shown.(43,44) Peripheral zone of the retina showing horizontal cells (brush-shaped cell) and axonic terminal at.(45) Complete cell shown by a montage: horizontal cell body hc and axon terminal at.(46,47) The clusters of knobs of the brush-shaped horizontal cell dendrites, shown in detail at two different magnification.(48) Axon terminals (Cajal's stellate cells) of two brush-shaped horizontal cells

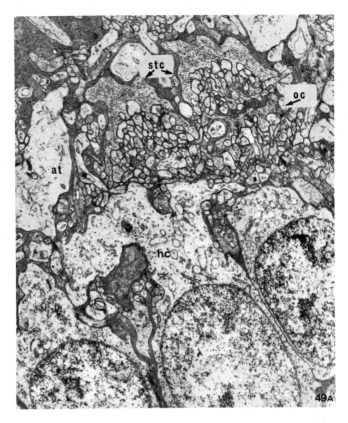

Fig. 49A. Electron micrograph of the chicken retina. Perpendicular section. Row of brush-shaped horizontal cells hc; axonic terminals at; oblique cone oc; straight cones stc

and structure give them the misleading appearance of a cell soma (Fig. 48). These axonic terminals ramify into thinner branches that spread out, giving origin to delicate endings in the form of small knobs directed towards the synaptic bodies of the photoreceptors (Figs. 44-48).

Reembedding of impregnated horizontal cells in whole flat mounted retinae and subsequent perpendicular sectioning reveals the location of the soma and axon terminals. The soma has the morphological features of the typical brush-shaped horizontal cells and is located in the inner or vitreal zone previously described, whereas the axon terminals are located in the external or sclerad zone of the outer plexiform layer and show the same morphology as the stellate horizontal cells described by Cajal (Fig. 48).

Further evidence on the location of the soma and axon terminals is revealed by electron microscopy studies. In perpendicular sections (Fig. 49A) the horizontal cell somata can be seen located in the external row of the inner granular layer and are closely packed. Between the somata of horizontal cells and the synaptic bodies of the photoreceptors thick cell processes are seen (Fig. 49B) obliquely

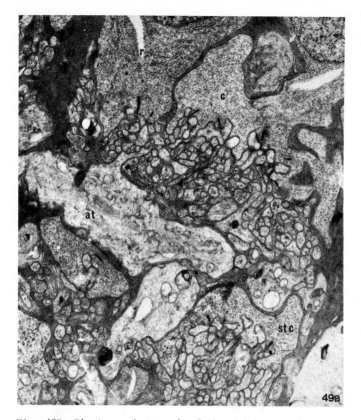

Fig. 49B. Electron micrograph of the chicken retina. Perpendicular ultrathin section showing one axon terminal at located between the synaptic bodies of the external row r: rod; c: cone, and the intermediate row stc: straight cones

orientated sections show these cellular processes intermingled with scattered photoreceptor synaptic bodies (Fig. 50). In transverse sections, these structures have the morphological aspect of cell bodies but in serial sections they show no nuclei and moreover their cytoplasm, filled with very thin neurofilaments, is not that of a cell soma. This evidence supports our identification of such structures with the axon terminals of the horizontal cells.

Nocturnal Birds. As in diurnal birds the retina of nocturnal animals has a single type of horizontal cell with axon. The general description of the horizontal cells previously made in the diurnal retina applies also to nocturnal birds with small differences in the morphological features of the soma, axon terminals and axon length. The location of the horizontal cell somata and of the axon terminals at different levels in the outer plexiform layer is similar to that found in diurnal birds retina; as observed in Golgi preparations and in electron microscopy analysis, the cell somata lie internally to the axon terminals.

In perpendicular sections of Golgi-impregnated retinas (Fig. 51), the horizontal cell soma has the typical appearance of the brush-shaped cell type described by Cajal and it closely resembles the correspond-

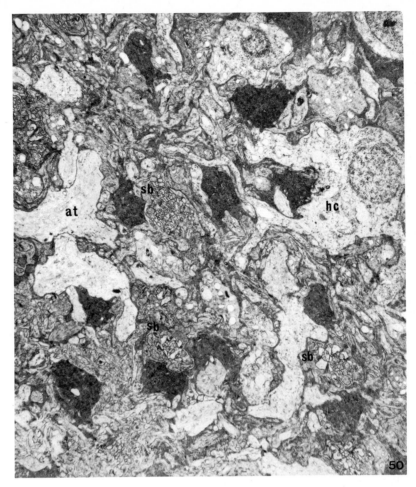

Fig. 50. Electron micrograph of the chicken retina. Oblique ultrathin section showing the cell body of a horizontal cell <u>hc</u> and an axon terminal <u>at</u>. Cone synaptic bodies <u>sb</u>

ing cell in diurnal birds' retina, except for its smaller diameter of 6 - 8 μ. The dendrites are directed vertically towards the photoreceptor synaptic bodies. Due to this arrangement the area covered by the dendrites is slightly smaller than that in the eagle retina. The axon terminals are somewhat larger and thicker in the "C. noctua" than in diurnal birds but the "A. flammeus" retinae show thinner axon terminals with a rich branching.

In whole flat-mounted retinae stained with the Golgi-Colonnier technique, both the somata and the axon terminals can be easily identified (Figs. 51-57). The closely packed dendrites are directed towards the photoreceptor synaptic bodies, and terminate in groups of two to four small knobs. These knobs do not form the well-defined cluster of the horizontal cell dendrites seen in diurnal birds; this might well be a characteristic of these cells or failure in the impregnation. The axon terminals show a wide enlargement of the axon which gives out several thick processes that extend very thin fibers, terminating in small isolated knobs. In two completely impregnated cell where the continuity of the soma, axon and axon terminal was convincing, the axon length was 800 and 900 μ.

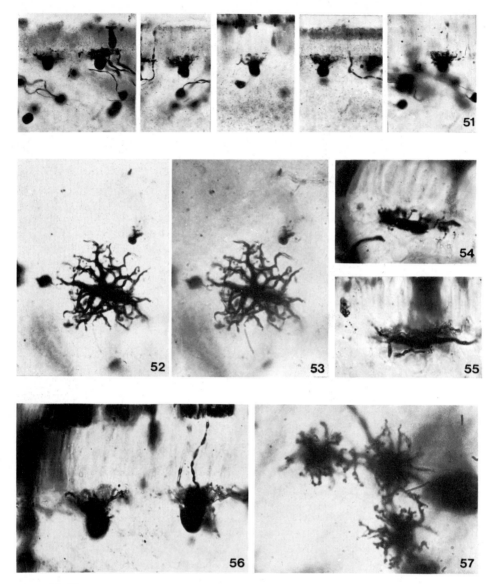

Figs. 51-57. Light micrographs of the owlet retina prepared by the Golgi-Colonnier method.(51) Perpendicular sections showing brush-shaped horizontal cells. (52,53) Horizontal cell axon terminal at two different focus levels. Flat-mounted retina.(54,55) Perpendicular sections of two horizontal cell axon terminals (Cajal's stellate cells).(56) Two brush-shaped horizontal cells as seen in a perpendicular section.(57) Three brush-shaped horizontal cell bodies as seen in a flat mounted preparation

In electron microscopy analysis of perpendicular sections (Fig. 58), the horizontal cell bodies form a well-defined row at the outer limit of the inner granular layer. Between the horizontal cell somata and the synaptic bodies of the photoreceptors large structures filled with very thin neurofilaments can be seen. Serial sectioning show no

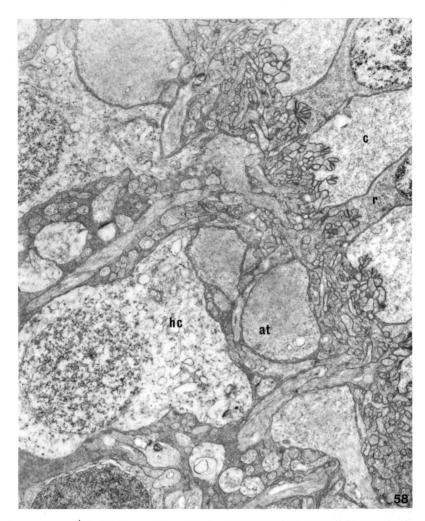

Fig. 58. Electron micrograph of the owlet retina. Perpendicular ultrathin section showing a brush-shaped horizontal cell body hc, axon terminals at and cone and rod pedicles c, r

nuclei in these structures, thus suggesting their identification with the horizontal cell axon terminals as in diurnal birds. Transversal sectioning reveals a plexus consisting of the axon terminals. This plexus is similar to that present in diurnal retinae, except for its large size and the fact that no synaptic photoreceptor bodies are seen at this level due to the absence in the nocturnal retina of straight and oblique cones which in diurnal birds are intermingled with the axon terminals (Fig. 59).

In the avian retina the relationship between horizontal cells and the synaptic bodies of the photoreceptors is difficult to study due to the existence of several types of visual cells (Fig. 60) and to the fact that in diurnal animals the photoreceptor synaptic bodies are located at three levels in the outer plexiform layer (Gallego et al., 1975b). In our study of the visual cells in the avian retina we have found that diurnal animals (chicken, eagle, and milano) have

52

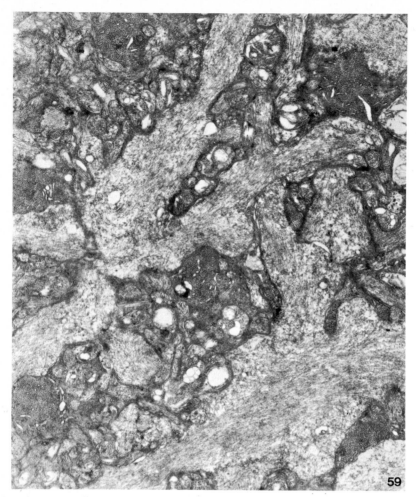

Fig. 59. Electron micrograph of the owlet retina. Transversal ultrathin section showing the axon terminals

rods with no oil droplet; double cones, the chief one with a yellow orange oil droplet and the accessory one with a greenish oil droplet, which confirm Meyer and Cooper's (1966) description of double cones; two types of single cones, the straight or cone II (Morris and Shorey classification, 1967) with a red oil droplet and the oblique or cone I with a green oil droplet. The photoreceptor synaptic bodies in diurnal birds are located at three different strata and they form three rows in the outer plexiform layer: the internal row is formed by the synaptic bodies of rods and double cones; the intermediate row consists of the synaptic bodies of single cones, straight or type II, and the internal row of the oblique cone or type I. In the nocturnal bird retina a simplified picture is found. These retinas lack both the straight and the oblique cone; they have rods with no oil droplet; the chief cone of the double cones displays a yellow-orange droplet, the accessory cone has no oil droplet; they have a unique type of single cone with a colorless oil droplet. On the other hand in the nocturnal retina all photoreceptor synaptic bodies are located at a single level in the outer plexiform layer.

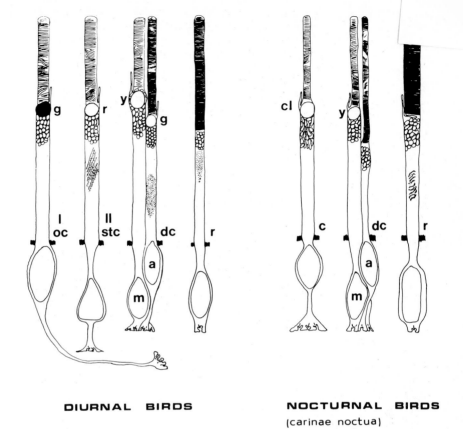

Fig. 60. Schematic drawings of the different types of photoreceptors in the avian retina. <u>Diurnal bird retinae</u>: oblique cone <u>oc</u> or <u>I</u>, straight cone <u>st</u> or <u>II</u>, double cone <u>dc</u>, consisting of the main <u>m</u> and the accessory <u>a</u> cones, and rod <u>r</u>.
<u>Nocturnal bird retinae</u>: single cone <u>c</u>, double cone <u>dc</u>, main <u>m</u> and accessory <u>a</u> member of the double cone and rod <u>r</u>. The oil droplets are green <u>g</u>, red <u>r</u>, yellow <u>y</u> and colorless <u>cl</u>

Somata and axon terminals of horizontal cells in both diurnal and nocturnal birds' retina impregnated with the Golgi-Colonnier method, have been reembedded for sectioning and electron microscopy observation. In nocturnal animals the dendrite terminals are seen incorporated to the cone triads of the single cones and of the chief member of the double cones. The axon terminals, are seen as isolated knobs, and are incorporated to the rod synaptic complex. This arrangement is, therefore, similar to that observed in mammal retinas. In diurnal retinas the dendrites of the horizontal cell contact both members of the double cone, the straight and the oblique cone: on the other hand the axon terminals contact the rods and also both members of the double cone. However, before reaching definitive conclusions, further investigations need to be done.

1.2.5. Discussion

It is of great interest to neurophysiologists to know the interrela-
tions established between the horizontal cells and the visual and bipolar cells, especially after the Svaetichin (1953) finding of the
"S" potentials and the subsequent demonstration of these potentials
being generated in the horizontal cells (McNichol and Svaetichin,
1958; Kaneko, 1970). Later a wealth of information emerged from the
investigation on the "S" potentials. To interpret these data correct-
ly the current confusion that exists concerning the horizontal cells
should be understood. Several reasons account for this confusion:

1. The different nomenclature used by investigators when referring to
   the horizontal cells.

2. The difficulty in cell-type identification when the horizontal
   cells are stained by different techniques and studied in different
   visualizing planes: whole flat mounts and perpendicular sections
   of the retina.

3. The tendency to generalize the data obtained in the study of retinae of an animal species to the retinae of other species of genus.

To illustrate the first point we should mention the several names
found in the literature that refer to the horizontal cells of terrestrial vetrebrates: "outer" and "inner", "brush-shaped" and "stellate"
horizontal cells (Cajal, 1893); silver-stained horizontal cells or
"amacrine cells of the outer plexiform layer" (Gallego, 1964); "large"
and "small" horizontal cells (Dowling et al., 1966); "A" and "B" horizontal cells in the primates retina (Boycott and Dowling, 1969); "A"
and "B" horizontal cells in the lower mammal retina (Fisher and
Boycott, 1974).

The results obtained in the present study strongly suggest that in
mammals and avian retinae there is only one type of horizontal cell
with axon, and not two types - external and internal - as was postulated
by Cajal. Moreover in some mammal retinae there is also a second type,
the axonless horizontal cell which was named in our earliest studies
"amacrine cell of the outer plexiform layer" (Fig. 61).

Subsequent to our description of the axonless horizontal cell in the
silver-impregnated retinas of lower mammals, two clearly different
types of horizontal cell were described in electron microscopy analysis of these retinas (Sobrino and Gallego, 1970; Gallego, 1971;
Fisher and Boycott, 1974) as well as in Golgi-impregnated sections
(Dowling et al., 1966; Gallego, 1971) and in Golgi flat mounts
(Boycott and Kolb, 1973; Gallego, 1975a) of the same retinas.

Dowling et al. (1966) identified the axonless horizontal cell, previously described in our 1964 publication, with what they call large
horizontal cell ("A" on their drawings). In later studies (Gallego,
1971), we also agreed with these authors as far as the identification
is concerned.

The horizontal cell with axon, or small horizontal cell according to
Dowling et al. (1966) is without any doubt, the classical type of
horizontal cell described by Cajal. Fisher and Boycott (1974) introduced a new nomenclature for the cat and rabbit horizontal cells impregnated by the Golgi method: A and B horizontal cells. The A cell
(Fisher and Boycott, 1974; Boycott, 1975) is identical with the axonless horizontal cell first described by us (Gallego, 1964) using

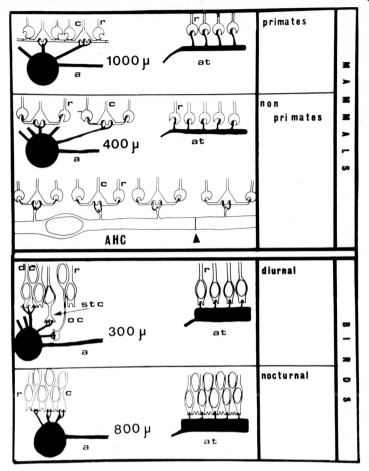

Fig. 61. Schematic drawings which summarize connections between horizontal cells and photoreceptors in the mammal and bird retinae. In black, short axon horizontal cell; AHC axonless horizontal cells; c cone; r rod; oc oblique cone; stc straight cone; dc double cone; a axon; at axon terminal. Arrowhead: tight junction. Average length of the axons is indicated in μ

neurofibrillar methods, thus confirming the previous identification made by Dowling et al. (1966) and by us (Gallego, 1971).

Due to the fact that in the terrestrial mammal retina as well as in the avian retina a single type of horizontal cell with axon seems to exist, the distinction of A and B horizontal cells in the primate retina (Boycott and Dowling, 1969) and the denominations of small and B cells given by Dowling et al. (1966) and Fisher and Boycott (1974) to the horizontal cell with axon in the cat and rabbit retina are superfluous and misleading. It would be more convenient to name this cell type "short axon horizontal cell" of "Cajal's horizontal cell".

The axonless horizontal cell of the lower mammal retina was first named (Gallego, 1964) "amacrine horizontal cell" (a, not + Gr. makros, long + is, inos, fiber), in agreement with the ethymology which led Cajal to name "amacrines" those cells without axon located at the sclerad row of the inner granular layer and whose cellular processes

spread through the inner plexiform layer. The amacrine horizontal cell
has also been named large horizontal cell (Dowling et al., 1966) and
A horizontal cell (Fisher and Boycott, 1974). If the resistance to
using the name amacrine horizontal cell is aimed at avoiding confusion
with the classical amacrine cells of the inner plexiform layer (see
Rodieck, 1973), perhaps it would be more convenient to name this cell
type <u>axonless horizontal cell</u>.

1.2.5.1 Short Axon Horizontal Cell (Cajal's Horizontal Cell)

The morphological differences that exist between the horizontal cells
located in the central and peripheral zones in mammal retina explain
Cajal's distinction of two types of horizontal cells: external and
internal horizontal cells. Cajal's investigations were made using
mainly the Golgi method, first fixing the retina by his enroullement
technique, with subsequent sectioning in several planes; this technique gave him only a partial view of the horizontal cells as seen in
perpendicular sections, which did not allow him to detect the morphological differences that exist between centrally and peripherally located horizontal cells; as a consequence he concluded that two types
of horizontal cell were present in the mammalian retina. Furthermore
he was influenced by the duplicity theory, and the existence of two
types of horizontal cells, one related to rods and the other to cones
was for him a very attractive idea. Credit must be given to Polyak
(1941) who, in the primate retina, first demonstrated that there is
a single type of horizontal cell, despite the fact that there are
great morphological differences between cells situated in central and
peripheral areas of the retina. The analysis of entire impregnated
retinas in flat-mounted preparations shows clearly that both in primate and in lower mammal retinae there is a single type of horizontal
cell with axon, as stated in the present study, in which the graded
morphological differences of those cells from the parafoveal zones to
the periphery are illustrated.

In the avian Golgi-impregnated retinae the peculiar morphology of the
axon terminals, which imitate a cell body, led Cajal to consider these
structures as a distinct cell type. Interestingly enough Cajal achieved,
even in retinal sections, a complete impregnation of the "brush-shaped"
cells, including soma, axon and axon terminals, due probably to the
shortness of their axon. In perpendicular sections of the avian retina
Cajal interpreted some of the axon terminals as stellate cells, in
agreement with his idea of the existence of two types of horizontal
cells in all vertebrate retinas.

Kolb (1970) confirmed the results of Polyak in the primate retina and
concluded, studying Boycott and Dowling Golgi-stained retinas, that
there is only one type of horizontal cell. In lower mammals a single
type of horizontal cell with axon has also been found.

In the avian retina there is only one type of horizontal cell, whose
axon terminals are large structures which were erroneously described
by Cajal as stellate cells. The length of the axon is shorter than
in mammalian retina (100 - 300 µ in diurnal birds and 800 - 900 µ in nocturnal ones).

1.2.5.2 Axonless Horizontal Cell ("Amacrine Horizontal Cell of the Outer Plexiform Layer" (Gallego, 1964)

The axonless horizontal cells described by us (Gallego, 1964) show a
clear differential characteristic from the horizontal cells with axon.

They form a plexus-like layer which spreads throughout the mammal retina, primates not included. The similarity of this plexus of horizontal cells with that formed by the first row of horizontal cells in
some fishes retina, particularly in the selachii "gynglimostoma cerratum", (Gallego, 1972, 1975b) is amazing. In the cat and rabbit retina,
in contrast with the short axon horizontal cells with axon which appear as isolated elements, the axonless horizontal cells establish between themselves wide membrane-to-membrane contacts, which suggest
functional electrical coupling (Sobrino and Gallego, 1970).

The axonless horizontal cell was not described either by Cajal (1893,
1909, 1933b) or Balbuena (1936), though both investigators have published micrographs and drawings of them from silver-reduced preparations (Cajal, 1909, Figs. 194 and 198; Balbuena in Rebslob, 1939,
Figs. 74-76); they consider such cells as "external" horizontal cells
with axon. However, Cajal's classical description of the horizontal
cells was made using the Golgi method. Is is understandable that Cajal
and Balbuena made erroneous interpretations, as the Golgi method very
rarely impregnates the axonless cell type (Dowling et al., 1966;
Gallego, 1971), whereas the reduced-silver methods only stain the
horizontal axonless cells which are very rich in neurofibrils (Gallego,
1964, 1965). Balbuena, who achieved very good stainings of the axonless horizontal cell, was convinced that it was Cajal's classical
horizontal cell with axon. On the other hand, Cajal's and Balbuena's
observations were carried out on small retinal zones sectioned in all
directions when using the enroullement technique or in perpendicular
section. However, using the technique of impregnating the entire retina (Gallego, 1953) which gives a panoramic view of the tissue, the
axonless horizontal type became manifest (Gallego, 1964, 1965, 1971,
1975a; Leicester and Stone, 1967; Honrubia and Elliot, 1969; Boycott,
1975).

1.2.5.3 Connections of the Horizontal Cells with the Photoreceptors

It has long been known, since Cajal's analysis of Golgi-impregnated
retinas, that the horizontal cells make contacts with the feet
(synaptic bodies) of the photoreceptors; Cajal (1911) thought that
the internal horizontal cells were related to cones, and the external
horizontal cells to rods, although in later studies (Cajal, 1933a) he
hesitated on this issue. The electron microscopy studies by Missotten
(1965) and the analysis made by Stell (1965) using the combined Golgi-
EM method have been conclusive in demonstrating that the horizontal
cell processes form the lateral components of the cone triads and that
they were also enclosed in the synaptic complex of the rods. However,
due to imperfect knowledge of the horizontal cell morphology and the
existence of a variety of visual cells in the vetrebrate retinas,
these findings were not sufficient to establish, from a functional
standpoint, the connections of the horizontal cells with the photoreceptors.

In the past five years, by using the Golgi-EM method, it has been
firmly established that in the mammalian retina the dendrites of the
horizontal cells with axon make contact with the cones and their axon
endings with the rods as was first shown by Kolb (1970) in the primate
retina. In our preparations we have observed groups of cones related
by way of the short axon horizontal cells to a large group of rods
400 μ distant in lower mammals and more than 1 mm away in the primates.

The variety of visual cell types that exist in the avian retina makes
it difficult to determine the synaptic relationship between those cells
and the dendrites and axon terminals of the horizontal cells. It is

particularly difficult in the diurnal birds' retina, where the synaptic bodies of the photoreceptors lie at different levels in the outer plexiform layer. However, in the nocturnal birds' retina, whose photoreceptor synaptic bodies lie at a single level in the outer plexiform layer, as shown in this study, it is possible to observe that the dendrites of the horizontal cells form the lateral components of the cone triads and that the axon endings penetrate the synaptic complex of the rods, identical organization to that found in mammals.

The axonless horizontal cells have been shown only in lower mammalian retina. In the cat and rabbit retina it has been demonstrated that the thin collateral branches of the axonless horizontal cells contact the cone synaptic bodies (Kolb, 1974). Our own results confirming this finding are shown in this work.

1.2.5.4 S-Potentials and Horizontal Cells

The functional role of both types of horizontal cells in lower mammalian retina could be determined by the analysis of the S-potentials, as has been done in the intact eye (Steinberg, 1969a,b,c, 1971) and in the isolated perfused eye of the cat (Niemeyer, 1973; Niemeyer and Gouras, 1973; Nelson et al., 1975). Steinberg's identification of the horizontal cells as a source of the recorded S-potentials has been discussed by Stell (1971) who doubted that they were obtained from horizontal cell recordings.

The existence of two types of S-potential in the cat retina has been claimed (Steinberg, 1971; Niemeyer and Gouras, 1973): S-potentials which are exclusively rod-dependent, and S-potentials which combine the action of both rods and cones. Nelson et al. (1975) recorded S-potentials from two different neuronal structures, which were identified using Procion Yellow injection from the recording electrode, as the cell body and the axon terminal arborization of the short axon horizontal cell. These investigators postulate that both axon terminal and the cell body receive input from rods and cones. The rods are responsible for 80% of the peak response recorded from the axon terminal and the cones for the remaining 20%, whereas the contribution to the peak response recorded from the cell body is 50% for both cones and rods.

The above-mentioned authors omit any clear reference to the existence of two different types of horizontal cells in the cat retina. According to our morphological data the two types of horizontal cells should behave functionally in a different way because of their mutual relationship and their different connections with the photoreceptors. The morphology of the horizontal cell with axon is that of a short axon neuron, whose dendrites make contact with cones and whose axon terminals make contact with rods; the axonless horizontal cells display wide membrane-to-membrane contacts between themselves, they form a plexus, and they make contact with cones only. The axonless horizontal cells, due to their size and the fact that they form a plexus throughout the retina, constitute the preferential element for microelectrode penetration. As a consequence, when recording from cat horizontal cells, the possibility that the microelectrode has impaled an axonless horizontal cell should be kept in mind. Such a possibility has not been taken into account by the above-mentioned investigators.

Nelson et al. (1975, Fig. 2a) record S-potentials from a cell which when stained with Procion Yellow was identified as a short axon horizontal cell. It seems to us that this cell shows morphological features that are characteristic of the axonless horizontal cell-type (Fig. 34);

namely, it shows no axon and the cell body extends five thick cell
processes which branch out dichotomously. While staining of the axon
would have been conclusive for identification purposes, failure to
stain it does not exclude its presence, due to the nature of intracellular techniques. On the other hand the short axon horizontal cell
shows typically thinner and more numerous dendrites (Fig. 22) than
the cell stained by Nelson et al. (1975).

Chances of blindly impaling the short axon horizontal cells are scarce
because of their dispersed distribution in the outer plexiform layer
and the small size of their cell body. Furthermore, S-potentials being generated in mammals' short axon horizontal cells are not reported
in the literature. Nevertheless, Nelson et al. have been able to record S-potentials from axon terminals, which are very fine structures
especially when compared to the cell body of the short axon horizontal
cells. Therefore to interpret such records with accuracy, consideration should be given to the existence of two types of horiontal cells
in the cat retina.

In our opinion Nelson et al. (1975) potentials have been recorded from
two different kinds of cell, the axon terminals of the short axon horizontal cell and the cell body of the axonless horizontal cell; as a
consequence their statement of functional independence between axon
terminals and cell bodies of the short axon horizontal cells seem to
have an erroneous basis. It remains to be shown how a group of cones
are functionally related to a larger group of rods through the horizontal cell axon, 400 µ long; the possibility of spikes being generated in the horizontal cell body and propagated along the axon to the
terminals has not been excluded.

Interestingly enough, the cat S-potentials show mixed components of
cones and rods; however, the cell structures generating those potentials make contact with single receptors: the axonless horizontal
cells with cones, the short axon horizontal cell dendrites with cones
and the axon terminals of the latter with rods. The records of Nelson
et al. (1975) from the body of one horizontal cell, whose dendrites
are connected only to cones, show a mixed rod-cone S-potential.
Niemeyer and Gouras (1973) also show that the cone contribution to
the missed rod-cone S-potential is much stronger than that of the
rods. The S-potentials recorded from the axon endings of a short axon
horizontal cell by Nelson et al. (1975) also show a mixed rod-cone
participation but in this case the rod contribution is stronger than
that of the cones. Such mixed rod-cone S-potentials recorded from
structures connected to a single type of photoreceptors strongly suggest a direct interaction between rod and cones.

Two kinds of contact between rods and cones have been described, the
well-known desmosome-like contacts and the contacts established by
the cone basal filaments which penetrate the synaptic complex of rods
(Gallego, 1971, 1975b, Fig. 16). The possibility exists of a rod-cone
interaction by way of these connections in such a way that the cones
signal transmitted to the dendrites of the horizontal cells could be
modified by the rods. The rod dominant response recorded from the
axon terminals, which are connected only to rods and also show a cone
participation, could be explained on the same basis.

The mammalian and avian short axon horizontal cell, whose dendrites
make contact with cones and whose axon terminals make contact with
rods, could be the element by which the interaction between both types
of visual cells is established as suggested by Brown and Murakami
(1968). Electrophysiological evidence for such interaction through

the short axon horizontal cell has not yet been shown: in the avian
retina not even S-potentials from the horizontal cells have been recorded.

Acknowledgements. This work was supported by the European Research
Office, Grant number DA-ERO-59-73-GOO43. We are grateful to Mrs. Rosa
Ayllón for technical assistance.

1.2.6. References

Balbuena, F.F.: Nota previa: Una fórmula para la aplicación del método de Cajal
a los cortes de retina. Trab. Lab. Invest. Biol. Univ. Madrid 20, 31-39 (1922)
Balbuena, F.F.: Conexiones de los conos y bastones al nivel de la capa plexiforme
externa. Arch. de Oft. Hisp. Amer. 35, 337 (1936)
Boycott, B.B.: The morphology of the horizontal cells of the retina of the domestic
cat and their comparison with those of other vertebrates. (Quoted by Fisher and
Boycott). Proc. R. Soc. London 186B, 317-331 (1974)
Boycott, B.B., Dowling, J.E.: Organization of the primate retina: light microscopy.
Phil. Trans. R. Soc. London 225B, 109-184 (1969)
Boycott, B.B., Kolb, H.: The horizontal cells of the rhesus monkey retina.
J. Comp. Neurol. 148, 115-140 (1973)
Brown, K.T., Murakami, M.: Rapid effect of light and dark adaptation upon the re-
ceptive fields organization of S-potentials and late receptor potentials.
Vis. Res. 8, 1145-1171 (1968)
Cajal, S. Ramon y: La rétine des vertebrés. La Cellule 9, 119-225 (1893)
Cajal, S. Ramon y: Textura del sistema nervioso del hombre y los vertebrados.
Moya, Madrid (1899-1904)
Cajal, S. Ramon y: Células nerviosas de cilindro-eje corto. Trab. Lab. Invest. Biol.
Univ. Madrid 1, 151-157 (1902)
Cajal, S. Ramon y: Un sencillo método de coloración selectiva del retículo proto-
plásmico y sus efectos en los diversos órganos nerviosos. Trab. Lab. Invest.
Biol. Univ. Madrid 2, 129-221 (1903)
Cajal, S. Ramon y: Histologie du système nerveux de l'homme et des vertebrés.
Maloine. Paris (1909-1911)
Cajal, S. Ramon y: Los problemas histofisiológicos de la retina. XIV Concilium
Ophtalmologicum. Madrid (1933a)
Cajal, S. Ramon y: La rétine des vertebrés. Trav. Lab. Rech. Biol. Univ. Madrid 28
(appendice) 1:144 (1933b)
Cajal, S. Ramon y, De Castro, F.: Elementos de técnica micrográfica del sistema
nervioso. Madrid: Tipografía Artîstica 1933, p. 135
Colonnier, M.: The tangential organization of the visual cortex. J. Anat. London 98,
327-344 (1964)
Crescitelli, F.: The visual cells and visual pigments of the vetrebrate eye.
In: Handbook of Physiology. Berlin: Springer 1972, Vol. VII/I
Dogiel, A.S.: Ueber die Retina des Menschen. Intern. Monatss. Anat. Physiol. 1,
143-151 (1884)
Dowling, J.E., Brown, J.E., Major, D.: Synapses of horizontal cells in rabbit and
cat retinas. Science 153, 1639-1641 (1966)
Fisher, S.K., Boycott, B.B.: Synaptic connexions made by horizontal cells within
the outer plexiform layer of the retina of the cat and the rabbit.
Proc. R. Soc. London 186B, 317-331 (1974)
Gallego, A.: Procedimiento de impregnación argéntica de la retina entera.
An. Inst. Farm. Esp. II, 171-176 (1953)
Gallego, A.: Description d'une nouvelle couche cellulaire dans la rétine des mammi-
fères et son rôle fonctionnel possible. Bull. Ass. Anat. 49, 624-631 (1964)
Gallego, A.: Connexions transversales au niveau des couches plexiformes de la
rétine. Actualités Neurophysiologiques, Paris: Masson 1965, Sér. 6, pp. 5-27
Gallego, A.: Horizontal and amacrine cells in the mammal's retina.
Vis. Res. 11 (3), 33-50 (1971)

Gallego, A.: Nota sobre las células horizontales de la retina del tiburón Gynglimostoma cerratum. Arch. Fac. Med. Madrid 21 (4), 237-245 (1972)

Gallego, A.: Las células horizontales de la retina de los mamíferos terrestres.
Trab. Inst. Cajal Invest. Biol. 65, 227-257 (1975a)

Gallego, A.: Las células horizontales de la retina de los vertebrados.
Real Acad. Nac. de Med., Instituto de España (1975b)

Gallego, A.: Horizontal cells of the terrestrian vertebrate retinae. Proc. 3rd
Intern. Symp. Structure of the Eye. Jap. Ophthalm. (suppl.) (1975c), in press

Gallego, A., Baron, M., Gayoso, M.: Horizontal cells of the avian retina.
Vis. Res. 15, 1029-1030 (1975a)

Gallego, A., Baron, M., Gayoso, M.: Organization of the outer plexiform layer of
the diurnal and nocturnal bird retinae. Vis. Res. 15, 1027-1028 (1975b)

Gallego, A., Perez Arroyo, M.: Photoreceptors and horizontal cells of the turtle's
retina. Proc. 3rd Intern. Symp. Structure of the Eye. Jap. J. Ophthalm. (suppl.),
(1975) in press

Gallego, A., Perez Arroyo, M.: Horizontal cells of the teleostei and selachii
retinae (1976, in prep.)

Gallego, A., Sobrino, J.A.: Horizontal cells of the monkey's retina.
Vis. Res. 15, 747-748 (1975)

Golgi, C., Manfredi, N.: Annotazioni istologiche sulla retina del cavallo.
R. Acad. di Med. di Torino V. Vercelino, Torino (1872)

Honrubia, F.M., Elliot, J.H.: Horizontal cells of the mammals retina.
Arch. Ophthal. New York 82, 98-104 (1969)

Kaneko, A.: Physiological and morphological identification of horizontal, bipolar
and amacrine cells in goldfish retina. J. Physiol. 207, 623-633 (1970)

Kolb, H.: Organization of the outer plexiform layer of the primate retina: Electron
microscopy of Golgi-impregnated cells. Phil. Trans. R. Soc. London 258B, 261-283
(1970)

Kolb, H.: The connections between horizontal cells and photoreceptors in the retina
of the cat: Electron microscopy of Golgi preparations. J. Comp. Neurol. 155,
1-14 (1974)

Kolliker, A.: Handbuch der Gewebelehre des Menschen. V. Auflage (1867)

Krause, W.: Die Retina. I. Die Membrana fenestrata der Retina. Intern. Monatss.
Anat. Physiol. 1, 225-254 (1884)

Lasansky, A.: Synaptic organization of cone cells in the turtle retina.
Phil. Trans. R. Soc. London 262B, 365-381 (1971)

Leicester, J., Stone, J.: Ganglion, amacrine and horizontal cells of the cat's
retina. Vis. Res. 7, 695-705 (1967)

Lorente de No, R.: Vestibulo ocular reflex arc. Arch. Neurol. Psych. 30, 245-291
(1933)

Marenghi, G.: Contributo alla fina organizzazione della retina. Memoria Reale
Accademia dei Lincei. Roma (1901)

McNichol, E.F., Svaetichin, G.: Electric responses from the isolated retinas of
fishes. Am. J. Opthal. 46, 26-40 (1958)

Merkel, F.: Ueber die menschliche Retina. Albrecht von Graefes Arch. Ophthal. 22 (4),
1-25 (1876)

Meyer, D.B., Cooper, T.G.: The visual cells of the chicken as revealed by phase
contrast microscopy. Am. J. Anat. 118, 723-734 (1966)

Meyer, D.B., May, H.C. Jr.: The topographical distribution of rods and cones in
the adult chicken retina. Exp. Eye Res. 17, 347-355 (1973)

Missotten, L.: The synapses in the human retina. In: II. Symp. über die Struktur
des Auges. Rohen, J.W. (ed.), Stuttgart: Schattauer 1965, pp. 17-28

Morris, Valerie B.: Symmetry in a receptor mosaic demonstrated in the chick from
the frequencies, spacing and arrangement of the types of retinal receptor.
J. Comp. Neurol. 140, 359-398 (1970)

Morris, V.B., Shorey, C.D.: An electron microscope study of types of receptor in
the chick retina. J. Comp. Neurol. 129, 313-339 (1967)

Müller, H.: Ueber sternförmige Zellen der Retina. Verhandl. d. Physik Med. Gess.
Würzburg 2, 216 (1851)

Müller, W.: Ueber die Stammesentwicklung des Sehorgans der Thiere. Leipzig
(1874-1876)

Muntz, W.R.A.: Inert absorbing and reflecting pigments. In: Handbook of Sensory
Physiology: Photochemistry of Vision. Berlin: Springer 1972, Vol. VII/I

Nelson, R., Lützow, A.v., Kolb, H., Gouras, P.: Horizontal cells in cat retina
with independent dendritic systems. Science 189, 137-139 (1975)
Niemeyer, G.: Intracellular recording from the isolated perfused mammalian eye.
Vis. Res. 13, 1613-1618 (1973)
Niemeyer, G., Gouras, P.: Rod and cone signals in S-potentials of the isolated
perfused cat eye. Vis. Res. 13, 1603-1612 (1973)
Ogden, T.E.: The morphology of retinal neurons of the owl monkey Aotes.
J. Comp. Neurol. 153, 399-428 (1974)
Orellana, J.M., Gallego, A.: Distribución topográfica de las células horizontales
de la retina. Actas Soc. Esp. Cienc. Fisiol. 5, 251-252 (1959)
Perez Arroyo, M., Gallego, A.: Horizontal cells of the reptiles and amphibian
retinae (1976, in prep.)
Polyak, S.L.: The Retina. Chicago: Univ. Chicago, 1941
Ranvier, L.: Traité technique d'Histologie, 6° fasc., F. Savy, Paris (1882)
Rebslob, E.: Histologie de la rétine. In: Traité d'Ophtalmologie I, 409-493.
Paris: Masson 1939
Reich, M.J.: Netzhaut des Hechtes. Jahrb. Anat. Physiol. 2, 228-229 (1875)
Retzius, G.: Beiträge zur Kenntniss der inneren Schichten der Netzhaut des Auges.
Biol. Untersuch. 1, 89-104 (1881)
Rivolta, S.: Delle cellule multipolari che formano lo strato intergranuloso o inter-
medio nella retina del cavallo. G. Anat. Fisiol. Patol. Animali, anno III,
1871, p. 185
Rochon-Duvigneaud, A.: Les yeux et la vision des vertebrés. Paris: Masson 1943
Rodieck, R.W.: The vertebrate retina. San Francisco: Freeman 1973
Schiefferdecker, P.: Studien zur vergleichenden Histologie der Retina.
Arch. Mkr. Anat. 28, 305-396 (1886)
Schultze, M.: Zur Anatomie und Physiologie der Retina. Arch. Mkr. Anat. 2, 175-286
(1866)
Schultze, M.: Üeber Stäbchen und Zapfen der Retina. Arch. Mkr. Anat. 3, 248 (1867)
Schwalbe, G.A.: Mikroskopische Anatomie des Sehnerven, der Netzhaut und des Glas-
körpers. In: Graefe-Saemish: Handb. ges. Augenheilkunde 1, 1874, p. 321
Sobrino, J.A., Gallego, A.: Células amacrinas de la capa plexiforme de la retina.
Actas Soc. Esp. Cienc. Fisiol. 12, 373-375 (1970)
Sobrino, J.A., Gallego, A.: Horizontal cells of the retina of Lampetra fluviatilis
(1976, in prep.)
Steinberg, R.H.: Rod and cone contribution to S-potentials from the cat retina.
Vis. Res. 9, 1319-1329 (1969a)
Steinberg, R.H.: Rod cone interaction in S-potentials from the cat retina.
Vis. Res. 9, 1330-1344 (1969b)
Steinberg, R.H.: The rod after effect in S-potentials from the cat retina.
Vis. Res. 9, 1345-1355 (1969c)
Steinberg, R.H.: The evidence that horizontal cells generate S-potentials in
the cat retina. Vis. Res. 11, 1029-1031 (1971)
Stell, W.K.: Correlation of retinal cytoarchitecture and ultrastructure in Golgi
preparations. Anat. Rec. 153, 389-397 (1965)
Stell, W.K.: The structure and relationship of horizontal cells and photoreceptor-
bipolar synaptic complexes in goldfish retina. Am. J. Anat. 120, 401-423 (1967)
Stell, W.K.: Comment on location of S-potentials in cat. Vis. Res. 11, 1027-1028
(1971)
Stell, W.K.: The morphological organization of the vertebrate retina. In: Handbook
of Sensory Physiology. Berlin: Springer 1972, Vol. VII/2
Stell, W.K.: Horizontal cell axons and axon terminals in Goldfish retina.
J. Comp. Neurol. 159, 503-519 (1975)
Svaetichin, G.: The cone action potential. Acta Physiol. 29 (106), 565-600 (1953)
Tartuferi, F.: Sull anatomia della retina. Intern. Monatss. Anat. Physiol. 4,
421 (1887)
Tomita, T.: Electrophysiological study of the mechanisms subserving colour coding
in the fish retina. Cold Spring Harb. Symp. Quant. Biol. 30, 559-566 (1966)
West, R.W.: Superficial warming of epoxy blocks for cutting of 25-150 um sections
to be resectioned in the 40-90 um Range. Stain Technol. 47, 201-204 (1972)
Yamada, E., Ishikawa, T.: The fine structure of the horizontal cells in some vertebrates retinae. Cold Spring Harb. Symp. Quant. Biol. 30, 383-392 (1965).

1.3 Neuronal Connections and Cellular Arrangement in the Fish Retina

J. H. SCHOLES

1.3.1 Introduction

Negative findings appear in print less frequently than positive ones, which must be a reason why so few workers (e.g. Kolmer, 1936; Leach 1963) have discussed the possibility of recognising, among cone cells of the mammalian retina, structural differences which might correspond with the three receptor functions in trichromatic vision. In fact, the three kinds of cone in the primate eye (Marks et al., 1964; Brown and Wald, 1964) have remained obstinately indistinguishable by their structure, so that almost nothing is known about the morphological basis of the neural colour channels in the advanced mammalian retina. Polyak (1941) must have felt this deficiency of knowledge particularly keenly as he applied black inks to his intricate diagrams of retinal neuronal architecture, unable to take account of the colour functions of the nerve cells he illustrated.

However, there is a curious evolutionary discontinuity in retinal structure between the known mammalian eyes and those of the other non-mammalian vertebrate groups, each of which possess conspicuously differentiated and easily recognisable colour cones. This convenience has made it possible to explore by direct means the patterns of synaptic connectivity which are involved in colour information processing in the first synaptic layer of the fish retina. My paper will review some of these findings, and also discuss the highly specific nature of retinal synapses, the numerical exactness of the retinal cell populations and the way in which they may be established during the continued growth of the retina.

1.3.2 Morphologically Differentiated Colour Cones in Non-Mammalian Vertebrates

Fish, amphibia, reptiles and birds all have retinal cone cells of more than one clearly differentiated structural type. Retinas from each of these vertebrate groups possess double cones, which are closely apposed pairs of receptors, one usually distinguishable from the other by its size, its position, or by the presence of a refractile or coloured oil droplet in its inner segment. In addition, there are usually two basic sorts of unpaired (single) cone populations, and in many species there may be as many as four. These receptors are often arranged in rather accurate mosaic patterns (Fig. 1).

The first applications of microspectrophotometry to retinal receptor cells showed that differently absorbing photopigments are segregated in the outer segments of these differently structured cones (Marks, 1965; Svaetichin et al., 1965). Subsequently, photopigment trichromacy has been shown to be surprisingly widespread, and distributed among the

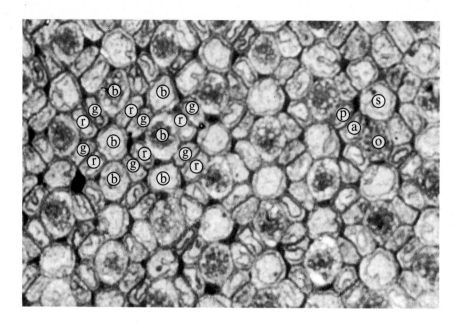

Fig. 1. Horizontal plane light micrograph of regularly arranged cone inner segments in a Cyprinid retina. A group of cones to the right of the micrograph are labelled according to their morphological identities; p: principal; a: accessory elements of double cones; s: single cone (or short single, Stell and Harosi, 1975); o: oblique cone (or miniature single, Stell and Harosi, 1975). Another group to the left are labelled according to their photo-pigment absorption maxima r: red; g: green; b: blue) as determined in goldfish and rudd by photographic densitometry (Scholes, 1975), microspectrophotometry (Stell and Harosi, 1975) and by a histological marker for spectral adaptation (Marc and Sperling, 1976). See text (x 3000)

same structural species of cone in often quite different animals (Liebmann, 1972). Double cones usually contain red-labile pigment in the outer segments of their larger (principal) elements, and a green sensitive one, similar to rhodopsin, in the smaller (accessory) ones. Blue sensitive pigment was early identified with at least one of the single cone populations.

This is the situation in the Cyprinid fish commonly studied (Marks, 1965; Scholes, 1975; Stell and Harosi, 1975), though in these fish, the single cone population is a complex and rather variable one. Each of them possesses unpaired cones located at the centre of squares formed by the double cone mosaic (where it is regular; cf. Fig. 1). These are referred to simply as single cones (Scholes, 1975) or short single cones (Stell and Lightfoot, 1975) and in goldfish and rudd they contain blue labile photopigment. A second population of unpaired cones appears, in the periphery of the retina if nowhere else, at the crosses of the double cone mosaic (cf. Fig. 1), called either oblique cones (Scholes, 1975) on account of the projection of their axons through the outer nuclear layer, or miniature single cones (Stell and Lightfoot, 1975). Microspectrophotometry suggests they also possess blue sensitive pigment in the goldfish (Stell and Harosi, 1975), and that has been confirmed by a recent study which used a histological indicator of the metabolic state of the cones in conjunction with selective spectral adaptation of the goldfish retina (Marc and Sperling, 1976). That technique also confirmed the identities of the other red, green and blue receptors given above.

In addition, some of the cyprinid retinas contain sporadically distributed larger single cones. In the rudd, they have the same structure and nervous connections as the principal elements of the double
cones, which strongly suggests they also contain red-sensitive pigment (Scholes, 1975). In goldfish, Stell and Harosi (1975) identified
two sorts of large single cones, containing either red or green labile
pigment. These large unpaired cones, of sporadic occurrence, therefore
complement numerically the chromatic function of the double cones in
the basic receptor mosaic of the fish retina. If they are absent, as
is often the case, and if it is true that oblique, as well as single
cones, contain blue sensitive pigment, then red, green and blue cones
are equally numerous in the receptor layer. Fig. 1 illustrates the
nomenclature for the different basic sorts of cone in the cyprinid
fish retina, and the chromatic identities attributed to them. It is
likely, however, that other fish have similar morphological cones with
different photopigments (Svaetichin et al., 1965).

1.3.3 Colour Coding of the Synaptic Connections of Receptors, Hori-
zontal Cells and Bipolars

Knowledge of the spectral functions of identifiable receptors in the
fish retina has been a stimulus to determine their patterns of connection with second-order nerve cells in the outer plexiform layer,
and recently significant steps have been made toward a comprehensive
description of the colour-coded synaptic architecture at this first
station for chromatic interaction in the visual pathway.

Stell (1967) was first to apply electron microscopy to Golgi-impregnated neurons in the retina, and he confirmed, in the fish, Cajal's
(1892) observation of separate bipolar cells for rods and cones. He
showed that the rod bipolars in fact also connect to cones, and confirmed that there are separable horizontal cells for rods and cones.
The method has since been used in the rudd to determine the connections between bipolar cells and the differently coloured cones, and
also the color coding of the direct synaptic connections among the
cones themselves (Scholes and Morris, 1973; Scholes, 1975). Furthermore, Stell and Lightfoot (1975) have demonstrated the colour specifi-
cities in the three populations of horizontal cells which are connected to cones in the goldfish. These findings will be briefly reviewed together here, for the two cyprinid species used are probably
substantially similar in their outer retinal organisation.

1.3.3.1 Direct Connections between Synaptic Terminals of Cones and Rods

Like those in the turtle (Lasansky, 1971), fish cones are directly
interconnected by a system of basal dendritic processes which ramify
in the plane of the outer plexiform layer, and invaginate the synaptic
terminals of neighbouring cone terminals (Witkovsky et al., 1974;
Scholes, 1975). Rods possess similar basal processes, but they never
make invaginating connections with other receptors; nor do the rod
spheres receive invaginating basal processes from cones.

Each red cone is connected in this way to between three and five of
the green cones in its neighbourhood, and the green cones each reci-
procate this relationship by sending basal processes to a similar number of adjacent red cone terminals. Single cones, which contain blue
sensitive pigment, direct their basal processes to invaginate the terminals of green cones, but that traffic is oneway only. Fig. 2 sum-

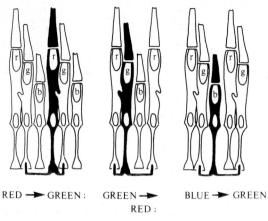

RED ➤ GREEN ;   GREEN ➤   BLUE ➤ GREEN
                 RED ;

Fig. 2. Summary diagram of colour coding of invaginating basal processes in the rudd retina. Basal processes ramify from their cone of origin over a field up to 10 μ in diameter, and invaginate other pedicles either en passant with small side branches, or with their terminal ends. Usually only one ending enters a given cone pedicle invagination, and that does not approach the synaptic ribbons there. Only a fraction of cone basal processes make invaginating connection with other cones; the remainder are restricted to the outer plexiform layer, and their connections are not yet known. Rod basal processes ramify at their own more sclerad level, and connect to one another with apposition junctions without invaginating any type of receptor terminal. Connection of oblique cone basal processes are not known

marizes the pattern of these connections, omitting oblique cones, whose basal process destinations are not yet known. Rod basal processes, also omitted, form an interconnected plexus via apposition junctions with one another (Witkovsky et al., 1974) and also sometimes with the external surfaces of the cone pedicles (Scholes, 1975).

There is no indication yet of what the physiological function of the invaginating basal process connections between (unlike) cones may be, though it seems sensible to envisage that it may be an antagonistic one. The range of the invaginating basal processes is less than the radii of the blur circles of the optics of moderately sized fish and turtle eyes (Scholes, 1975), and is about the same as the range over which cones of similar spectral sensitivity electrically pool their excitation in the turtle retina (Baylor et al., 1970). The anatomical basis of that interaction is not yet known; certainly the close membrane appositions between pedicles and their basal processes seem strong candidates in principle, but the difficulty is that, in the fish retina at least, they are formed non-selectively with respect to the chromatic functions of the cones (Scholes, 1975).

1.3.3.2 Connections between Receptor and Horizontal Cells

Horizontal cells are arranged in separate, rather regular, populations in fish, and they all receive hyperpolarising input, electrotonically pooled over wide retinal areas, from some, and in certain cases all, of the major classes of colour cones (Naka and Rushton, 1966c). Some of them receive in addition depolarising input from one or more classes of cones (Naka and Rushton, 1966a,b). There is a separate type of horizontal cell connected directly only to rods (Stell, 1967). In addition to being synaptically driven by the hyperpolarising receptor

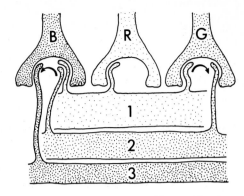

Fig. 3. Stell and Lightfoot's summary diagram (1975) of the colour coding of horizontal cell connections with the cones in the goldfish retina. Only one type of
horizontal cell is selective for just one colour channel; the vitread cells connect
to short and miniature single cones, both of which are thought to contain blue sensitive pigment. Two more sclerad layers of cells connect to combinations of colour
cones; the intermediate one synapses with green and blue cones, and the outer one
connects to all three types of cones. The arrows indicate the direction of synaptic
traffic between horizontal cells which Stell and Lightfoot (1975) suggest could
explain the origin of chromatic S-potential responses in these cells. According to
their scheme, H1 cells would have luminosity responses, while H2 and H3 cells would
have chromatic S-potentials, biphasic and triphasic respectively. This model remains
to be tested by recording and Procion marking, but it is not inconsistent with findings by those means in the turtle retina. All of these horizontal cells have axons
which terminate in large sinuous swellings at the outer margin of the inner nuclear
layer, but a fourth class of cells, connected exclusively to rods, apparently does
not. Reproduced with the permission of the Journal of Comparative Neurology

potentials of the cones to which they connect, horizontal cells, in the
turtle at least (Fuortes, 1973; Fuortes and Simon, 1974), exert a depolarising synaptic feedback onto the cones, which is of significant
effect only when large areas of the retina are illuminated. The ultrastructural basis of this two-way physiological traffic, however, is
not yet clear.

Stell and Lightfoot (1975) showed that the structural colour coding of
the horizontal cells in fish is complex (Fig. 3). Of those connected
to cones, external horizontal cells send dendritic processes to invaginate the terminals of red, green and blue receptors. An intermediate
layer of cells connects in the same way with green and blue cones only,
and a more widely ramifying set of vitread horizontal cells synapses
with the blue cones alone. Stell and Lightfoot (1975) argued that the
responses observed in the horizontal cells are only explicable in these
structural terms if certain of them are influenced synaptically by the
others, in addition to the direct input they receive from the cones
they invaginate (Fig. 3). They suggested, for example, that the blue
connecting cell is a type which generates spectrally biphasic S-potentials by virtue of occupying the blue cone synaptic invagination in
common with processes from the intermediate horizontal cells connected also to green cones. That interaction could in principle be directly mediated, or result from horizontal cell feedback on to cones
(cf. Fuortes et al., 1973).

1.3.3.3 Connections between Receptors and Bipolar Cells

The bipolar cells are the most numerous neurons in the fish retina and different types can be reliably distinguished not only by the patterns of connection they make with different receptors but also by the destinations to which they direct their axons in the stratified neuropile of the inner plexiform layer (Scholes and Morris, 1973; Scholes, 1975). Seven types have been analysed comprehensively, but there are likely to be as many as fifteen in all, if not more. Those which have been studied fall into three rather arbitrary, hopefully functional, classes (Fig. 4).

1. Rod bipolars are cells with massive processes and a vitread address in the inner plexiform layer. Their outer plexiform dendrites invaginate the spherules of rods, and also the pedicles of the red cones, where their counterparts in the primate retina connect exclusively to rods. There are two distinct morphological classes of such cells (Fig. 4, left). And the physiology of one, at least, is rather well known (see Fig. 5 and page 72).

2. Selective cone bipolars are filamentous neurons with wide dendritic fields, which, as Cajal observed long ago (1892; cf. Stell, 1967), connect only to cones. In the event, different types synapse exclusively with one or other of the colour channels, though variants have only been seen for green or for blue cones, and not for red cones (Fig. 4, centre).

3. Mixed cone bipolars connect predominantly to cones, though some invaginate the rod spherules as well. They are morphologically distinct from the rod bipolars proper, and, unlike them, connect to combinations of different colour cones. The cell type most frequently encountered synapsed with red and green cones, and with rods, and the majority of such cells directed their axons to the most sclerad strata of the inner plexiform layer (Fig. 4, right). Other sorts have been observed which connect to green and blue cones only, or to all three types of cones (Scholes, 1975).

Details of these findings encourage only very cautious comparison with the organisation of the outer plexiform layer in primates, and emphasize the likelihood that the evolutionary step between lower vertebrates and advanced mammal retinas involves more profound matters than the structural differentiation or nondifferentiation of the colour cones. Invaginating basal process connections have not been observed between primate cones, and there is not yet any sign that their horizontal cells are differentiated to receive input from distinct population of colour cones. A large fraction of their bipolar cells, connecting in the central area to individual cones only (Polyak, 1941; Boycott and Dowling, 1969) clearly do not directly sample combinations of colour receptors in the way characteristic of a likely majority of fish bipolars (Fig. 4). Perhaps primates, which have less complicated neuronal feature-detecting arrangements at the retinal level than appear in the inner plexiform layer of many non-mammalian vertebrates, can also afford to postpone some of the colour interactions, elsewhere undertaken in the outer plexiform layer, until a higher level in their visual pathway.

In what has been described there has been no mention of the exact ultrastructural characteristics of the receptor synapses, and it has been possible to gloss over that complex subject because fish cone terminals are invaginated by processes from other receptors, from horizontal cells and from bipolars, and those contacts are accordingly easily and clearly definable at a macroscopic level (Scholes, 1975; Stell and Lightfoot, 1975). However, all the cells involved may be

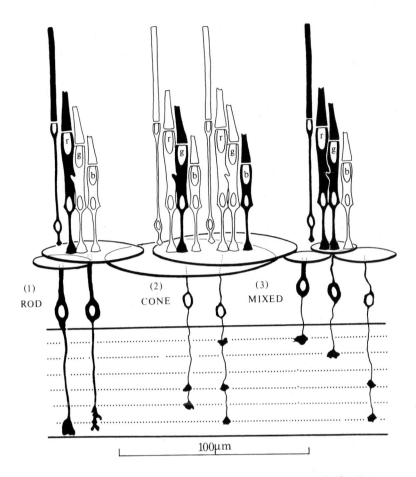

Fig. 4. Summary diagram (from Scholes, 1975, reproduced by permission of the Roy. Soc. London) of the colour coding of connections between bipolar cells and different receptors in the rudd retina. Some bipolars, filamentous cells with wide dendritic trees (centre group), connect selectively with different types of cones (though a red coded type has not been found). Other cell types, with smaller, more densely branched fields, connect to combinations of receptors. One class, particularly easily identified by light microscopy on account of their massive processes and vitread axon terminals in the inner plexiform layer (Cajal, 1892; Stell, 1967; Parthe, 1972), synapse with rods and with red cones only (left-hand group). Another more varied class of bipolars (right-hand group) connect to combinations of colour cones (e.g. red and green; blue and green) and some of them connect with rods as well. These cells generally have the smallest outer plexiform dendritic fields, their axons predominantly address the more sclerad strata of the inner plexiform layer.

The grouping suggested by the diagram is somewhat arbitrary, but designed to respect the light microscopic nomenclature (Cajal, 1892) for fish bipolars. Likewise, no account is taken of the detailed ultrastructure of the connections involved

either hyperpolarised or depolarised by light, both directly and indirectly, and it has been well established that vertebrate retinal neurons make a variety of ultrastructurally distinct contacts with the receptor terminals (e.g. Dowling and Boycott, 1966; Missotten, 1975; Lasansky, 1972).

Stell (pers. comm.) has urged that fish bipolars make very different sorts of synapses inside the cone terminals they invaginate, and this encouraged me to examine some of them with this in mind. But before describing the results, it may be useful to outline the more general organisation of the synapses of vertebrate retinal rods and cones.

1.3.4 The Comparative Organisation of Retinal Receptor Synapses

In the primate retina, different sorts of bipolar cells make their synaptic connections with the pedicles of cones in two quite distinct ways. Flat bipolars send dendritic processes to abut against, or slightly indent, the vitread surfaces of the cone terminals (Missotten, 1965; Dowling and Boycott, 1966). At these points of contact, there is a dense specialisation of the receptor membrane, and an accurate cleft separating receptor and bipolar plasma membranes contains a weakly opaque matrix. There are no presynaptic aggregations of vesicles, and the location of the junctions is remote from the presynaptic ribbon complexes of the pedicles.

Processes of invaginating bipolars, which are the second sort, penetrate the cone terminals, and arrange themselves, along with the horizontal cell dendrites that accompany them as triad complexes in the pedicle cavities they occupy, around the apical specialisations of the receptor presynaptic ribbons (Missotten, 1965; Dowling and Boycott, 1966). They form central process profiles in the triads, opposite the synaptic ribbons, and they are flanked by lateral processes belonging to the horizontal cells (Dowling, 1965). A typical cone pedicle contains some 30 of these complexes, and most of them are occupied by the multiple processes of one bipolar cell (Boycott and Dowling, 1969; Kolb, 1970). Although there is a clear distinction between cells of this sort, whose processes associate with the ribbons, and flat bipolar cells whose processes do not, the invaginating bipolar dendrites frequently participate in joint membrane specialisations with mammalian receptor terminals en route to the ribbons (Kolb, 1974), and those junctions are similar to the flat bipolar's only contact with the cones. Bipolar processes which connect to rods in primates end in invaginated triad centre processes, and there are apparently no flat contacts with those receptors (Missotten, 1965; Dowling and Boycott, 1966).

In non-mammalian vertebrate retinas, this distinction between the flat and invaginating contacts of bipolars with cones is often less clear. The overall topology of the cone pedicles varies in a confusing way between different animals (Lasansky, 1971, 1973; Witkovsky et al., 1974), and the same, rather than different bipolar cells may be involved in both sorts of junction with different receptors (Lasansky, 1973). The locations of bipolar and horizontal cell process profiles around the ribbon apical specialisations is less clearly defined, and bipolar processes sometimes appear at stations occupied by the horizontal cells in primate pedicles, and vice versa. In the turtle retina (Lasansky, 1971), some dendritic processes which probably belong to bipolars occupy the centre station at cone pedicle triads, and like their counterparts in the primate retina, can make apposition junctions with the terminals on their way there. But they are relatively rare, while basal junctions similar in size and position to the flat contacts between primate flat bipolars and cones, are prominent and numerous.

Fish cone pedicles are numerically more complex than those in the other vertebrate groups studied (Witkovsky et al., 1974; Scholes, 1975), and most of the bipolar processes connected to them are very small, which makes the distinction between ribbon-directed and simple apposition contacts less clear to see. It describes the fish cone pedicles quite well to say that their three-dimensional arrangement is as if the entire vitread surface of a primate (or turtle) cone pedicle, with its triad invaginations around multiple synaptic ribbons, and its separate basal junctions with bipolar processes, were folded in to shape a single large secondary invagination, occupied in common by all the neuron processes in synaptic contact with the cone.

Accordingly, the synaptic ribbons (longer than their primate counterparts) are arranged around the lining of a common synaptic cavity, flanked in the usual way by horizontal cell dendrites. Stell (pers. comm.) has convinced me that only a very small minority of the hundreds of bipolar processes which enter fish cone synaptic invaginations make their way to the ribbons. Likewise, only one of the processes which enter rod spherules does so (see below). The majority appose themselves in a simple way to projections from the pedicle cavity lining, which mingle with the bipolar processes as they enter the synaptic terminal cavity. These junctions are not conspicuous in low resolution Golgi-EM material, and often no more striking than many other membrane adjacencies between other components of the synaptic cavity. Nevertheless, there is a distinction between the minority of bipolars whose processes associate with the ribbons of rod and cone terminals and the majority whose dendrites do not.

Now, electrical recordings from mammalian bipolar cells have not yet been described, but the responses of these neurons are well documented in amphibian, fish and reptile retinas. According to type, they are, as Werblin and Dowling (1969) first showed in Necturus, either depolarised or hyperpolarised when the receptors directly connected to them are illuminated. Bearing these two sorts of post-synaptic activity in mind, there are three good reasons for examining the details of the synaptic contact between certain identified fish retinal bipolar cell and the rods and cones.

1. Cyprinids have both on- and off-centre bipolars. Kaneko (1970) has marked examples of each in goldfish by Procion injection, and one of his cells is clearly to be identified with a type encountered in Golgi impregnations, and known to connect to rods and to red cones (Scholes, 1975). Toyoda (1973) has shown that both the rods (in dark adaptation) and the cones (in the light) depolarise these bipolars in the carp, so it is of interest to enquire whether they make the same contacts, ribbon associated or apposition, with both classes of receptors.

2. Some fish have more than one type of bipolar connected in common with red cones and rods (Fig. 4, left), and it would be interesting if the ultrastructure of their connections with those receptors differs, for it has been claimed (Toyoda, 1973) that while the majority of bipolars, connected to rods and accessible to recording, are depolarised by light, some are hyperpolarised.

3. Whenever the convergence of rod and cone signals has been studied in fish retinae (at the ganglion cell level) (Raynauld, 1972; Beauchamp and Daw, 1972), it has always entailed the same sort of interaction. The rod and red cone mechanisms work synergistically (but separately, according to the state of light adaptation) to excite or inhibit ganglion cell discharge, while green cones work antagonistically to both of them. It is possible that this conver-

(a)

Fig. 5A and B. A) Camera lucida sketch
of a small field rod bipolar from the
rudd retina (Scholes, 1975; reproduced
by permission of the Roy. Soc. London).
It is suggested that these cells are
morphologically and functionally homologous in the two closely related fish.
That identity enables a characterisa-
tion of the detailed synaptic structures which mediate their depolarising
responses to illumination of their receptive field centres (see text)
B) Procion injected on-centre bipolar
cell from goldfish retina (Kaneko,
1970; reproduced by permission of the
J. Physiol.). Golgi-impregnated

gence of rod, red and green cone signals, measured at the ganglion
cell level, is mediated by a particular class of bipolars, mixed
cone bipolars, which connect directly to just that combination of
receptors (Fig. 3, right). Do they perhaps make different contacts
with the green cones than with the red cones and the rods?

1.3.5 The Detailed Contacts between Bipolars and Receptors in Fish

1.3.5.1 Rod Bipolars

Two sorts of bipolar cells comprise the most conspicuous vertical pathway for rod signals in the Cyprinid retina (Scholes, 1975), and similar
ones are found in other quite different fish retinas (Parthe, 1972).
Both have the massive aspect which Cajal (1892) first described for
the rod-connecting bipolars in the fish, and they both synapse with
red cones as well as with rods. They differ in the width of their dendritic trees in the outer plexiform layer (Fig. 3, left), and in the
exact appearance of their connections with receptors there (Figs 6, 7).

The vertical profile of one of these cells, a small field rod bipolar,
drawn by camera lucida from a Golgi impregnation in the rudd retina,
is compared in Fig. 5 with Kaneko's (1970) micrograph of a Procion-
marked depolarising bipolar from the goldfish retina. There is a great
resemblance in detailed morphology between these two neurons, and their
identity across Cyprinid species is attested by Toyoda's more recent
(1973) demonstration in the carp retina that they are depolarised by
the activity of both the rods and the red cones.

Dendrites of the large field rod bipolars embrace more receptors in a wider retinal area, but connect, nevertheless, to the same curious combination of red cones and rods. Even at the level of the light microscope, however, the way in which their terminal processes synapse with these receptors is quite different. The processes of small field rod bipolars terminate in rod spherules as quite distinctive question mark figures (Fig. 7a, arrows), while the finer endings of large field rod bipolars remain straight and unelaborate where they invaginate these rod synapses (Fig. 7c). Similarly, where the small field cells connect with red cone pedicles they do so with ten or so dendritic processes, which, where they invaginate the terminals, have a characteristic compact and clustered aspect in the light microscope (Fig. 6a). Those of the large field cells, on the contrary, are appreciably finer and appear in a more dispersed circlet closer to the lateral outlines of the red cone pedicles (Fig. 6b). That circular arrangement in the pedicles suggests they are related to the presynaptic ribbons there, which are similarly disposed in the synaptic terminal cytoplasm.

The retinal areas shown in Figs. 6a and b were thin-sectioned for electron microscopy, and the cone terminals indicated by a circle or by boxes in the light micrographs, are shown at the resultant higher magnification in Figs. 6c,d, and e. In the thin section plane of Fig. 6c a cluster of processes from the small field rod bipolar occupies the synaptic cavity of a red cone pedicle, but none of them approach the presynaptic ribbons, and simply appose, at intervals, the receptor plasma membrane lining the invagination (arrow). This was true in the remaining sections of this sequence, and in a number of other cells of this type which were examined. It is sensibly non-commital to call these junctions apposition contacts, rather than to identify them explicitly with the flat or basal junctions of bipolar cells with cone pedicles elsewhere (Missotten, 1965; Dowling and Boycott, 1966; Lasansky, 1971, 1972), which they resemble but which are properly defined at a resolution this Golgi material did not permit.

These unspecialized apposition contacts with the pedicle plasma membrane, therefore, appear to be the morphological substrate for the depolarising action (Kaneko, 1970) of red cone activity upon the small field rod bipolar cells.

By contrast, the dendritic processes of the large field rod bipolars to direct themselves into the proximity of the red cone presynaptic ribbon specialisations, occupying 'triad' centre process locations over short distances of the lengths of the apical ridges in the pedicle cavities (Fig. 6d,e). They are the only bipolar processes consistently seen in Golgi impregnations to do this, in red, or in any other type of cone, though in most of the section planes inspected there were additional fine processes of unknown origin interposed between them and the ribbon apical specialisations (Fig. 6d,e).

What of the connections of these two kinds of cells with rod spherules? The little question mark cyphers characteristic of the small field rod bipolar connections with rods (Fig. 7a) are loops, usually in the plane of the retina, low down in each innervated spherule, which provide a wide area of apposed membrane contact with the lining of its synaptic invagination (Fig. 7b). The loops are quite separate from the rod sphere ribbons, and never participate in the triadic process profiles around them. In this case, therefore, connections with the rod spheres are similar to (if more prominent than) those made with the red cones (Fig. 6c), and at first sight that is a gratifying agreement with the synergy of synaptic control the two sorts of receptors are known to exert upon these bipolars (Toyoda, 1973).

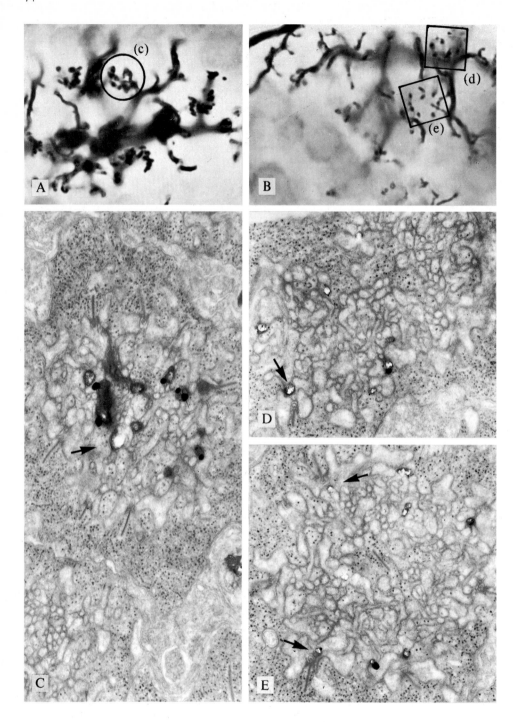

By contrast, the connections of <u>large field</u> rod bipolars with rod
spheres are different from those they make with the red cones. Their
processes do not penetrate the spherules to the level of the ribbon
apparatus, and have never been seen at the triad centre profile location. Their small size and the quality of the material meant that
their contacts could not be seen with the clarity possible in the
case of the small field cells, and they typically appear, as in
Fig. 7d, in limited contacts with spherule plasma membrane and perhaps also with horizontal cell processes. This bipolar type, then,
has ribbon contacts with red cones, but not with the rods it synapses
with.

1.3.5.2 Mixed Cone Bipolar Cells

The synaptic ribbon apparatus in rod spherules has just one bipolar
cell process apposed to its apical specialisation along the greater
part of its length, and that 'triad' centre process does not belong
to either of the two sorts of cells designated rod bipolars above.
It originates in a class of cells intermediate in their morphology
between rod and cone bipolars proper, called mixed cone bipolars
(Scholes, 1975; Fig. 3, right), which synapse with rods in addition
to red and green cones. The profile morphology of one such bipolar
is shown in Fig. 8c, and the thin section in Fig. 8b transects the
rod connections of another example, showing the destinations of its
dendritic processes at the rod synaptic ribbons. One ribbon is cut
sufficiently obliquely to suggest the way the silvered processes occupy long lengths of the apical specialisations.

However, the dendrites of these cells, where they invaginate the pedicles of the red and green cones, do not approach their synaptic
ribbons, and simply appose the receptor plasma membranes in a way
similar to the connections of the small field rod bipolars with cones.
This is illustrated in Fig. 8a, where processes of the same bipolar
shown in Fig. 8b occupy adjacent red and green cone pedicles.

The varied connections of these three cell types with receptors is
not consistent in any simple way with the suggestion that ribbon and
apposition contacts might be the separate bases of the depolarising

---

◁ <u>Fig. 6A-E.</u> Rod bipolar connections with red cones. A) Horizontal view of the outer
plexiform dendritic tree of a <u>small field</u> rod bipolar cell in the rudd retina. The
focal plane is in the layer of cone pedicles, and group of dendritic processes invaginating one of them is outlined with a circle. There are three other such synaptic sites in the picture, and it is characteristic of these cells that they connect
in this way only with red cones (Fig. 3) (x 4000). B) Similar picture of part of
the dendritic field of a <u>large field</u> rod bipolar from close by in the same retina.
Two of the sites at which it sends multiple dendritic processes to invaginate red
cone terminals are outlined with squares (x 4000). C) The dendritic tree shown by
light microscopy in (A) was sectioned for electron microscopy, and this figure shows
an horizontal section through the encircled synaptic complex shown in the light
micrograph, at a level close to the focal plane chosen there. Dendritic processes
enter the invagination of the red cone pedicle, and make apposition contacts
(e.g. <u>arrow</u>) with its plasm membrane: they do not approach the synaptic ribbons
(e.g. <u>sr</u>) (x 20,000). D, E) In the same way, the synaptic complexes of the large
field rod bipolar, outlined by squares in the light micrograph (B), are shown at
higher magnification in thin E.M. section. The cone terminals invaginated by the
silvered processes belong to red cones, and in these section planes, some of the
processes (<u>arrows</u>) appose themselves to the apical specialisations of the synaptic
ribbons of those terminals (x 20,000).

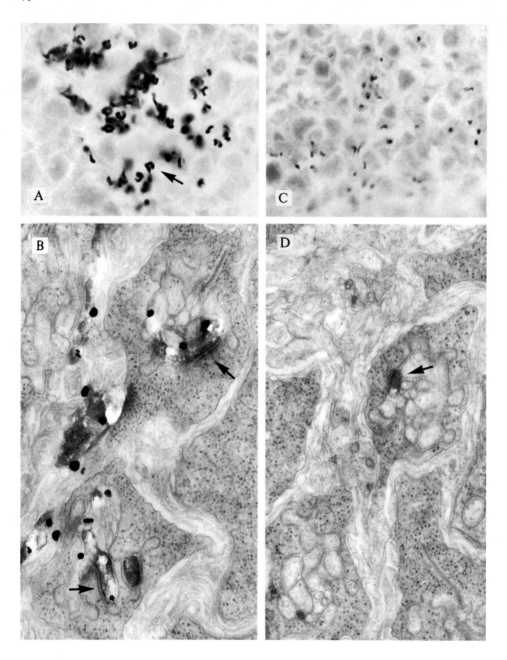

and hyperpolarising responses of different bipolar cells, and they admit only one clear comparison with the physiological responses of those neurons in the fish retina.

If it is correct to identify, as in Fig. 5, <u>small field</u> rod bipolars in rudd with the on-centre neurons in goldfish and carp of Kaneko (1970) and Toyoda (1973), it follows in their case if in no other,

◁ Fig. 7A-D. Rod bipolar connections with rods. A) Light micrograph of the dendritic field of a small field rod bipolar cell, taken with the focal plane at a more sclerad level than that shown in Fig. 6A, to show the endings of its dendritic processes in the rod spheres. They take the form of little incomplete loops (e.g. arrow) (x 4000). B) By electron microscopy, at this level in the retina, the endings are seen to appose the rod spherule plasma membranes (arrows), and, like their counterparts in the red cones (Fig. 6C) they do not approach the spherule synaptic ribbons (x 29,000). C) Light micrographs of some of the contacts of a large field rod bipolar with rod spherules. D) Electron micrograph of large field rod bipolar contacts with rod spheres. The processes involved are finer than those of the small field bipolar cells (cf. A and C), but like them appose the rod spherule membrane (arrow), without approaching the synaptic ribbon apparatus. Thus they differ from the corresponding contacts the large field cells make with red cones (Figs. 6D,E) (x 29,000)

that depolarising responses to centre field illumination are conveyed by apposition contacts with the rods and the red cones. Were ribbon and apposition contacts to mediate different forms of synaptic activity, however, the large field rod bipolars would be driven one way by their apposition contacts with rods in dark adaptation, and in another when their ribbon connections with red cones are effective at higher levels of illumination. Again to find rod spherule ribbon synapses converging, on the dendritic processes of the mixed cone bipolar cells (Fig. 8), with apposition synapses to red and green cone pedicles, is inconsistent in these terms with the widespread synergy between rod and red cone signals, and with their antagonism to green cone signals, which are frequently observed in fish retinal ganglion cell responses (Raynauld, 1972; Beauchamp and Daw, 1972).

Perhaps some of these difficulties might be removed were it possible to take account of horizontal cell involvement at the synapse, and were the findings assimilated with the long-standing but neglected evidence (Evans, 1966), that rod and cone synaptic vesicles are ultrastructurally quite different. Hints accumulate of differences in the synaptic metabolism of these receptors, for example in their structural responses to abnormal regimes of light and dark (Wagner, 1973).

In the context of the title of this volume, it is worth remarking that in the outer plexiform layer of lower vertebrate retinas one is confronted with a high degree of complexity in the synaptic wiring of the neuronal colour channels. Pathways between cones seem to be duplicated by their convergence onto common horizontal cells and on to bipolar cells (Figs. 2, 4). The difficulty is that one has insufficient knowledge of the functional roles these connections, and presumably many others, play in the analysis of patterns of coloured light falling on the retina. In this sense, our descriptions of the arrangement of the cog wheels and springs of vision have begun to outstrip, in a way hard to justify, our apprehension of the function of the mechanism they comprise.

1.3.6 The Accuracy of Neuronal Connections in the Outer Plexiform Layer

Although the fish retina is impressive for the degree of differentiation of its various types of nerve cell, and for the correlated specificity of their synaptic connections with the receptors, bipolars have been noticed from time to time which make odd contacts with receptors other than those which provide their dominant synaptic input (Scholes, 1975). They are designated error connections on the grounds that,

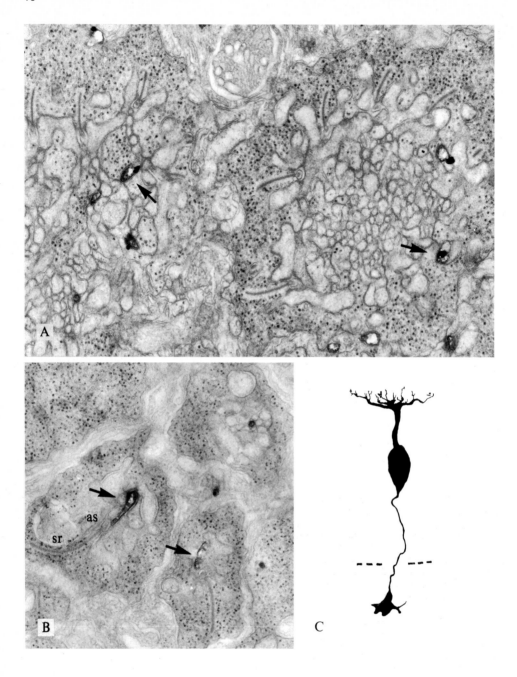

within a given bipolar cell class, their occurrence is sporadic, and
because it is unlikely that they could contribute significant functional synaptic activity in relation to the much more numerous and
consistent 'correct' connections. They probably account for between
1 and 5% of contacts between bipolars and all species of cones, and
less than that figure for rod connections.

Fig. 8A-C. Mixed cone bipolar connections with red and green cones, and with rods. A) Horizontal plane electron micrograph of the pedicles of a red r and an adjacent green g cone in the rudd retina. Golgi-impregnated processes from a mixed cone bipolar cell invaginate both of them to appose the plasma membrane lining the synaptic cavities (arrows), but they do not approach the synaptic ribbon complexes sr (x 28,000). B) By contrast, processes from the same impregnated bipolar cell which penetrate the rod spherules at their more sclerad level, conspicuously terminate (arrows) close to the apical specialisations as of their synaptic ribbons sr (x 28,000). C) Vertical profile of one of these mixed cone bipolar cells, reproduced from Scholes (1975). They are the only cells found to have ribbon synapses with the rod spheres, though their connections with cones are quite different (Fig. 8A). Their axons terminate in the more sclerad strata of the inner plexiform layer (x 1300)

An example of these minority synapses is shown in Fig. 9, where part of the dendritic field of a bipolar cell specific for blue single cones connects to all eleven of the receptors of that type within range of its major dendrites. But three small branches from one dendrite leave the focal plane in Fig. 9a without approaching nearby cone pedicles, and terminate by invaginating three spherules at their different level in the retina (Fig. 9b). At that level, a fourth rod contact is also visible some distance to the right. These rod contacts, similar to that in Fig. 7c, were verified by electron microscopy, and the rod spherules they involved were in no visible way abnormal.

From the point of view of the neuron concerned, the error connections comprise some 5% of all synaptic contacts with the receptors, bearing in mind that multiple processes are directed to most of the blue cones, and only single ones to the four rod spherules. Conversely the four spherules innervated represent only a very small fraction of the rods within the outline of the bipolar dendritic field, where connectivity with the blue cones in the area was almost comprehensive.

The likelihood is that error connections occur systematically according to bipolar cell type. It must be acknowledged cautiously however, because given their rarity, the error connections are not easily quantitatively inventoried in Golgi-EM material, but the issue seems important enough to list the fragmentary observations that suggest it. Among four blue cone connecting bipolars analysed, two of them synapsed with a total of six rods among a total of 55 blue cone connections retrieved in the electron microscope. Two green cone selective cells contacted 43 green cones between them and also three blue cones, while no errors were found in the connections of more than 15 small field rod bipolars. Of five large field rod bipolars, two were found to make single error connections with green cones, but otherwise synapsed according to pattern with red cones and rods. No error connections were noticed in the fields of six mixed cone bipolars, specific for red and green cones and rods.

The mechanisms which shape selective synaptic connections in the nervous system are little understood. Had they to do, for example, with specific cell surface affinities (e.g. Roth, 1973) amplified data consistent with that outlined above might suggest such things as that the blue cone membrane surface phenotype has more in common with that of rods than of any other receptor type, or that there are systematically different degrees of accuracy with which bipolar cell surfaces may match those of the receptors with which they predominantly synapse.

The frequency and circumstances of the error connections can profitably be compared with the situation in the fly's eye, where the absolute accuracy of receptor synapses with second-order neurons has been com-

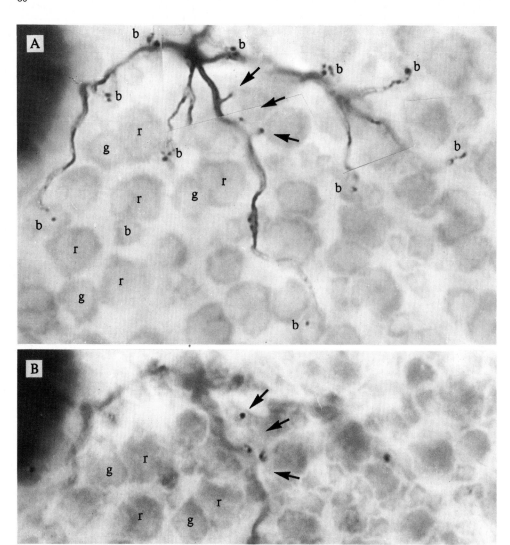

Fig. 9A and B. A) Horizontal plane light micrograph of part of the dendritic field
of a selective Golgi impregnated cone bipolar connected to blue cones. Its connections with those receptors appear usually as small clusters of process, and each
of them is labelled b. The pale profiles of other cone pedicles in the vicinity
are visible, and some of them are labelled according to their colours (r: red;
g: green). Three small dendritic branches (arrows) skirt neighbouring cones and
leave the focal plane in the sclerad direction. B) The same retinal field, but the
focal plane has been adjusted to lie in the layer rod spherules (visible as small
angular profiles). The 'error' dendrites (arrows) terminate in rod spheres (x 4000)

prehensively analysed. There, accuracy is complete (Horridge and
Meinertzhagen, 1970) except where there are imperfections in the geometry of facets in the eye (Meinertzhagen, 1972). The nature of the
errors of connectivity in these circumstances seems to be determined
by spatial considerations of a sort unlikely to apply in the fish
retina.

Be that as it may, a degree of outer plexiform synaptic convergence
as high as 30-fold in the case of green cone connections with the bipolars specific for them and the roughly equal degree of divergence
in the synapsis of red cones with bipolar cells (Scholes, 1975), in
the fish retina, suggests a greater numerical scope for mistakes there.

1.3.7 Regularities in the Retinal Receptor Mosaic

Fish retinas are well known for the frequently almost crystalline orderliness of their receptor mosaics (Fig. 1), and in some cases the
periodicities in the packing of the different cones in the receptor
layers are clearly reflected by similar periodicities in the packing
of nerve cells and their dendritic processes below (Hibbard, 1971).

The following construction is one way of describing the typical teleost cone mosaic. Arrange oblique cones in a hexagonal pattern whose
three-fold axes run at approximately +60°, -60°, and parallel to the
local radius of the entire retinal disc (Fig. 10a). Take another hexagonal pattern of single (blue) cones of the same periodicity and orientation, and intersperse them in the oblique cone lattice by first
superimposing them then shifting them half a mosaic period along the
radius-parallel axis of the pattern (Fig. 10b). Now, bearing in mind
that double cones are twice as numerous as the single cone populations,
take first a hexagonal pattern of double cones, again with the same
periodicity and orientation, and with each double oriented along one
of 60° the hexagon axes of the mosaic. Superimpose it again on the
oblique cone lattice and shift it half a period, this time along a
60° axis (Fig. 10c). Finally take another identical lattice of double
cones, reflect it in the plane of the paper so that its elements now
align with remaining diagonal axis, and intersperse it in the same
way as before along that axis, to complete the mosaic.

Viewed in this way, the distinctive pattern of the mosaic can clearly
be seen as an economically space-filling and even way of packing the
four different populations of cone cells together, given the simple
integral ratios that exist between their numbers. Those exactly integral ratios attest to a degree of precision in the proliferation and
differentiation of cones in the developing retina, and in that context, irregular mosaics, like that of the rudd are as interesting as
regular ones.

The basic outline of the regular teleost mosaic can easily be discerned
in certain areas of the rudd retina, but in others the lattice is distorted almost out of recognition by the apparently haphazard insertion
of significant numbers of additional cones of various sorts. The interesting thing is that these insertions, although sporadic in occurrence,
obey certain quite fixed rules when they do appear. One example must
suffice here to illustrate this principle. Frequently, extra double
cones appear in units of the mosaic, always oriented along the radii
of the retina, and located close to the blue single cones. Less frequently, there are extra single cones as well, always in their neighbourhood, and such extra single cones are never found unaccompanied
by an extra double. That implies a conditional sequence to the production, in development, of these extra cones, in which the different
steps have different probabilities of completion. Occasionally in the
rudd, each repeating unit in the basic lattice is occupied with the
full complement of these insertions, and the packing mosaic again becomes regular, but now yet more complex and aesthetic (Fig. 11), and
the ratios of red, green and blue receptors are quite changed.

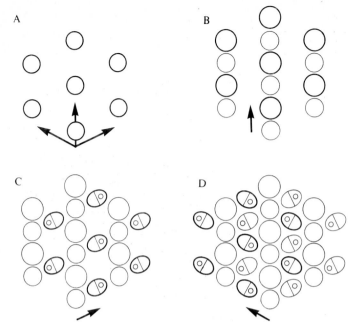

Fig. 1OA-D. Illustrates, by a construction, the hexagonal packing of the different cone populations in the regular receptor mosaic of teleost fish (cf. Fig. 1).
A) Hexagon pattern of oblique cones, with the three packing axes indicated by arrows, the vertical one pointing toward the retinal margin. B) A similar lattice of single cones (thick outlines) has been interpolated by a half period shift (arrow) along the vertical axis. C) Another lattice of double cones (thick oval outlines) is interpolated by shifting a half period along a diagonal axis (arrow). Notice, however, that the orientations of the component red and green elements of the double cones (circled and blank profiles, respectively) are determined by a rather complicated rule. D) The same pattern of double cones is reflected about the vertical axis of the mosaic and interpolated by half period shift along the remaining diagonal axis (arrow), to complete the mosaic pattern

Now the numerical precision with which the different populations of receptors (and nerve cells for that matter) are established in development here, the order of these insertions of extra cones, and even the mirror symmetries in the regular receptor mosaic (Fig. 1Oc,d), are all suggestive of regular and iterative programmes of mitosis among the developmental precursors of these differential retinal cells. The fish retina grows in area throughout life from a narrow zone of undifferentiated and continually dividing neuroblasts arranged around its periphery (Glücksmann, 1940; Müller, 1952; Gaze and Watson, 1968; Hollyfield, 1972) are different cells which differentiate out of the growth zone filially related by fixed and repetitive lineages of cell division there?

1.3.8 Cell Proliferation in the Retina

Retinal cell division cycles have mainly been studied in embryonic material, using tritiated thymidine as an autoradiographic tag for newly replicated DNA. In early embryos, cell proliferation proceeds

Fig. 11. Diagram of the way in which extra double and single cones (thick outline) inserted into each unit of the
regular teleost mosaic (upper part of
diagram) can generate a more exotic
cone pattern (lower part) sometimes
encountered in the rudd retina

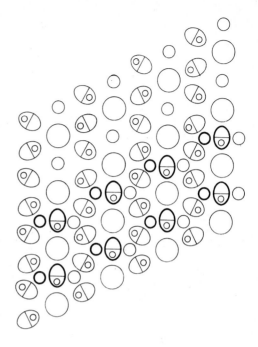

in parallel over wide areas of the partially differentiated retinal
disc, at a maximum rate before the tissue stratifies into its three
main cellular layers (Sidman, 1961; Fujita and Horii, 1963). It has
repeatedly been shown that the precursors of the different cell classes complete their terminal proliferation at significantly different,
if overlapping, times in embryonic development. Ganglion cell precursors invariably finish their division first, and in fish (Hollyfield,
1972) there is an orderly progression outward through the retina, with
receptor cell precursors finishing their mitosis last. In other animals the order may be different (Fujita and Horii, 1963; Kahn, 1974)
with inner nuclear layer cells last.

In the general way, these findings support the notion that scheduled
programmes of cell division in the retina commit their products to
the differentiation of the distinct cellular phenotypes which comprise
the tissue, and that is important in view of the evidence accumulated
from genetically mosaic eyes in insects, that the identities of the
receptor cells in single ommatidia are quite unrelated to their cell
lineage in embryonic development (Benzer, 1973). It is tempting also
to use them as support for a suggestion of Jacobson (1970) that functionally related mosaic units of receptors and nerve cells in the depth
of the retina might arise in development as parallel clones of definite and fixed lineage. Perhaps large numbers of such units proceed in
rough synchrony through a routine of proliferation which produces ganglion cells first, and the other retinal cells in their order.

The difficulty is that there is no evidence that proliferation in one
cell class is contingent upon its earlier completion in another, and
no indication that the separation in time of peaks of proliferative
activity in the precursors of the different retinal layers is related
in any way to cell cycle duration. Furthermore, one reason why different cell classes complete proliferation at different times might be

that they achieve different final numbers: one suspects this factor might have something to do with the late completion of inner nuclear layer cells in the chick retina (Kahn, 1974).

That consideration does not apply in one instance where the cell types comprising a retinal layer have been distinguished from one another in autoradiographs. Morris (1973) found that the rods, double cones and single cones in the chick retina all complete their terminal S phases at slightly different times in embryonic development, though again there was no relation between their sequence and the estimated duration of the cell cycle, and considerable temporal overlap between them. She interpreted this to mean that different receptors in units of the chick retinal mosaic are not filially related through common mother cells, in ways which might relate to their local arrangement there.

The present author has begun similar observations in the growing juvenile fish eye, using brood mates of the Black Molly, a viviparous Cyprinid fish. ^{3}H-Tdr was administered post-natally in small intra-peritoneal doses (1µCi), whose half-time of availability for incorporation was estimated by scintillation counting to be about 40 min.

A sufficient period was allowed for labelled cells to take their place and be recognisable in the differentiated retina, before preparation for autoradiography. Then identified quadrants of the retinas were dissected out and embedded in plastic for thin sectioning (1µm) close to the growth zone in a plane tangent to the receptor surface there. Sections were prepared for light microscope autoradiography by emulsion dipping, and exposures were usually short enough that grains did not obscure the nuclei they marked, while at the same time adequate quantitative data could be gathered by following the same cells through serial sections and pooling grain counts from them.

Growth rate in these fish retinas amounted to roughly one unit, or row of cones, in the mosaic in 24 h, and labelled division products distributed themselves in accurate narrow fronts, parallel to the growth zone, through the depth of the tissue (Fig. 12a). Once cell division is completed, therefore, the mobility of differentiating cells in the retina is highly restricted, unlike the rod precursors in Rana pipiens tadpoles which are free to migrate over great retinal distances before differentiating (Hollyfield, 1968). In the receptor layer, the main axis of the cone mosaic is not quite perpendicular to the margin of the retina, so fronts of labelled receptors are never restricted to particular mosaic rows over more than small distances round the retinal circumference. Nevertheless, the distribution of grains between the labelled fronts and the retinal periphery (Fig. 12a) approximates to that expected if cells which successively differentiate after administration of a pulse label contain geometrically diluted quantities of isotope (cf. Hollyfield, 1972).

Looking in detail among the cells of a labelled front in the receptor mosaic (Fig. 12) one often sees jointly labelled neighbour cone nuclei of the same and of different sorts. However, these groupings appear to be random. For example, in Fig. 13, both members of a double cone are labelled, but in several other cases, one element carries label and the other does not. Two labelled single cones have quite different patterns of labelled neighbours surrounding them. This condition is typical of the wide areas of labelled retina which have been scrutinised, where neighbouring cells were tagged apparently at random, and where labelled cells would frequently be seen with no others within distances very appreciable besides the limited ones (Fig. 11) over which the receptors are free to move once having taken their place in the mosaic. If there are orderly programmes of cell division giving rise to differ-

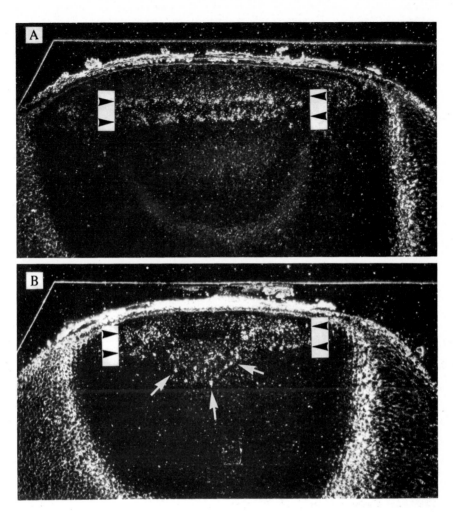

Fig. 12 A and B. A) This is a dark field light micrograph of an autoradiograph thin section (1μm) taken in a plane tangent to the retinal surface of the margin of the eye. The retinal growth zone is at the top, and the plane of the section goes through the curvature of eye cup to intersect the inner nuclear layer in the centre, grazing receptor outer segments near sides of the frame. The fish received intraperitoneal doses of ^{3}H-Tdr on its first and fifth post natal days, and was allowed to grow for a further nine days before preparation for autoradiography. Two precise labelled bands of nuclei show through the depth of the retina, parallel to the growth zone, and separated by about four units in the receptor mosaic B) Similar preparation at a more peripheral level in the same retina. The two fronts of label show, less precisely, among the receptor nuclei, but in addition, heavily labelled rod nuclei appear at a more central location in the retina. See page (x 150)

ent retinal cells, this significant point suggests that when a labelled division gives rise to a cone as one daughter product, the other daughter must go through sufficient rounds of division, generating retinal cells outside the receptor layer, that the label it carries be diluted beyond detectability before delivering a second cone of any species into the mosaic. That argument presupposes that label ac-

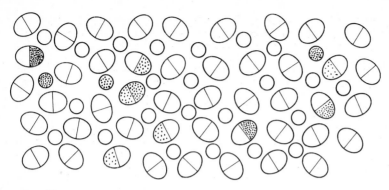

Fig. 13. Quantitative distribution of autoradiographic grains over nuclei of different kinds of cones in a small strip of the receptor mosaic. Dot: grain counts above
a criterion of one per nucleus per section in a serial sequence through the outer
nuclear layer. See text

quired during S-phase is divided equally at each subsequent round of
division, but there are uncertainties about the way newly replicated
DNA is partitioned between the progeny of cell division which deserve
outline before it can be accepted.

In many systems of proliferating eukaryotic cells, label applied during
one round of DNA replication dilutes geometrically into progeny at each
subsequent round of division in cold precursor. The distribution of
grains peripheral to a front of heavily labelled cells in the growing
thymidine pulsed retina (Fig. 12a) suggest strongly that the same happens in the tissue as a whole (cf. Hollyfield, 1972, and Fig. 16). One
explanation of this is that chromosomes segregate randomly into daughter cells during mitosis. However, there are many firm indications from
a number of types of proliferating cells (e.g. Lark et al., 1966), that
chromosomes may segregate non-randomly according to the age, in cell
cycles, of their component DNA strands, and that experimentally observed randomness of segregation may be due to the effects of premitotic
sister chromatid exchanges (Taylor, 1958), known to be encouraged by
radiation (Gibson and Prescott, 1972), and possibly occurring at an
artificially high rate in isotopically labelled chromosomes.

For present purposes, however, it does not matter much what balance
exists, in the labelled fish retina, between these different modes of
partition of DNA into mitotic daughter cells, since the difference between them can only be detected in the second and subsequent cell cycles after a pulse label. At the second division if chromosomes segregate according to the age of their DNA strands (Lark et al., 1966), label
acquired from a pulse during grandparent S-phase is concentrated in
two of four granddaughter cells and remains in two cells among an increasing unlabelled population of progeny at each subsequent round of
division. Thus only in the second and later cell cycles after an isotope pulse can one of two directly filially related cells carry label
and the other not. The overall distribution of label in the fish retina (e.g. Figs. 12, 16) suggests random partition of labelled chromatin at is, but if a minority of segregations are non-random, one cannot invoke them to explain instances, like that shown in Fig. 13, of
labelled cells isolated among unlabelled ones in the receptor layer,
because many of them are heavily labelled, and therefore have only
undergone one division since incorporation of labelled nucleoside.

Normally, groups of labelled rods and cones, or isolated labelled
cells, differentiate out of the growth zone apparently at random with
no evidence of lineal relationship between close neighbors in the mosaic. They assume a position in a tree ring appropriate to the point
in time at which the ^{3}H-Tdr pulse administered, and the accurate front
of the tree ring extends through to other cell types in the depth of
the retina (Fig. 12a). However, there is one exception to this pattern
of accretion of cells at the extending margin of the retina.

Depending upon the time of administration of a thymidine pulse, additional heavily labelled rod nuclei appear scattered in a more diffuse
band closer to the central area than the normal tree ring of activity
(Fig. 12b). Also, very infrequent mitotic figures can be seen in the
outer nuclear layer at the approximate position those labelled rods
would have occupied at the time of administration of the pulse, a position where the rods have not begun to differentiate along with the
other retinal cell types, but where, nevertheless, their nuclei are
recognisable by their dense chromatin.

Evidently, therefore, the precursors of rods are the only cells in the
retina to divide outside the growth zone, and they do so some time after they have been generated by mitosis among the stem cells there.
This second wave of cell division results only in an approximate dou-
bling of the number of rods (2) per unit of the cone mosaic provided
by normal mitoses in the growth zone. In agreement, there is a much
narrower spectrum of (high) grain counts per nucleus among the rods
involved than there is, as a result presumably of <u>repeated</u> divisions
after incorporation, among the daughters of premitotic synthesis phases
pulsed in the growth zone proper.

In many of the autoradiographic preparations, rod nuclei were not labelled in this way, and carried isotope only in the restricted retinal area of the normal growth ring, even though their numbers doubled
between the extreme periphery and more central areas of the retina.
That observation suggests that this secondary proliferation of rods
is conducted intermittently, and prompts the question whether proliferation of other retinal cells is episodic in the same way.

Although labelled receptor cells of all types differentiate at random
out of the growth zone following a thymidine pulse (e.g. Fig. 13),
there is usually a preponderance of one receptor type or another, depending upon the preparation, extending over long distances round the
margin of the eye. Fig. 14 shows histograms, compiled from different
animals, of the fraction of total autoradiographic grain counts which
were distributed over the nuclei of the different cone types in the
peripheral retinal mosaic. The data come from serial section autoradiographs through the depth of the outer nuclear layer, extending along
between 50 and 100 <u>files</u> of the cone mosaic at the dorsal eye margin.
They are weighted, in each case, to take account of the relative frequency of each cone type in the mosaic (e.g. there are about twice as
many double cones as singles and obliques); each contains 1000-2000
grain counts from 75-100 labelled cells.

The histogram of Fig. 14a shows that, at the time of ^{3}H-Tdr administration, synthetic activity among precursor cells for oblique cones
exceeded by a considerable and significant factor that in precursors
for the remaining cone species. In the retina of Fig. 14b single cone
precursor synthesis predominated, particularly at the expense of oblique cones, while in Fig. 14c accessory cones predominate at the expense of principal elements and single and oblique cones are roughly
equally labelled.

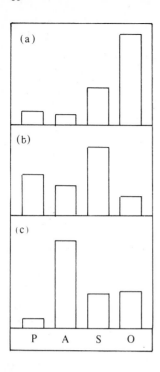

Fig. 14. Histograms compiled from labelled fronts in the receptor mosaic at the dorsal margins of eyes from different broodmate fishes, pulsed with ^{3}H-Tdr at different times (a,b,c) in their first 36 post-natal h, showing that precursors of the different kinds of cone cells undergo terminal mitotic activity episodically in the growth zone at the retinal margin. The height of each bar indicates the proportion of the total label in all cones, distributed among receptors of the types indicated, and is weighted inversely according to the frequency with which that cone population is distributed in the receptor mosaic (cf. Fig. 1). Rods behave in a similar manner to cones, but they are omitted from the histogram because their chromatin is more compact, and their autographic grain counts need rather complex correction before they are comparable with the figures for cones. Principal elements P; Accessory cones A; Single cones S; Oblique cones O

These data are examples selected from the same (dorsal) retinal areas in different fish, part of a larger experimental group of broodmate juveniles who received single ^{3}H-Tdr pulses at different times on a three hour schedule throughout their first 36 h post-natal life and were prepared for autoradiography 10 days later. They show clearly that there is episodic production of the different receptor populations by mitosis in the growth zone of the retina, loosely synchronised over circumferential distances as great as a quarter of the margin of the eye, if not right round it. The experiment was intended to show whether there is a diurnal rhythm (cf. Scheving and Chiakulas, 1965) to cone production in the growing retina, and whether terminal divisions generating the different cone populations follow one another in a definite sequence. That it broadly failed to do, largely, one suspects, because of great variability in the growth rate of individual fish.

Fig. 15 illustrates a further point, generally, though not always, true in most of the labelled retinas examined. It is a small part of a retina where oblique cone nuclei carried the most activity, and shows that though some cones of other types were labelled, they individually carried less isotope and were located closer to the retinal margin than the oblique cones. This is shown quantitatively in the histograms of Fig. 16, where grain counts along an extended strip of the retinal margin are related to functional cone types in successive ranks of the cone mosaic (lower sketch), in the direction of retinal growth (abscissal scale). Labelled oblique cones are concentrated in a front farthest away from the growth zone, while less heavily labelled cones of other types, fewer in number, are located in more peripheral mosaic ranks. Both their position and their activity suggest they originate from cell lines which have gone through more than one round of division since their precursors incorporated labelled nucleoside from the pulse. Although some of these weakly labelled second

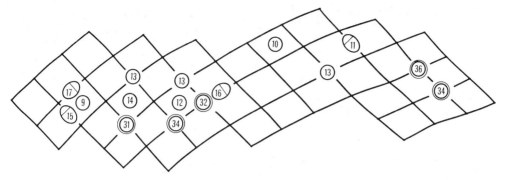

Fig. 15. Numbers indicate autoradiographic counts for the cone nuclei indicated in this diagram of a strip of labelled receptor mosaic. The labelled cone outlines only are drawn, and the periodicity of the mosaic indicated by lines. Oblique cones, predominantly labelled, are located further from the growth zone (parallel to, and in the direction of, the top of the diagram) than less heavily labelled nuclei of other cone types. See also Fig. 16

generation cones were located directly peripheral to heavily labelled cones (Fig. 15), the majority of them were uncorrelated in this way, and therefore not directly related through their lineage to the cones which differentiated first.

To summarize, two points arise from these autoradiographic observations which should be accommodated by any other theory of the way in which retinal cells differentiate in growth. First of all, despite the orderly arrangement of cells in the receptor layer (Fig. 1), no two neighbours in the mosaic, not even the paired elements of the double cones, regularly arise either as mitotic daughters of common mother cells, or as the progeny of two successive stem cell cleavages. Since differentiation keeps pace with cell division at the margin of the growing retina, cell proliferation must proceed according to a stem cell mode of mitotic lineage. Because labelled cones regularly appear in the autoradiographs quite isolated in the receptor layer, the stem cells must, immediately after they have delivered a division product to differentiate as a cone, then proceed as a matter of rule to deliver at least a few cells into the nervous layers of the retina before again adding to the receptor population.

The second point is that although the stem cell divisions which deliver daughters into the receptor layer are often synchronised and, over quite large retinal areas, involve cells predominantly of just one functional class (Figs. 14, 15), they nevertheless represent only a fraction of the mitoses in the growth zone that are proceeding at any time. The majority are engaged in delivering cells to the two nervous layers of the retina (Fig. 12). Following a thymidine pulse, first-generation cones which differentiate into the receptor layer and the second and later generation receptors (of different types) which follow them (Fig. 16) are not lineally related in any direct way. Therefore, the later generation cells must be the progeny of stem cells whose divisions at the time of administration of labelled nucleoside were involved with the nervous layers of the retina.

These two points suggest, but do not prove, that individual stem cells at the retinal margin proceed through repeated and regular routines of cleavage, producing single receptor cells of particular types, followed

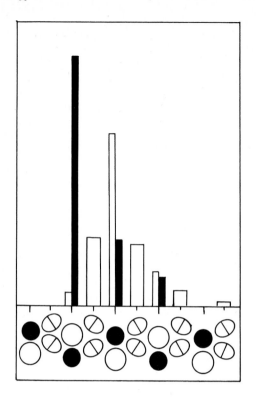

Fig. 16. The histogram is compiled from grain counts along a hundred files of cones
along the retinal margin, a small area of which is shown in Fig. 15. Individual oblique cones at the front of the strip of label were treated as though they all were
located in the same coherent rank in the cone mosaic, and the positions, relative
to them, of other labelled cones in their neighbourhood were preserved by shifting
them into common register along the growth axis (abscissa and sketch) of the cone
mosaic. Oblique cones are marked black in the mosaic drawing and narrow black bars
in the histogram are grain counts referring to them. Narrow white bars: single cone
counts, and broad white bars: pooled grain counts for principal and accessory elements of the double cones

by the more numerous retinal nerve cells associated with them in the
differentiated tissue. Divisions associated with particular functional
types of receptors tend to synchronise over quite long distances around
the retinal margin. But while it is clear that mitoses which produce
cones are scheduled to the extent that they are both preceded and followed by divisions involving nerve cells, it is not yet possible to
specify their order, nor to say whether the same stem cell lines produce all of the receptor types and their attendant neurons in turn,
or there are separate lines for each functional system.

Now, the neural connectivity of the different receptors in the fish
retina suggests that strict numerical relations are established in
growth or embryonic development between the colour cone populations
and the different neuronal systems which selectively synapse with
them. In the rudd retina, just as there are some 30% more red than
green cones, so bipolar cells which connect to red cones outnumber
those connected to green by just the same factor (Scholes, 1975).
Moreover, in retinas where the orderly receptor mosaic is reflected

by a congruent neuronal lattice below (Hibbard, 1971), there is an exact numerical correspondence between individual receptors and nerve cells in the tissue, even in retinal areas (unpublished observations) where there are insertions or other irregularities (Sect. 6 of this Chap.) in the basic receptor mosaic, and even though the nerve cells involved synapse with quite large and usually somewhat loosely speci- fied numbers of receptors. It is important to establish whether these interspersed populations of different cellular phenotypes owe their numerical exactness directly to regular programmes of cell prolifera- tion in development and growth, or to intervening regulative organis- ing processes acting upon essentially uncommitted products of cell division.

1.3.9. References

Baylor, D.A., Fuortes, M.G.F., O'Bryan, P.M.: Receptive fields of cones in the retina of the turtle. J. Physiol. 214, 265-294 (1970)

Beauchamp, R.D., Daw, N.W.: Rod and cone input to single goldfish optic nerve fibres. Vis. Res. 12, 1201-1212 (1972)

Benzer, S.: Genetic dissection of behaviour. Sci. Am. 229 (12), 24-37 (1973)

Boycott, B.B., Dowling, J.E.: Organisation of the primate retina: light microscopy. Phil. Trans. R. Soc. London 255B, 109-184 (1969)

Brown, P.K., Wald, G.: Visual pigments in single rods and cones of the human retina. Science 144, 45-52 (1964)

Cajal, S.R.: La retine des vertebrés. Cellule 9, 119-259 (1892). Translated by S.A. Thorpe and M. Glickstein. In: The structure of the retina. Springfield, Ill., Thomas and by R.W. Rodieck. In: The Vertebrate Retina. San Francisco: W.H. Freeman 1973

Dowling, J.E.: Foveal receptors of the monkey retina: fine structure. Science 147, 57-59 (1965)

Dowling, J.E., Boycott, B.B.: Organisation of the primate retina: electron microscopy, Proc. R. Soc. London 166B, 80-111 (1966)

Evans, E.M.: On the ultrastructure of the synaptic region of visual receptors in certain vertebrates. Zeit. Zellforsch. 71, 499-516 (1966)

Fujita, S., Horii, N.: Analysis of cytogenesis in chick retina by tritiated thymidine autoradiography. Arch. Histol. Japan. 23, 359-366 (1963)

Fuortes, M.G.F., Schwartz, E.A., Simon, E.J.: Colour dependence of cone responses in the turtle retina. J. Physiol. 234, 199-216 (1973)

Fuortes, M.G.F., Simon, E.J.: Interactions leading to horizontal cell responses in the turtle retina. J. Physiol. 240, 177-198 (1974)

Gaze, R.M., Watson, W.E.: Cell division and migration in the brain after optic nerve lesion. In: Growth of Nervous System. Wolstenbolme, O'Connor (eds.). Ciba Foundation Symp. London: Churchill 1968, pp. 53-67

Gibson, D.A., Prescott, D.M.: Induction of sister chromatid exchanges in chromo- somes of rat kangaroo cells by tritium incorporated into DNA. Exp. Cell. Res. 74, 397-402 (1972)

Glücksmann, A.: Development and differentiation of the tadpole eye. Brit. J. Ophthalmol. 24, 152-178 (1940)

Hibbard, E.: Grid patterns in the retinal organisation of the cichlid fish Astronotus ocellatus. Exp. Eye. Res. 12, 175-180 (1971)

Hollyfield, J.G.: Differential addition of cells to the retina in Rana pipiens tadpoles. Dev. Biol. 18, 163-179 (1968)

Hollyfield, J.G.: Histogenesis of the retina in the killifish Fundulus hetero- clitus. J. Comp. Neurol. 148, 373-380 (1972)

Horridge, G.A., Meinertzhagen, I.A.: The accuracy of the patterns of connexions of the first- and second-order neurons of the visual system of Calliphora. Proc. R. Soc. London 175B, 69-82 (1970)

Jacobson, M.: Developmental Neurobiology. New York: Holt 1970

Kahn, A.J.: An autoradiographic analysis of the time of appearance of neurons in the developing chick neural retina. Develop. Biol. 38, 30-40 (1974)

Kaneko, A.: Physiological and morphological identification of horizontal, bipolar and amacrine cells in goldfish retina. J. Physiol. 207, 623-633 (1970)

Kolb, H.: Organisation of the outer plexiform layer of the primate retina: electron microscopy of Golgi-impregnated cells. Phil. Trans. R. Soc. London 258B, 261-283 (1970)

Kolb, H.: The connections between horizontal cells and photoreceptors in the retina of the cat: Electronmicroscopy of Golgi preparations. J. comp. Neurol. 155, 1-14 (1974)

Kolmer, W.: Die Netzhaut. In: Handbuch der Mikroskopischen Anatomie des Menschen. v. Möllendorf, W. (ed.). Berlin: Springer 1963, Vol. III, pp. 295-468

Lark, K.G., Consigli, R.A., Minocha, H.C.: Segregation of sister chromatids in mammalian cells. Science 154, 1202-1204 (1966)

Lasansky, A.: Synaptic organisation of cone cells in the turtle retina. Phil. Trans. R. Soc. London 262B, 365-381 (1971)

Lasansky, A.: Cell junctions at the outer synaptic layer of the retina. Invest. Ophthalmol. 11, 265-274 (1972)

Lasansky, A.: Organisation of the outer synaptic layer in the retina of the larval tiger salamander. Phil. Trans. R. Soc. London 265B, 471-489 (1973)

Leach, E.H.: On the structure of the retina of man and monkey. J. Roy. Micros. Soc. 82, 135-143 (1963)

Liebman, P.A.: Microspectrophotometry of photoreceptors. In: Handbook of Sensory Physiology. Berlin: Springer 1972, Vol. VII/1, pp. 482-528

Marc, R.E., Sperling, H.G.: Colour receptor identities of goldfish cones. Sciences 191, 487-489 (1976)

Marks, W.B.: Visual pigments of single goldfish cones. J. Physiol. 178, 14-32 (1965)

Marks, W.B., Dobelle, W.H., MacNichol, E.F.: Visual pigments of single primate cones. Science 143, 1181-1183 (1964)

Meinertzhagen, I.A.: Erroneous projection of retinula axons beneath a dislocation in the retinal equator of Calliphona. Brain Res. 41, 39-49 (1972)

Missotten, L.: The Ultrastructure of the Human Retina. Brussels: Arscia Uitgaven 1965

Morris, V.B.: Time differences in the formation of the receptor types in the developing chick retina. J. Comp. Neurol. 151, 323-330 (1973)

Müller, H.: Bau und Wachstum der Netzhaut des Guppy (Lebistes reticulatus). Zool. Jahrb., Abtl. F. Allg. Zool. Physiol. 63, 275-322 (1952)

Naka, K.I., Rushton, W.A.H.: S-potentials from colour units in the retina of fish (Cyprinidae). J. Physiol. 185, 536-555 (1966a)

Naka, K.I., Rushton, W.A.H.: An attempt to analyse colour reception by electrophysiology. J. Physiol. 185, 556-586 (1966b)

Naka, K.I., Rushton, W.A.H.: S-potentials from luminosity units in the retina of fish. J. Physiol. 185, 587-599 (1966c)

Parthe, V.: Horizontal, bipolar and oligopolar cells in the teleost retina. Vis. Res. 12, 395-406 (1972)

Polyak, S.L.: The Retina. Chicago: Chicago Univ. 1941

Raynauld, J.P.: Goldfish retina: sign of the rod input in opponent colour ganglion cells. Science 177, 84-85 (1972)

Roth, S.: A molecular model for cell interactions. Quart. Rev. Biol. 48, 541-563 (1973)

Scheving, L.E., Chiakulas, J.J.: Twenty four hour periodicity in the uptake of tritiated thymidine and its relation to mitotic rate in urodele larval epidermis. Exp. Cell. Res. 39, 161-169 (1965)

Scholes, J.H., Morris, J.: Receptor-bipolar connectivity patterns in the fish retina. Nature 241, 52-54 (1973)

Scholes, J.H.: Colour receptors, and their synaptic connexions in the retina of a cyprinid fish. Phil. Trans. R. Soc. London 270B, 61-118 (1975)

Sidman, R.L.: Histogenesis of mouse retina studied with thymidine - H^3. In: Structure of the Eye. Smelser, G.K. (ed.). New York: Academic Press 1961, pp. 487-505

Stell, W.K.: The structure and relationships of horizontal cells and photoreceptor - bipolar synaptic complexes in goldfish retina. Am. J. Anat. 121, 401-423 (1967)

Stell, W.K., Lightfoot, T.: Colour specific Interconnexions of Cones and Horizontal
    Cells in the Retina of the goldfish. J. Comp. Neurol. 159, 473-502 (1975)
Stell, W.K., Harosi, F.: (1975, in prep.)
Svaetichin, G., Negishi, K., Fatehchand: Cellular mechanisms of a Young-Hering
    visual system. In: Ciba Foundation Symposium on Colour Vision. DeReuck, A.V.S.,
    Knight, J. (eds.). London, Churchill 1965, pp. 179-203
Taylor, J.H.: Sister chromatid exchanges in tritium labelled chromosomes,
    Genetics 43, 515-529 (1958)
Toyoda, J.I.: Membrane resistance changes underlying the bipolar cell response
    in the carp retina. Vis. Res. 13, 283-291 (1973)
Wagner, H.J.: Darkness induced reduction in the number of synaptic ribbons in
    fish retina. Nature 246, 53-55 (1973)
Werblin, F.S., Dowling, J.E.: Organisation of the retina of the mudpuppy
    Necturus maculosus. J. Neurophysiol. 32, 339-355 (1969)
Witkovsky, P., Shakib, M., Ripps, H.: Interreceptoral junctions in the teleost
    retina. Invest. Ophthalmol. 13, 996-1009 (1974).

1.4 Golgi, Procion, Kernels and Current Injection

K. Naka

1.4.1 Introduction

The study of the retina, in an attempt to determine how visual information is processed, comprises primarily two interrelated fields:
biochemistry, to learn how energy carried by the incident photons generates a polarization of the receptor membrane; and functional morphology, to discover how such changes in the receptor membranes are transformed into a series of spike discharges. During the last eight years
our group at the California Institute of Technology has been working
exclusively on the second phase, to ascertain how a signal is processed within the retina.

From the outset of this venture, we had two important motivations:
(1) We wished to focus the many and varied capacities of several different disciplines into a systematic procedure to solve a given problem. This multifaceted methodology is somewhat different from the traditional single-avenued approach in which an isolated phenomenon is
probed in great depth in various preparations but with a limited number of techniques; (2) we tried in all instances to correlate our observations with the end product of all the information processing in
the retina, the ganglion cell discharges, as a reference.

This multidisciplinary approach has been time-consuming and expensive,
not so much in terms of money but more in terms of effort expended.
We had to develop and modify our procedures empirically, and serially
rather than in parallel, as we proceeded. Clearly, the decision most
crucial for ultimate success was the initial selection of the preparation itself, since any judgment as to the wisdom of the choice had to
wait until the entire project had made considerable progress.

Experience has indicated that the retina of the channel catfish,
Ictalurus punctatus, was a good choice, based on several factors:
(1) the fish is readily available in Pasadena, and (2) is stable the
year round in that it does not experience seasonal variations in physiological activity as do the frog and turtle. (3) Its retina lacks
the C-type (or chromatic) horizontal cells and (4) its classical ganglion cells have typically a concentric (or biphasic) receptive field
organization. Furthermore, almost all neuron types Cajal (1893) saw
in other vertebrate retinas can be found in the catfish retina. Double and/or accidental staining, a serious problem in which Procion
dye appears in more than the one cell recorded from, does not occur
in the catfish retina as it does in other densely packed retinas such
as the frog's. The disadvantages we have encountered - the small receptors and the large dendritic fields of the proximal neurons which
make their complete structural description difficult - are relatively
insignificant when compared with the problems we encountered in the
frog, mudpuppy and turtle retinas (Matsumoto and Naka, 1972; Ohtsuka,
pers. comm.). Much more effort might have been required to enable us
to reach a similar level of understanding had we chosen a different
retina with a more complicated functional organization.

The catfish retina is a horizontal cell-preponderated retina, like
that of the skate or dogfish shark. It has multiple papillae which
serve as convenient landmarks for the location of dye-injected cells.
Curiously, no thorough physiological or structural study of this retina has yet been reported, except for a few attempts at the level of
the receptors (Walls, 1967; Adomian, 1972; Witkovsky et al., 1974).

From the outset our work was categorized into three main areas:
(1) structural definition of the various retinal cell types, (2) their
functional characterization, and (3) determination of the pathways and
mechanisms of signal transmission within the retina. Structural identification on a gross level was based on neuronal shapes: the merits
of this method are best exemplified by the work early in this century
of Cajal whose classification scheme is still being honored. Our functional identification sought to assign specific response parameters to
specific classes of neurons through a systems analysis approach. Although a popular pursuit in engineering, this methodology has not been
given much attention in neurophysiology. Identification of signal transmission involved a search both for synapses among cells and for transfer functionals for given neuron chains.

In this paper we will elaborate on these three identification procedures to show how we developed our view on the processing of signals
in the catfish retina.

1.4.2 Structural Identification

Morphology has as its aim the organized categorization of neurons
based on their overall shapes and substructures; it has preceeded its
functional counterpart by almost a hundred years because of its lack
of dependence on modern technology. It involves analyzing cells not
only for their fine structural detail, as revealed by various staining procedures, but also for their place in a comprehensive classification scheme.

Staining itself is a very artificial process and is bound to produce
artifacts. Accordingly, in the catfish retina the morphological data
we obtained on almost every neuron were fragmentary, since not all
structural features of a cell were revealed by any single staining
method. The most preferred technique is not a stain at all, but an impregnation procedure - the Golgi Rapid Silver Method. It is known to
be capricious, especially so in fish retinas, but it has an undisputable advantage in that it produces cell silhouettes of high contrast because of the dark crystalline silver precipitate within each impregnated
neuron. For reliable identification of smaller processes this is very
important. On the other hand in the catfish retina, at least, this
method was not enough to identify axons unequivocally, either because
they were not impregnated or because they could not be distinguished.
from other look-alike processes. As Gallego (1971) has observed: "The
interpretation of the structure of the retina based only on knowledge
acquired by the Golgi method can be misleading."

This dilemma was remedied in part by the use of the methylene-blue
vital stain which easily stained the soma and principal dendrites
of a large number or neurons and clearly revealed the axons as they
joined the optic fiber bundles, although it produced only the outlines
of the major dendritic expansions. Insofar as ganglion cells can most
easily be distinguished structurally from amacrine cells by their
axons, methylene blue played a significant role in our classification

studies. Furthermore, the contrast between the stained material and the background was much like that in Procion dye-injected cells, making correlation of neuron images in the two preparations more relevant than when comparisons were made with the starker Golgi images.

The intracellular dye-injection technique represents the single most important advance in functional morphology, as it readily provides a means of correlating structure with function for any given class of neurons. The first convincing and esthetically pleasing examples of dye-injected retinal cells were published by Kaneko (1970) using the then newly discovered Procion yellow dye. However, this dye has a serious failing in that the contrast with the surrounding material, especially when in flat mount, is very poor, so that dye-injected ganglion cells, for example, appear to have rather sparse dendritic fields, whereas similar neurons in Golgi preparations are revealed to have densely criss-crossed dendritic arborizations. Furthermore, the dye also often fails to penetrate axons; in our experience usually only one in eight ganglion cells, identifiable also otherwise as such - functionally, by their spiking discharges - appeared to have axons (Chan, 1975; Naka and Ohtsuka, 1975). Due to these difficulties, at least in the catfish, reliable structural identification had to be made by correlating the images seen in Golgi, methylene-blue, and Procion preparations. However, this three-phased identification procedure provided a means for comparing preparations produced by the three staining methods and thereby exposed the artifacts peculiar to each.

Structural analysis of retinal neurons has traditionally been based on their side views, even though they exist in life as three-dimensional entities. Cells seen in tangential (or flat-mount) as well as in radial (or side) view can, therefore, be visualized more realistically, since identification of a class of neurons based only on one of these two aspects can be misleading. In this respect the series of studies on the dogfish shark retina by Stell and Witkovsky (1973) and Witkovsky and Stell (1973) deserves recognition; without their effort we would not have devised flat mounting for our Procion preparations. In the catfish retina, then, neurons were always first studied as flat-mount preparations and then were radially sectioned to reveal their side views (Naka and Ohtsuka, 1975). In this way (1) a large number of Procion dye-injection experiments could be performed in one retina, (2) the results of which could immediately be seen without involving the tedious process of sectioning and (3) all neurons into which dye had been injected could be examined to exclude the possibility of double staining, a significant consideration because in radial sections detection of double staining was very difficult. Also without initial flat mounting we would never have realized that some neurons extended processes more than 1 mm across the retina or been able to verify a ganglion cell's identity because its axon could be seen to enter the optic fiber bundle.

A most difficult phase in structural identification is the registration of neuron images and their objective classification. Neurons, like people, come in all sizes and shapes; no two neurons, except the photoreceptors, are exactly alike. Because the structures of the various retinal neurons constitute a continuum rather than discrete stages it is imperative to obtain a statistically significant number of examples with which to correlate the function of a particular class of neurons with its structure. Preconceptions can be treacherously deceptive. For example, although the distinction between a ganglion and an amacrine cell should be unequivocal insofar as one possesses an axon whereas the other does not, our study of the catfish retina suggests that to distinguish between these two cell types, based on fragmentary observations, is extremely difficult. Furthermore, there may be some intraretinal ganglion cells whose axons terminate within the synaptic layer.

Future efforts will be directed toward a coherent quantitative and objective description and classification of the various neuron shapes; three-dimensional computerized reconstruction of neuron images will be the first step. A thorough study of the ultrastructure of these cells by electron microscope will have to be carried out to locate morphological counterparts for functional phenomena, to add ultrastructural information to our structural description, and to confirm pathways of information processing already suggested by white-noise analysis and current injection studies.

1.4.3 Functional Identification

Structural categorization of neurons in the retina is naturally paralleled by an attempt to define its functional components. This functional identification relies on the input-output relationship of the system - in which a known input is given to a system to evoke some resultant output from it - to classify neurons into types which should correspond to classes organized on a strictly morphological basis. Although the term "functional identification" has not been used much by neurophysiologists, they do in fact spend most of their time sorting out neuron types according to their response patterns and assigning them to specific neuron types. In the past, confusion has arisen because stimulus configurations have almost always consisted of step inputs or pulses of light and the resulting responses have been described more or less qualitatively or subjectively. An example is the classification of amacrine cells as either sustained or transient types which our analysis in the catfish reveals are really very distinct neuron classes with quite dissimilar structural and response properties (Naka et al., 1975). Obviously the previously accepted classification criterion is arbitrary and oversimplified. Naka and Ohtsuka (1975) have shown that some sustained responses are more transient than others and vice versa. Some ganglion cells produce a sustained depolarization to one mode of photic input (a spot of light) and a transient depolarization to a different input configuration (an annulus of light). Therefore, the type of responses a given class of neurons produces is just as dependent on the pattern of stimulus used as it is on the type of neurons from which responses are recorded. This is also to be expected from an analysis of ganglion cell spike data (Naka and Nye, 1970). Clearly this is not a satisfactory procedure by which to develop a comprehensive functional identification scheme. It therefore becomes necessary to find those inputs which will enable us to define responses in a more quantitative fashion.

The most commonly used stimulus for quantitative analysis is the sinusoidal input; responses at each frequency can be obtained and the gain and phase relationships plotted. This is the most powerful as well as the most traditional input in linear analysis and has been used extensively in the study of the signal processing in the vertebrate retina (Hughes and Maffei, 1966; Spekreijse, 1969; Spekreijse and Norton, 1970). It was also this stimulus that Toyoda used to define the dynamics of the intracellular responses from the carp retina (Toyoda, 1974). An attempt was made to extend the utility of this sinusoidal input by adding noise to it, a process known as linearization of the system (Spekreijse, 1969). However, sinusoidal inputs have several drawbacks when applied to study the intracellular responses from the vertebrate neurons: (1) The fact that they are repetitive becomes a serious disadvantage in recording from smaller neurons where time is of prime value. (2) The process of collecting and analyzing data for several frequencies is also time consuming and therefore inefficient. (3) For highly nonlinear responses, such as the transient on-off depolarizations from the presumed amacrine cells (Werblin and Dowling, 1969; Kaneko, 1970), measurements of gain and phase themselves are problematical.

To overcome these difficulties inherent in the traditional forms of inputs Marmarelis and I, during the past five years, have explored the use of white-noise analysis as a probe to identify neurons functionally in the vertebrate retina (Marmarelis and Naka, 1972a, 1973). In 1958 Wiener had proposed that any nonlinear (as well as linear) system could be completely characterized by determining its response to a white-noise input which, in possessing a flat power spectrum over the entire bandwidth of the system and amplitudes of Gaussian distribution, would, effectively, test the system with all possible stimulus combinations. Credit must go to Marmarelis (1971) for the first comprehensive effort to apply Wiener's nonlinear analysis technique to biological systems, in particular, to the study of vertebrate retinal neurons. Our most recent work is in fact a revived version of an earlier series of studies (Rushton and Naka) undertaken to analyze the S-potential (now known as the horizontal cell response). There the task was to describe the potential within a reasonable framework of logic rather than to define its particular parameters.

In the vertebrate retina our group has experimented with three white-noise stimulus inputs: a spot or an annulus of light whose intensity was modulated in white-noise fashion (Naka et al., 1975), a thin ring of light whose diameter was modulated in white-noise fashion (Chan, 1975), and current injected into neurons with white-noise amplitude modulation (Marmarelis and Naka, 1972a). Each stimulus mode has its own problems, the least complex being the current input.

In all cases the input and output signals are cross-correlated after several steps of "pre-conditioning" (Marmarelis and Naka, 1973) to remove the slow drifts accompanying the intracellular recording and also to adjust the power level of the output signals, since the various intracellular recordings have signals of varying amplitudes. The first-order cross-correlation gives rise to the first-order kernels, the best linear approximation of the system response and coincidentally the source of the most significant information. Similarly the second-order correlation produces the second-order kernels. Our experience indicates that computation of kernels up to the second order is enough to predict the responses of most retinal neuron types with reasonable accuracy, with the exception of the type C neurons, which require additional third-order kernels to model their responses effectively (Naka et al., 1975; Chan, 1975). While prediction of very sharp changes of potential from baseline would require a much larger assembly of kernels, the signals we record from retinal neurons lack such abrupt fluctuations (except for spike discharges) and are almost always noisy. The fact that only a small set of kernels is sufficient to model the system responses is apparently surprising to engineers but in reality it reflects the intrinsic noisiness of both biological signals and our recording techniques.

This use of the white-noise stimulation and cross-correlation technique to obtain a set of kernels has many advantages over the traditional methods of testing a neuron by step or sinusoidal inputs: (1) White-noise is a rich input which maximizes the information gathering within any given span of time, a critical consideration in defining the

response characteristics of smaller retinal neurons in which both recording and intracellular staining must be made within a brief period
of stability. (2) The process of cross-correlation eliminates most un-
wanted (or uncorrelated) noise at the output, a further important factor in the study of smaller retinal neurons in which drifts or noise
from electrodes contaminate the response. (3) The resulting kernels
are a statistical average over the entire experimental period and thus
provide a more reliable estimate of the system characteristics.
(4) Since they are expressed in canonical, or mathematically compact
and consistent form, kernels from different neurons or varying experimental conditions can be compared quantitatively. (5) The process of
cross-correlation allows responses evoked by multiple-inputs to be
analyzed into separate components as if each were derived individually
(Marmarelis and Naka, 1974).

The last-mentioned point is the most unique feature of the white-noise
analysis technique and deserves further comment. Neurons in the vertebrate retina possess biphasic (or concentric) receptive field organi-
zations in which the configuration of the light input, whether a spot
(stimulating the center) or an annulus (stimulating the surround),
plays the critical role. Two-input white-noise analysis (in which the
intensities of both the spot and annulus are modulated independently)
performed on these neurons has enabled us to separate their response
patterns into two components, each due either to the spot or to the
annular stimulus, to determine the relative contribution from and the
mutual interaction between each input. Without the analysis procedure
such a separation of receptive field components under dynamic conditions is not possible (Marmarelis and Naka, 1974). Furthermore, this
method can easily be extended to study color processing by separating
responses into components corresponding to each chromatic input and,
in the spatial domain, to define the spatial dynamics of the various
retinal neurons (S. Yasui, pers. comm.).

For a successful execution of white-noise analysis one must have a
fair knowledge of the system to which it is applied because the method
itself provides a quantitative rather than a qualitative description
of that system. For example, the input and output have to be specified
together with a rough estimate of their respective bandwidths. For traditional biologists definition of a system input and output may be only
one of a number of compounding problems and may well restrict the general application of white-noise analysis, particularly as so much of
biology is devoted to little more than qualitative descriptions of ob-
servable phenomena.

As for problems involved in its application, the use of the white-noise
analysis technique to analyze the light-evoked responses from the retinal neurons involves a serious drawback in that there is no antithet-
ical stimulus mode to photic inputs: a light signal has only a positive
value; negative light does not exist. Thus the white-noise signal, instead of being modulated around a zero value, as originally prescribed,
is set to modulate in Gaussian fashion around a mean (or average) intensity value; this input represents what Rushton called the "field
adaptation" (1965). It is a dynamic adjustment of retinal sensitivity
according to the level of average light intensity of the environment
and if often also called, reasonably enough, light adaptation. We are
yet to agree as to how the retina controls its gain automatically but
the fact it does undergo this field adaptation is undeniable.

Therefore, a Gaussian-modulated white-noise stimulus is exactly the
kind of input the "automatic gain control mechanism" in the retina is
best equipped to handle. When stimulated by a white-noise signal, the
system of neurons within the retina quickly computes the average light

intensity level and sets up a new gain setting. Thus a white-noise input directly affects the state of the retina.

The most basic - and most serious - limitation of the applicability of Wiener's analysis is the lack of stationarity of the system itself. If we restrict our observations only to a specific state at a given mean intensity level, the system can be considered roughly, but not exactly, stationary. Thus a small set of kernels, while capable of mimicking the system response with a considerable degree of accuracy, is nevertheless valid only for that particular input intensity level at which the responses were recorded. We have had to reconcile ourselves to the fact that a "complete description" of a system can therefore be so only in a very narrow sense, since each set of kernels will exhibit different gain and dynamic characteristics for each spatial configuration and mean intensity level of photic input. It seems, then, that the application of the Wiener-Marmarelis analysis method is not as straightforward as we once thought. Nevertheless, it is fair to say that it still represents the most sophisticated, even if qualified, approach to evaluating functionally any biological system, especially as there are a number of ingenious ways in which its inherent restrictions can be minimized, if not avoided altogether, and its power and scope exploited to the fullest.

1.4.4 Identification of Neuron Interconnections

A natural sequel to the classification of neurons based on morphological and functional traits is an attempt to establish their interconnections through a procedure of three phases: (1) morphological evidence, (2) direct functional indications and (3) indirect functional inferences.

1. Morphological Evidence. The most conventional approach is to seek patterns of connections using an electron microscope, a process routine to the point of being tedious but certain of revealing neural pathways where they exist.

2. Direct Functional Indications. Connectivity between pairs of neurons can be determined by injecting test signals into one of them and recording the resulting responses from the other. This is the most direct and straightforward approach and has been used to establish the existence of functional contacts among receptors (Baylor et al., 1971), between external and intermediate horizontal cells and receptors, and between horizontal and ganglion cells (Maksimova, 1970; Naka and Nye, 1971). The injected current is a somewhat unnatural input, however, in that in life no stimulus configuration approaches its size as a point source, nor are single units ever excited in such total isolation. Furthermore, some minute cell pairs may defy being impaled, so that their contribution to information processing in the retina might be overlooked.

3. Indirect Functional Inferences. An alternative means to establish patterns of neuron connections involves mathematical analysis; a classical example is the identification of the laminas formed by rows of horizontal cells as the S-spaces (Naka and Rushton, 1967). Recently Chan (1975) suggested an indirect method of establishing neuron connectivity using nonlinear analysis to derive mathematically the transfer functionals between pairs of neurons by knowing the overall transfer functionals between the common input (light, in the case of the retina) and the individual outputs. If the transfer functionals between

101

any two stages are physically unrealizable, no connections can exist
between the neurons they represent; if the transfer functionals are
physically realizable a connection between the two stages is thereby
possible. Although the development of this technique is still preliminary this seems to be the only practical method to resolve functionally complex patterns of neuronal connection.

1.4.5 Catfish Retina

We have been applying these three identification procedures to the
cat fish retina to establish not only the functional morphology but
also the interconnections of its component neurons. Our results can
be briefly summarized in Figs. 1 and 2.

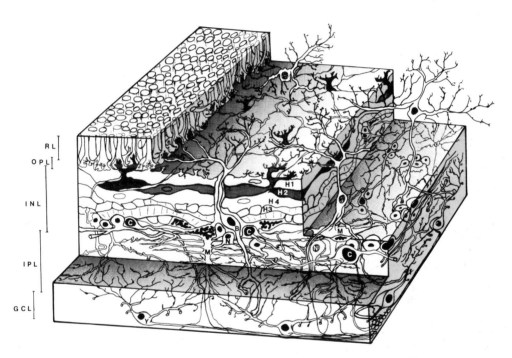

Fig. 1. A three-dimensional view of the channel catfish (Ictalurus punctatus) retina, illustrating the structural relationship of the various layers and their component neurons. The receptor layer: RL; the outer synaptic layer: OPL; the inner
nuclear layer: INL containing the layers of external H1, intermediate H2, striated
or internal H3 and nucleated H4 horizontal cells interspersed by the cell bodies
and main vertical processes from bipolar cells B and Müller fibers M; inner synaptic
layer IPL; types N and C cells (labeled N and C respectively); ganglion cell layer
GCL with ganglion or type Y cells. This drawing was made based on observations made
on a large number of Golgi-, methylene-blue, and Procion yellow-stained neurons

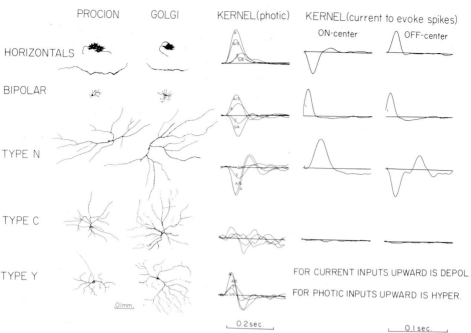

PROCION   GOLGI   KERNEL(photic)   KERNEL(current to evoke spikes)

Fig. 2. Morphological and functional traits of the catfish retinal neurons. In Procion and Golgi preparations neurons are seen in flat mount; pairs of each preparation were chosen arbitrarily to match each other. The first-order kernels from
light-evoked responses are for external horizontal, on-center bipolar, type NA,
type C and type YB (or off-center ganglion) cells. S and A are for single-input
experiments in which intensity of a spot or an annulus of light was modulated in
white-noise fashion whereas S/A and A/S are for two-input experiments in which intensities of the spot and annular inputs were modulated by independent white-noise
signals. S/A is for the spot component in the presence of an annular input whereas
A/S is for annular component in the presence of a spot input. Experimental parameters are similar to those described in Naka et al. (1975). Upward deflection is
for an hyperpolarization of the cell membrane. Notice that in the presence of the
complimentary member of two receptive field components the dynamic gain of both
the spot and annular components in the bipolar cell responses showed a marked increase. In the type N (depolarizing variant of type NA) neuron the presence of an
annular component depressed the spot component. The first order kernels to produce
ganglion cell discharges of current injected into various neurons are shown for
both on-center (type A) and off-center (type B) ganglion cells. For the on-center
ganglion cells a hyperpolarization of a horizontal cell or a depolarization of a
type NA neuron produced discharges whereas for the off-center ganglion cell the
relationship was reversed. Note the longer peak time of the kernels for the type
NA neuron. Depolarization of bipolar cells, regardless of ganglion cell types, always produced discharges from the latter neuron. Current injected into type C neurons never excited or depressed the nearby ganglion cells. The electrode for injecting current was placed a few hundred microns away from the spike recording
electrode. Linear kernels were computed by cross-correlating the white-noise current signal against normalized spike discharges. Durations of white-noise current
injection were about 60 s during which dye was also deposited. Upward deflection
is for depolarization of the presynaptic neurons. Raymond Y. Chan kindly provided
us with the Procion drawings

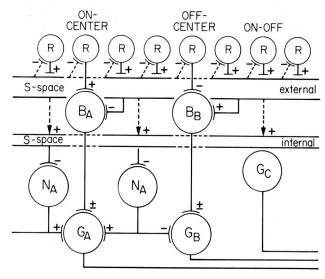

Fig. 3. Summary diagram of the functional and morphological organization of the catfish retina to produce a biphasic (or concentric) receptive response. Type A pairs
form the on-center fields and type B pairs the off-center fields, whereas the type C
neurons (the presumed displaced ganglion cells) are inferred to form on-off fields.
Although no information is yet available as to the types of neurons which feed signals to these type C neurons, it is not unreasonable to assume, from their highly
nonlinear characteristics, that they receive inputs from all other neuron types,
except, obviously, the type Y or classical ganglion cells. The basic patterns of
types A and B fields are formed by three neuron types, receptor, horizontal and
bipolar cells. Type NA neurons seem to act as modifiers of the basic field patterns.
We do not know exactly how the internal horizontal cells act on the bipolar cells.
At "synapses" identified by + signs signals are transmitted without polarity reversal (i.e. a presynaptic hyperpolarization produces a post-synaptic hyperpolarization) whereas at "synapses" identified by - signs signals are transmitted accompanied by their polarity reversal (i.e. a presynaptic depolarization produces a
postsynaptic hyperpolarization or vice versa). At synapses identified by ± signs
a depolarization of presynaptic neurons always leads to an excitation of the postsynaptic neurons and an hyperpolarization always to an inhibition or post-inhibitory
rebound in the postsynaptic element

1.4.5.1 Receptors

We have not studied the catfish receptors to any extent, although
Chan (1975) has noted that light induces only hyperpolarizing responses in cones.

1.4.5.2 Horizontal Cells

There are at least three classes of horizontal cells, the external
(H1 in Fig. 1), intermediate (H2 in Fig. 1) and internal horizontal
cells (H3 in Fig. 1). H4 in Fig. 1 represents a nucleated internal
horizontal cell (Naka and Carraway, unpubl. data). The external horizontal cells' distal processes not only invaginate the cone terminals
(probably to receive signals) but also make conventional synapses
back onto the receptor terminals (Naka and Carraway, unpubl. observation), presumably to improve their frequency response (Naka et al.,
1975), and onto bipolar cell processes as well. By use of a thin ring

Note added in proof: Polarity of some signs in Fig. 3 is reversed by mistake.

of light whose diameter was varied in white-noise fashion Chan (1975) found that for a large diameter modulation (mean diameter about 600 μ) the first order kernels indicated a depolarization of the membrane potential, possibly due to feedback of horizontal cell potentials to the receptors. The intermediate horizontal cells receive inputs from the rods and probably from synapses similar to those of the external horizontal cells back to the receptors and onto bipolar cells processes. The "internal horizontal" cells are those structures which normally lack nuclei and which have previously been thought to be axons from the external cells (Stell, 1975). In the catfish retina, however, in both Procion and EM preparations some of these cells definitely had nuclei and, in contrast to the smooth contours of the unnucleated type, had more angular outlines. The uncertainty of classification of these neurons, which initially seem very simply organized, indicates the degree of complexity involved in establishing any functional morphology among the retinal neurons.

Both the external (probably intermediate) and internal horizontal cells form individual laminas, or S-spaces (Fig. 3), in which current spreads uniformly in all directions among cells of the same type (Marmarelis and Naka, 1972b). The space constant of the external cells is far smaller than that of the internal cells; in the external cells the potential appears to be generated by an increase in the source resistance (probably produced by the transmitter release from the receptors), while in the internal cells the potential is due to an increase in the source current.

Within the range of modulation used the horizontal cells produced nearly linear responses and formed monophasic receptive fields. However, recently Chan (1975) discovered that for some modes of inputs the external horizontal cells are capable of displaying very complex response patterns. He attributed this phenomenon to the apparent feedback of horizontal cell potential back onto the receptors. Furthermore, extrinsic polarization of the horizontal cells produced ganglion cell discharges (Naka and Nye, 1971). Depolarizations elicited the same ganglion cell response as that evoked by a spot of light, whereas hyperpolarizations mimicked the action of an annulus of light. The first order kernels of the ganglion cell discharges evoked by current injected into the external cells have shorter latencies and peak response times than do those produced by current injected into the internal cells. This evidence strongly suggests that the two horizontal cell types are activating the ganglion cells via different pathways (Fig. 3). Further functional (Naka et al., 1975) as well as morphological evidence (Naka and Carraway, unpubl. observation) shows that the internal horizontal cells are communicating with processes in the inner synaptic layer, probably from type N (or true amacrine) cells, and as such constitute a vertical transmission line in addition to the classical one formed by the bipolar cells (Fig. 3).

The role of the horizontal cells is to produce integrating (or global) signals representative of the average intensity level of the environment (Naka and Nye, 1970; Schwartz, 1973); thus they are instrumental in the mechanisms of field adaptation rather than in bleaching adaptation.

1.4.5.3 Bipolar Cells

The distal processes of bipolar cells (labeled B in Fig. 1) extend into the outer synaptic layer to receive synaptic input from the (external and intermediate) horizontal cells and to become not only the central elements but, like the horizontal cell processes, the lateral elements as well within receptor terminals (Naka and Carraway, unpubl. observation). Bipolar cells are thus both secondary neurons after the receptor cells and third order neurons after the horizontal cells. The evidence in the catfish retina suggests that bipolar cells act as summing points or common mode rejecting devices in which the large dc components in the signals from both receptors (local signal) and hori- zontal cells (integrating signal) are cancelled against one another (Fig. 3), (Naka et al., 1975).

The proximal processes of bipolar cells are simply organized in the inner synaptic layer and receive a large number of conventional synap- ses while making only a relatively few ribbon synapses. Catfish bipolar cells form biphasic (or concentric) receptive fields in which a spot and an annulus of light give rise to responses of opposing polarity (Naka et al., 1975): they are fairly linear and have bandpass frequen- cy response characteristics.

The first order kernels of the ganglion cell discharges produced by an extrinsic polarization of all bipolar cell types are fast depolariza- tions (Fig. 2), indicative that whenever a bipolar cell depolarizes to a sufficient degree ganglion cells produce spike discharges; single bi- polar cells activate more than one ganglion cell and a single ganglion cell is activated by more than one bipolar cell. So far, these results are consistent with the view that type A bipolar cells (on-center) are communicating with type A ganglion cells (on-center) whereas type B bipolar cells (off-center) are communicating with type B ganglion cells (off-center) (Fig. 3). Synapses formed by bipolar cells are the only classical synapses found in the catfish retina in the sense that pre- synaptic depolarizations always result in a generation of ganglion cell spike discharges and hyperpolarizations always produce a suppression of ganglion discharges which rebound at the offset of current injection (Fig. 3). By use of a thin ring of light whose diameter was modulated in white-noise fashion Chan (1975) showed that the center of the cat- fish bipolar cell receptive field had an average diameter of 500 μ.

1.4.5.4 Proximal Neurons in the Catfish Retina

We found it was extremely difficult to classify neurons in the proximal layer of the retina according to their morphology or physiology alone (Naka and Carraway, 1975; Naka et al., 1975). In a series of papers we developed several independent categorizations based on different clas- sification approaches. In this paper we will adopt the identification scheme based on the degree of nonlinearity of light-evoked responses, thereby grouping the proximal neurons into three types, N, C, and Y. However, we stress the fact that this arrangement, although far more objective and consistent than those based only on response waveforms evoked by steps of light, is nevertheless neither unique nor definitive but is arbitrarily organized according to a set of particular response parameters. Correspondence between the functional and morphological classes is on a statistical rather than a rigid one-to-one basis.

1.4.5.5 Type N Neuron

The soma of the typical type N neuron (N in Fig. 1) is fairly small and lies in the proximal inner nuclear layer or in the distal inner synap- tic layer (Naka and Carraway, 1975). Its arborizations are simply orga- nized, tending to radiate laterally in the inner synaptic layer from the more or less centered soma which often is displaced to the side of a principal process by a "neck" of varying size. Some of these cells have very large fields, extending to 1 mm diameter in some instances.

These neurons are the only kind which fit the accepted definition of an amacrine cell (in that they lack axons) and correspond to Cajal's unistratified amacrine cells. Preliminary electron microscope studies suggest the internal horizontal cells make synaptic contact with these cells.

Type N neurons form monophasic receptive fields in which any form of light input always depolarizes (type NA) or hyperpolarizes (NB) the cell (Naka et al., 1975). So far we have not found any correlation between the two functional types and any conspicuous morphological characteristic. To steps of light NA neurons produce sustained depolarizations whereas NB neurons produce sustained hyperpolarizations (Naka and Ohtsuka, 1975). Type N neurons are linear within the modulation range used and give rise to highly underdamped (or differentiating) first order kernels (Fig. 2).

Discharges from types A and B ganglion cells are differentially evoked by extrinsic polarization of single type NA neurons whose mode of action on the ganglion cells is complimentary to that exerted by the horizontal cells (Fig. 2). The fact that the polarities of the first order kernels modeling ganglion cell discharges are independent of the kind of input from the type N neurons indicates the dual nature of their signals (Fig. 3). Functional as well as morphological evidence thus suggests that these type N neurons are receiving (at least some) synaptic inputs from internal horizontal cells (Fig. 3).

The functional traits of the type N neurons are much like those of the horizontal cells; Gallego (1971) has even deemed some neurons which are structurally similar to the catfish type N neurons to be the "amacrine cells of the outer synaptic layer".

1.4.5.6 Type C Neurons

The somata of the type C neurons (labeled C, Fig. 1) are typically spindle-shaped and lie adjacent to the somata of the type N neurons (Naka and Carraway, 1975; Naka and Ohtsuka, 1975; Naka et al., 1975). The bistratified dendritic fields of the type C neurons proliferate in the inner synaptic layer. In a few Procion preparations axons have been seen to issue from these neurons and in the methylene-blue preparations cells of this type always appear to possess axons, so that in Golgi preparations they are classified according to their characteristic shape of soma and dendritic field as a variation of ganglion cells.

Functionally, the type C cells, like the type N cells, form monophasic receptive fields but, unlike the type N cells, are highly nonlinear, requiring third-order kernels to predict their responses (Fig. 2); this suggests not only that their spike-like depolarizations are produced by some sort of threshold process but also that these neurons act like frequency-doublers (electronic devices which produce on-off events). In fact, responses from type C neurons to steps of light are on-off transients which remain unchanged in spite of variations in the stimulus parameters (Naka and Ohtsuka, 1975). Thus the power content of their responses is almost independent of the power content of the input signals (Kaneko and Hashimoto, 1969; Chan, 1975). Extrinsic current injected into type C neurons has always consistently failed to influence discharges from nearby ganglion cells (Fig. 2).

Thus, although types N and C neurons are referred to as sustained and transient amacrine cells in other retinas (Kaneko, 1973), our analysis suggests that these two types are really two different types of neurons altogether, a conclusion which could only have been reached through our multi-faceted approach. In the catfish retina, a more proper classifi-

cation would be to refer to the type N or sustained amacrine cell as
the true amacrine cell and the type C or transient amacrine cell as
a displaced ganglion cell (Fig. 3). Possibly type C neurons form a
third receptive field (or on-off receptive field) type which is specialized to detect only changes in light inputs, temporal as well as
spatial (Chan, 1975).

1.4.5.7 Type Y Neurons

The morphological and functional parameters of the type Y neurons -
or classical ganglion cells - vary considerably ($\underline{Y}$, Fig. 1). Their
somata and dendritic fields display a continuous spectrum of size and
shape. Naka and Carraway (1975) have proposed a scheme of classification similar to that devised by Ramón-Moliner (1962) in order to cate-
gorize these cells into seven (arbitrary) groups; one-, two-, three-,
four-, and multipolar, kite and spindle cells (the last-mentioned
type corresponds to the type C neuron). All the type Y neurons cross-
correlated in methylene-blue, Golgi- and Procion-preparations fall to
the simple end of Ramón-Moliner's classification progression and correlate with either his leptodendritic ("living fossils" within the
mammalian nervous system) or radiating (the prototype from which all
other dendritic patterns evolved) types. That the retina of the catfish represents such a basic, primitive organization may bear relevant implications in the study of more highly evolved visual systems.

Functionally there are two types of Y cells: the type A (on-center)
and the B (off-center) cells. Both form biphasic receptive fields. A
variety of intracellular responses have been observed from these neurons, some being transient and others sustained, some accompanied by
the clear spike discharges which others lack. These widely differing
functional traits reflect either the complex behavior patterns of the
type Y neurons, the varied membrane characteristics of different portions of the neurons from which the responses were recorded, or a combination of both circumstances.

The type Y neurons are moderately nonlinear, the most prominent non-
linearity being that of half-wave rectification, reflecting the nature
of the synaptic transmission between the bipolar and type Y neurons.
Using a thin ring of light whose diameter was modulated in white-noise
fashion Chan determined the outer diameter of the center of the receptive field to be about 700 μ, a value considerably larger than that of
the bipolar cell field center (Chan, 1975).

1.4.5.8 Organization of the Catfish Retina

The catfish retina comprises three distinct systems: the receptors,
and the vertical and the horizontal networks (Fig. 3). The receptors
interact with photons to produce certain electrical changes on which
we have little data in the catfish. The vertical networks include the
bipolar cell-ganglion cell pairs which are segregated into three independent channels: The types A (on-center) and B (off-center) receptive fields, organized by the type Y or classical ganglion cells, and
the type C (on-off) receptive field, involving probably the type C or
displaced ganglion cells. The horizontal and type N neurons constitute
the horizontal networks; their signals are colorless in the sense that
they are neither excitatory nor inhibitory but rather represent an on-
going integrating or lateral mechanism: It is incumbent upon the vertical system, in sharing the inputs from the horizontal networks, to
determine the (functional) polarity of these incoming signals. In this
sense lateral inhibition does not literally lead to a depression of

spike discharges but should ·be interpreted as a surround or integrating signal which opposes the action of the central or local signal.

1.4.6 Conclusion

The retina is a very intricate system. Any strategy to decipher the
processes of information transformation within it has to be equally
intricate. To define retinal neurons only according to their shapes
or responses to simpler inputs overlooks the basic complexity of the
retinal neurons themselves and their interrelationships. Even the
major (morphological) classifications of retinal neurons which looked
so firmly and unalterably established at the turn of this century must
be questioned.

There are many sources of ambiguity in categorizing cells. Each staining procedure we used emphasized a different facet of each cell's morphology, just as the various stimulus configurations elicited sometimes apparently inconsistent response patterns from the same neuron.
For example, Chan (1975) reported that sustained (type N) neurons responded transiently to spatially modulated inputs.

One of the most disappointing aspects of our venture to date is that,
despite our massive effort in which literally hundreds of neurons were
Procion dye-injected, Golgi silver-impregnated, or methylene-blue
stained and in which diversified inputs such as step, sinusoidal, temporally and spatially modulated white-noise stimuli were used, we are
not yet able to distinguish among the functional subtypes of any neuron type by their structure alone. The on-center bipolar cells appear
identical to the off-center bipolar cells, the two kinds of type N
cells look just alike, and there is no apparent functional difference
among the type Y (classical ganglion) neurons which can be correlated
to their structural variants. This illustrates the limitations of our
current techniques, as does the still speculative classification of
the horizontal cells which originally looked so straightforward: it
was not until we examined the horizontal cell layers under the electron microscope that we discovered the nucleated internal horizontal
cells (H4 in Fig. 1), after which we began to recognize this cell type
also in Procion preparations. Undoubtedly there remain other classes
of neurons (see Dowling and Ehinger, 1975) yet to be found in the catfish retina; we will benignly neglect their possible existence until
one day we happen across them by accident. Nevertheless, within the
limitations of our approach lies, paradoxically, its power. We would
possibly never have come to realize the very existence of the nucleated internal horizontal cell or the distinction between the type N
(true amacrine) cell and the type C (displaced ganglion) cell had we
not been applying a number of independent identification techniques
and cross-referencing the results. Furthermore, although the retina
outwits the white-noise input by adjusting its gain to match the mean
intensity level of the stimulus, white-noise analysis still has an
undisputable edge over other such conventional forms of stimulation
as step or sinusoidal inputs.

In other retinas, the (morphological) subtypes of each neuron are
easily recognizable, such as the bipolar cell variants in the dogfish
or primate retinas. The obstacle to our developing a more detailed
description of each subclass of neuron may derive from the inherent
simplicity of the catfish retinal neurons, a systematic deficiency
in our approach, a basic flaw in our philosophy which seeks to emphasize the similarities rather than the differences among neurons or
their subtypes, or a combination of all these considerations.

The many previous studies on the vertebrate retina have been limited in scope and piecemeal in approach: very complicated experiments performed to answer only a specific question, such as to classify the ganglion cells based on just their discharge patterns. However, it is obvious that the neural organization underlying the biphasic (concentric) receptive field cannot be elucidated reasonably through detailed descriptions of such an isolated phenomenon as the ganglion cell discharges alone. Neuron networks and processes leading to the production of these discharges have to be studied to answer even this simple question, so that the situation must be that much more complex for clarification of the neuron mechanisms responsible for such involved behavior as directional selectivity. I believe any approach must be systematic as well as comprehensive to make a lasting contribution to this field. I do not think that there are many preparations which are amenable to the multi-avenued approach; therefore, the selection of a suitable preparation is the very first problem to be faced. In this respect our choice of the catfish was a very fortunate one. It was also a very fortuitous circumstance that we happened to be with the Information Science Department and not in a traditional biology division, as our emphasis has concerned systematic acquisition and analysis of information rather than anecdotal descriptions of animal behavior.

To summarize this paper let it suffice to say that our approach has been experimental, to expand to its limits an analytical study of the vertebrate retinal organization. So far, we believe, our attempt has achieved its modest goal but it remains to be seen whether such will still be the case when we finally put together the various pieces of information to make a plausible model of the retina.

Acknowledgements. The author thanks the colleagues, past and present, in the Biosystems Group, Information Science Department, California Institute of Technology for their unfailing support and inspiring discussions. I also thank N. Franceschine for making useful comments. Research cited in this paper has been supported by PHS Grants NS 10629, NB 03627 and EY 00898.

1.4.7 References

Adomian, G.: The ultrastructural basis of the photomechanical response in the visual receptors of the catfish. Thesis, Univ. Calif. Los Angeles, 1972

Baylor, D.A., Fuortes, M.G.F., O'Bryan, P.M.: Receptive fields of single cones in the retina of the turtle. J. Physiol. (London) 214, 265-294 (1971)

Cajal, S.R.: La rétine des vertébrés. La cellule 9, 17-257 (1893)

Chan, R.Y.: Spatial dynamics of vertebrate retinal neurons. Thesis, Calif. Inst. Tech. Pasadena, 1975

Dowling, J.D., Ehinger, B.: Synaptic organization of the amine-containing interplexiform cells of the goldfish and cebus monkey retinas. Science 188, 270-273 (1975)

Gallego, A.: Horizontal and amacrine cells in the mammal's retina. Vis. Res. 3, 33-50 (1971)

Hughes, G.W., Maffei, L.: Retinal ganglion cell response to sinusoidal light stimulation. J. Neurophysiol. 39, 333-352 (1966)

Kaneko, A., Hashimoto, H.: Electrophysiological study of single neurons in the inner nuclear layer of the carp retina. Vis. Res. 9, 37-55 (1969)

Kaneko, A.: Physiological and morphological identification of horizontal, bipolar and amacrine cells in goldfish retina. J. Physiol. (London) 207, 623-633 (1970)

Kaneko, A.: Receptive field organization of bipolar and amacrine cells in the goldfish retina. J. Physiol. (London) 235, 133-163 (1973)

Maksimova, Y.M.: Effects of intracellular polarization of horizontal cells on the
activity of the ganglion cells of the retina of fish. Biophys. J. 14, 570-577
(1970)

Marmarelis, P.Z.: Nonlinear Dynamic Transfer Functions of Certain Retinal Neuronal
Systems. Thesis, Calif. Inst. Tech. Pasadena, 1971

Marmarelis, P.Z., Naka, K.I.: White-noise analysis of a neuron chain: An application of Wiener's theory. Science 175, 1276-1278 (1972a)

Marmarelis, P.Z., Naka, K.I.: Spatial distribution of potential in a flat cell:
Application to the catfish horizontal cell layers. Biophys. J. 12, 1515-1532
(1972b)

Marmarelis, P.Z., Naka, K.I.: Nonlinear analysis and synthesis of receptive-field
responses in the catfish retina. I. Horizontal cell to ganglion cell chain.
J. Neurophysiol. 36, 605-618 (1973)

Marmarelis, P.Z., Naka, K.I.: Identification of multi-input biological systems.
IEEE Trans. Bio.-Med. J. 21, 88-101 (1974)

Matsumoto, N., Naka, K.I.: Identification of intracellular responses in the frog
retina. Brain Res. 42, 59-71 (1972)

Naka, K.I., Carraway, N.R.G.: Morphological and functional identifications of catfish retinal neurons. I. Classical morphology. J. Neurophysiol. 38, 53-71 (1975)

Naka, K.I., Marmarelis, P.Z., Chan, R.Y.: Morphological and functional identifications of catfish retinal neurons. III. Functional identification.
J. Neurophysiol. 38, 92-131 (1975)

Naka, K.I., Nye, P.W.: Receptive-field organization of the catfish retina: Are at
least two lateral mechanisms involved? J. Neurophysiol. 33, 625-642 (1970)

Naka, K.I., Nye, P.W.: Role of horizontal cells in organization of the catfish
retinal receptive fields. J. Neurophysiol. 34, 785-801 (1971)

Naka, K.I., Ohtsuka, T.: Morphological and functional identifications of catfish
retinal neurons. II. Morphological identification.
J. Neurophysiol. 38, 72-91 (1975)

Naka, K.I., Rushton, W.A.H.: The generation and spread of S-potentials in fish
(Cyprinidae). J. Physiol. (London) 192, 437-461 (1967)

Ramón-Moliner, E.: An attempt at classifying nerve cells on the basis of their
dendritic patterns. J. Comp. Neurol. 119, 211-227 (1962)

Rushton, W.A.H.: The Ferrier Lecture, 1962. Visual Adaptation. Proc. R. Soc. London
162B, 20-46 (1965)

Schwartz, E.A.: Organization of on-off cells in the retina of the turtle.
J. Physiol. (London) 230, 1-14 (1973)

Spekreijse, H.: Rectification in the goldfish retina: Analysis by sinusoidal and
auxiliary stimulation. Vis. Res. 9, 1461-1472 (1969)

Spekreijse, H.: Norton, A.L.: The dynamic characteristics of color-coded S-potentials. J. Gen. Physiol. 56, 1-15 (1970)

Stell, W.K.: Horizontal cell axons and axon terminals in goldfish retina.
J. Comp. Neurol. 159, 503-520 (1975)

Stell, W.K., Witkovsky, P.: Retinal structure in the smooth dogfish. Mustelus
canis; general description and light microscopy of giant ganglion cells.
J. Comp. Neurol. 148, 1-32 (1973)

Toyoda, J.-I.: Frequency characteristics of retinal neurons in the carp.
J. Gen. Physiol. 63, 214-234 (1974)

Walls, G.L.: The Vertebrate Eye and Its Adaptive Radiation. New York: Hefner 1967

Werblin, F.S.: Dowling, J.E.: Organization of the retina of the mudpuppy Necturus
maculosus. II. Intracellular recording. J. Neurophysiol. 32, 339-355 (1969)

Witkovsky, P., Stell, W.K.: Retinal structure in the smooth dogfish, Mustelus canis.
Light microscopy of bipolar cells. J. Comp. Neurol. 148, 47-60 (1973)

Witkovsky, P., Shakib, M., Ripps, H.: Interreceptoral junctions in the teleost
retina. Invest. Ophtal. 13, 996-1009 (1974).

1.5 Electrophysiological and Histological Studies of the Carp Retina

R. Weiler and F. Zettler

1.5.1 Functional Analogies

The comparative approach today offers special promise in the study of the retina, as it did in the days of Cajal, for there are, of course, a number of differently evolved and highly analogous visual systems. By comparing various types of eye one hopes to distinguish the general principles of neuronal organization from special adaptations. Mechanisms common to molluscan, arthropod and vertebrate eyes are indeed general ones; their separate evolutions have been shaped by the same physical parameter, light. And in fact, the morphology of these eyes alone reveals a high degree of correspondence. In all cases many receptors are associated to form an organ, and there is convergence - over one to several intermediate stages - upon a relatively small number of neurons linking the eye to the CNS.

Recent research on the retina has led to a massive accumulation of empirical data - so vast, in fact, that it is hardly impossible to separate the general principles underlying these data from special adaptations without the comparative view, a method which has after all been essential to all great biological theories.

Cajal, whose work was done during a comparable phase of explosively rapid accumulation of histological data, made clear the anatomical analogy in his general remarks concluding his 1915 book. The point was illustrated by the drawings in Fig. 1. The central drawing is a modification of the accurately represented insect retina on the left, with the cell bodies of the monopolar neurons displaced downward to emphasize the analogy with the vertebrate retina shown on the right.

With the above considerations in mind, and wishing to carry Cajal's analogy further at the level of function, we extended our visual studies - first focussed upon the fly retina (2.4 of this vol.) - to the retina of the carp. Specifically, our experiments were designed to permit comparison among the receptors, of the bipolars with the monopolars, and among the various horizontal cells of the retina.

We first found that intracellular recording from neurons of the vertebrate retina was more difficult than our experience with the insect retina had led us to expect. The penetration of single neurons presented a particular problem, even though these cells are larger than those in the insect retina. For this reason we developed a "magnetostrictive hammer"; with this we have been able to penetrate the cells relatively easily and obtain stable intracellular recordings. Fig. 2 shows diagrammatically the construction of the hammer. A ferrite rod is contracted by magnetostriction when a magnetic field is applied; when the coil current is interrupted it expands very rapidly, with an acceleration of as much as 1500 g. The length of the hammer stroke is continuously adjustable up to a maximum of 2 μm. The hammer stroke is transmitted to the electrode in an axial direction, and thus facilitates penetration.

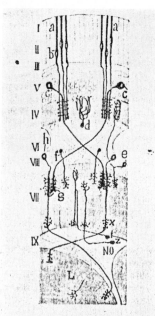
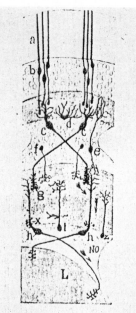
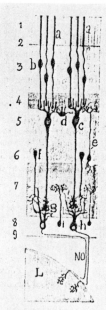

Fig. 81. — Esquema de las tres empaliza-das neuronales de la retina de un insec-to. Los somas aparecen en su posición natural. — *a*, bastoncito; *b*, núcleo del bastón; *c*, segunda neurona visual; *h*, tercera neurona visual; NO, nervio óptico; L, lóbulo óptico.

Fig. 82.—Esquema de las tres empaliza-das neuronales de la retina de un insec-to. Para que la comparación con la re-tina de los vertebrados sea más fácil se ha rectificado la posición de los núcleos, emplazándolos en los parajes donde ha-bitan en los vertebrados. Las letras marcan los mismos elementos que en la figura precedente.

Fig. 83.—Esquema de las prin-cipales capas y anillos neuro-nales de la retina de los ver-tebrados. Compárese con la retina de los insectos repre-sentada en las figuras 81 y 82. Los números señalan las mis-mas zonas retinianas; NO, nervio óptico; L, lóbulo óp-tico; *a*, bastoncitos; *b*, núcleo del bastoncito; *d*, célula ho-rizontal; *c*, corpúsculo bipo-lar; *f*, amacrina ordinaria; *i*, amacrina dislocada.

Fig. 1. Vertical sections through the insect retina (left) and the vertebrate ret-ina (right), as they appear after staining by the Golgi method. From Cajal (1915). In the center, the insect retina has been redrawn so as to emphasize the analogy between it and the vertebrate retina

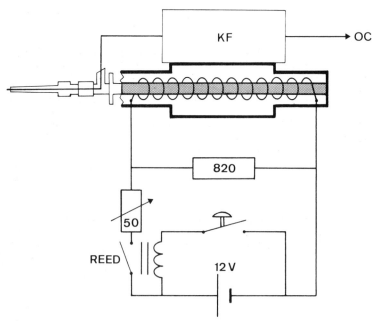

Fig. 2. Diagram of the magnetostrictive hammer. Inside the coil is a ferrite rod (10 cm long). KF: first stage of the cathode follower; OC: connection to oscillo-scope; REED: reed relay

Fig. 3. Comparison of bipolar and monopolar potentials. The vertical scale, at the left, represents 10 mV; the horizontal scales represent 200 ms, which is the duration of the point light stimulus. The two recordings on the right show responses to an intensity greater by a factor of 10^2 than that on the left

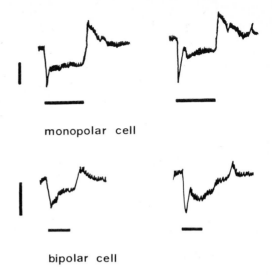

monopolar cell

bipolar cell

Fig. 3 shows a very preliminary comparison of the carp and fly; the potentials are recorded from a rod bipolar of the carp and the analogous neuron in the insect retina, an L2 monopolar. Both cell types respond to increase in the intensity of the point light source - which necessarily involves increased stimulation of the adjacent cells by scattered light - by a corresponding increase in the depolarizing component and an enhancement of the on-off behavior. Although very few ' measurements have yet been made in support of this comparison, Fig. 3 indicates clearly that functional analogies can exist in these two cells.

1.5.2 Horizontal Cells

The classical (Cajal, 1909) anatomical classification of horizontal cells in the fish retina - in terms of external, intermediate and internal - on the basis of their position has recently been the subject of renewed discussion. Stell (1975) has raised a particular challenge, in asserting that the internal horizontal cells are clearly identifiable as terminal structures of the axonal processes of external horizontal cells. One must say, of course, that there is a risk in classifying a structure as belonging to an axonal ending on the basis of Golgi staining which distinctly reveals the connection in a few cases. Other structures, perhaps quite different in function, which are incompletely stained may be lumped together with the unambiguously identified end structures, because of morphological similarities. On the other hand, in basing a functional analysis on the marking of cells with dye one must reckon with the possibility that the dye may not diffuse into all parts of the cell.

Our experiments, combining intracellular recording in the carp retina with cell-marking and Golgi staining, have suggested a provisional classification of the horizontal cells. Functionally, we can clearly distinguish two different L-potentials; there are preliminary indications that a third exists as well, and this potential will be discussed in connection with the internal horizontal cells. With respect to

the anatomical origin of these potentials, we can say the following:
L-potentials with spectral sensitivity maximal at 520 nm, with a slow
time course, and with a small receptive field can definitely be ascribed to the median horizontal cells. This result is consistent with
those of Kaneko and Yamada (1972). The fine dendrites of the intermediate horizontal cells extend through gaps in the layer of external
horizontal cells, making synaptic contact only with rods (Stell, 1967).
These fine dendrites are always easily discernable in cells marked
with Procion yellow, and facilitate identification of the cells. The
second type of L-potential, with a maximum response at 620 nm and a
rapid time course, could be assigned to the external horizontal cells
and to cells lying proximal to the intermediate horizontal cells. As
will be shown later, these structures are not in fact separate cells,
but rather are the axon terminals of the external horizontal cells.

These L-potentials exhibit marked spatial summation (Naka and Rushton,
1967; Norton et al., 1968), ascribable to an electrical coupling like
that demonstrated by Kaneko (1971) in the dogfish retina.

If the structure at the end of the axon coming from an external horizontal cell is not presynaptic but a postsynaptic element, which is
not yet clear, one would expect its geometry to be reflected in the
measured receptive field of this cell. As a test of this possibility,
the receptive field was measured with a point source of light (20 µm
in diameter on the retina) moved as much as 1500 µm off-center in
the x and y directions. (The center was the spot at which the electrode penetrated the tissue.) In all experiments whole-eye preparations were used in which the sclera, the chorioid and the pigmented
epithelium on the optic-nerve side had been removed. The orientation
of the eyes in the apparatus was random, so that measurements made
only along the x and y axes are sufficient to establish an asymmetry
of the receptive field.

Fig. 4a, an example of many such measurements, confirms that the field
is symmetrical. In no case was it possible to demonstrate an asymmetry
correlated with the direction of the axon.

Of the horizontal cells with a C-type potential, our experiments so
far (in contrast to those of Norton et al., 1968) have revealed only
the red/green type - that is, only cells with a depolarizing response
in the long-wavelength region and with a hyperpolarizing response in
the green region. It seems remarkable that in some recordings the
depolarizing response suddenly changes into a hyperpolarizing response, independently of the wavelength of the stimulus light and with
no change in the recording conditions. Such an effect has been observed before, by Saito et al. (1974) in their study of the turtle
retina.

All the C-type potentials observed were recorded from the external
or internal horizontal-cell layer. Just as Kaneko found in 1970
(Kaneko, 1970), there was no clear association of the L- and C-type
potentials with either of these two layers. The receptive fields of
cells with C potentials were measured as described above, in order
to test whether the depolarizing and hyperpolarizing responses could
be distinguished spatially (Fig. 4b). However, in no experiment did
either the depolarizing or the hyperpolarizing component of the response change in any way other than would be expected given symmetrical spatial summation. The inference is that C-potentials evidently
arise in cells with electrical coupling, as do the L-potentials with
maximum response at 620 nm.

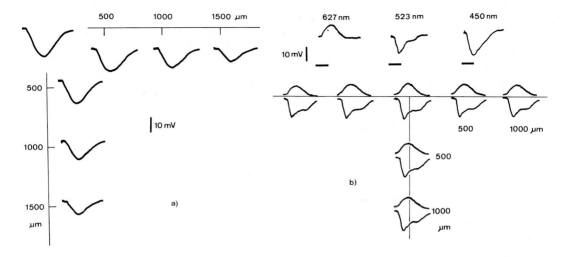

Fig. 4a and b. a) Receptive field of an L-type horizontal cell, measured with a point light stimulus 20 μm in diameter on the retina. b) Receptive field of an external C-type horizontal cell, measured as in (a) but at two wavelengths, 523 nm and 627 nm

The internal horizontal cells in the teleost retina, which have been given a number of different names since their discovery (see 1.2 of this vol.), represent the terminal structures of the axons of the external horizontal cells, according to the most recent results from the goldfish retina (Stell, 1975). A contradictory result was obtained by Wagner (see 1.1 of this vol.) in the fish Nannacara. There, the structure appearing to be an axonal process in Golgi-stained preparations is not in fact an axon, but rather is the dendrite of a bipolar cell.

The combination of Golgi staining and intracellular marking allows us to confirm, for the carp retina, the presence of external horizontal cells with axons and terminal structures. It is striking that the axon, which connects the external cell to the terminal structure, has knobby swellings at regular intervals (Fig. 5). Recordings from the terminal structures, as mentioned above, were L-potentials with maximum response at 620 nm, and in a few cases C-potentials; that is, apart from the presence of spatial summation (Kaneko, 1970) these recordings correspond to those obtained from the cell bodies in the external layer. Whether these potentials are propagated as graded potentials along the axon from the external horizontal cells to the terminal structures, and whether the knobby swellings routinely found (Fig. 5) may function as amplification stations, is not known.

In addition to these terminal structures, which correspond to the internal horizontal cells described by Kaneko (1970), we found a second internal-horizontal-cell type, which we shall term the "true internal horizontal cell" to distinguish it from the first. Morphologically, this cell differs from the terminal structures in four respects: (1) its position is different, (2) axonal processes were never observed, (3) no spines were detectable on its surface, and (4) it has a nucleus (the terminal structures are also referred to as anucleate cells). Fig. 6 shows a comparison between the terminal structure and

116

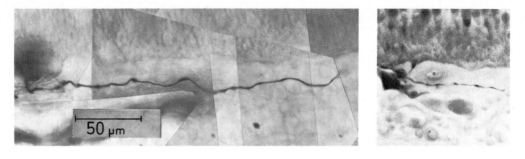

Fig. 5. Axons of external horizontal cells. The preparations on the left: stained by the Golgi method; the cells on the right: marked by intracellular injection of Procion yellow. In both cases the regularly spaced knobby swellings are revealed

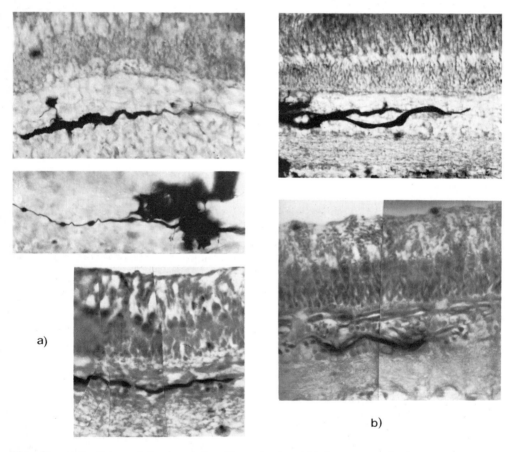

a)

b)

Fig. 6a and b. Internal horizontal cells: a) classified as a terminal structure of the axon of an external horizontal cell. Top and middle: Golgi stain; bottom: intra-cellular marking with Procion yellow. b) classified as a "true horizontal cell". Top: Golgi stain; bottom: marking with Procion yellow

the "true horizontal cell". The lower photographs are montages of cells marked with Procion yellow.

Whereas the terminal structures in general run parallel to the other horizontal-cell layers, the true horizontal cell follows an arc, centripetally from the level of the intermediate horizontal cells to the proximal end of the second nuclear region - i.e. to the inner synaptic layer. Naka has found similar cell types in the catfish retina (Naka and Carraway, 1975). He does not rule out the possibility that these internal horizontal cells may participate directly in the inner synaptic layer (Naka et al., 1975).

None of the cells designated true horizontal cells showed even the slightest hint of an axon, with either Golgi staining or intracellular marking (in contrast to the preparations of Fig. 6a). The absence of spines was just as convincing. The electron micrographs necessary to demonstrate conclusively the existence of a nucleus have not yet been made. However, nuclei are usually more deeply stained by Procion-yellow injection than the rest of the cell, and can be identified by this criterion. On this basis, nuclei are clearly present in several marked cells, and comparisons with Bodian preparations corroborate this identification.

The third type of L-potential, mentioned at the outset, arises in these true horizontal cells. These are slow, hyperpolarizing potentials, the maximal amplitude of which is half that of the other S-potentials, with a depolarizing inflection during the plateau. The electrophysiological data, however, are not yet adequate to provide a functional description of the cell. It remains to be clarified whether this cell is actually directly involved in the synaptic plexus of the inner region. Should this be the case, a second direct centripetal pathway, in addition to the bipolars, would be available.

1.5.3 Interactions in the First Synaptic Layer

The notion that the hyperpolarizing potentials of the horizontal cells not only affect the centripetal transmission of information, but also can act in the reverse direction, upon the receptors, was considered by Naka and Rushton (1966). That this reciprocity in fact applies was demonstrated for the turtle retina (Baylor et al., 1971; Fuortes, 1972; O'Bryan, 1973).

We have now found evidence of such a centrifugal effect in the carp retina. Fig. 7 summarizes the relevant measurements. The site of origin of these potentials has not yet been definitely established. The recording site was in the region of transition from receptors to horizontal cells, possibly directly in the triad; this situation could account for the fact that the centrifugal effect is so pronounced in these records. Recordings from the receptors of the carp retina - from cones, specifically - have been published previously only by Tomita et al. (1967) and Kaneko and Hashimoto (1967). In their publication, the cone response is compared with the S-potential. The latencies of the two potential types do not differ conspicuously, but there are clear differences in spatial summation. This result is consistent with our findings in marked cones and horizontal cells. With respect to the rods, the evidence is inconclusive, since an identification was possible in only one case.

118

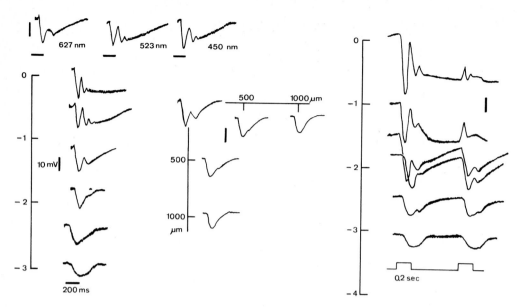

Fig. 7. Potentials resulting from feedback between horizontal cells and receptors. The potentials were recorded in the first synaptic layer. Left-hand column: dependence upon intensity of a point light stimulus 200 ms in duration; row of potentials above this: wavelength dependence of the depolarization. Column on the right: effect of paired stimuli. Center: a two-dimensional array of responses showing the effect of displacement of the stimulus from the center

To bring out the centrifugal effect more strongly, we increased the intensity of our point stimulus, so as to increase stimulation of the surrounding cells by stray light.

The column of traces on the left in Fig. 7 shows that the depolarizing deflection of the hyperpolarizing response is dependent on intensity; at the highest intensities (least attenuation) the potential can even acquire the character of a spike. The depolarizing effect is more pronounced in the short-wavelength region than at long wavelengths, as the row of three traces at the upper left shows. We cannot yet say definitely whether the wavelength dependence reflects differential involvement of the rods.

The columns on the right side of Fig. 7 show the responses of a cell to a sequence of two point-light stimuli, the second at the same intensity and position as the first, but following it 0.6 s later. Even when there is a prolonged hyperpolarization following the first stimulus, the subsequent stimulus produces a depolarization; an explanation is that the horizontal cells have returned to a potential near the resting level and thus respond to the second stimulus with a renewed hyperpolarization, which in turn depolarizes the receptor.

If the light stimulus strikes the retina at some distance from the point of electrode penetration, the cell at the electrode is still excited by scattered light; however, the depolarizing effect must be reduced according to the spatial-summation properties of the horizontal cell, as is demonstrated in the central set of recordings in Fig. 7.

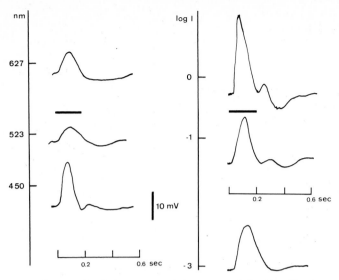

Fig. 8. Depolarizing potential measured in the first synaptic layer. Left: wavelength dependence; right: intensity dependence

These first results are far from sufficient to elucidate the mechanism operating in the first synaptic layer of the carp retina. In particular, we do not yet even know all the types of cells that are involved. For example, the potentials shown in Fig. 8 were also recorded in the first synaptic layer. This cell is depolarized in response to stimuli at all wavelengths, and the response is graded over the entire physiological range of intensities. The cell has not yet been classified morphologically.

Acknowledgments. This work was supported by the Deutsche Forschungsgemeinschaft. We would like to thank Dr. Marguerite A. Biederman-Thorson for translating the German text into English and Miss Christa Weyh for technical assistance.

1.5.4 References

Baylor, D.A., Fuortes, M.G.F., O'Bryan, P.M.: Receptive fields of cones in the retina of the turtle. J. Physiol. 214, 265-294 (1971)
Cajal, S.R.: Histologie du système nerveux de l'homme et des vertébrés. II. Madrid: Instituto Ramon y Cajal, 1909 (reprint 1955)
Cajal, S.R., Sanchez, D.: Contribucíon al conocimiento de los centros nerviosos de los insectos. Trabajo del lab. invest. biol. Univ. de Madrid, 13, 145-150 (1915)
Fuortes, M.G.F.: Responses of cones and horizontal cells in the retina of the turtle. Invest. Ophthalmol. 11, 275-284 (1972)
Kaneko, A.: Physiological and morphological identification of horizontal, bipolar and amacrine cells in goldfish retina. J. Physiol. 207, 623-633 (1970)

Kaneko, A.: Electrical connexions between horizontal cells in the dogfish retina. J. Physiol. 213, 95-105 (1971)

Kaneko, A., Hashimoto, H.: Recording site of the single cone response determined by an electrode marking technique. Vis. Res. 7, 874-851 (1967)

Kaneko, A., Yamada, M.: S-potentials in the dark adapted retina of the carp. J. Physiol. 227, 261-274 (1972)

Naka, K.I., Rushton, W.A.H.: S-potentials from L-units in the retina of fish (Cyprinidae). J. Physiol. 185, 587-599 (1966)

Naka, K.I., Rushton, W.A.H.: The generation and spread of S-potentials in fish (Cyprinidae). J. Physiol. 192, 437-461 (1967)

Naka, K.I., Carraway, N.R.G.: Morphological and functional identifications of catfish retinal neurons. I. Classical morphology. J. Neurophysiol. 38, 53-71 (1975)

Naka, K.I., Marmarelis, P.Z., Chan, R.Y.: Morphological and functional identifications of catfish retinal neurons. III. Functional identification. J. Neurophysiol. 38, 92-131 (1975)

Norton, A.L., Spekreijse, H., Wagner, H.G., Wolbarsht, M.L.: Receptive field organization of the S-potential. Science 160, 1021-1022 (1968)

O'Bryan, P.M.: Properties of the depolarizing synaptic potential evoked by peripheral illumination in cones of the turtle retina. J. Physiol. 235, 207-224 (1973)

Saito, T., Miller, W.H., Tomita, T.: C- and L-type horizontal cells in the turtle retina. Vis. Res. 14, 119-121 (1974)

Stell, W.K.: The structure and relationship of horizontal cells and photoreceptor-bipolar synaptic complexes in goldfish retina. Am. J. Anat. 121, 401-424 (1967)

Stell, W.K.: Horizontal cell axons and axon terminals in goldfish retina. J. Comp. Neurol. 159, 503-520 (1975)

Tomita, T., Kaneko, A., Murakami, M., Pautler, E.L.: Spectral response curves of single cones in the carp. Vis. Res. 7, 519-531 (1967).

1.6 Interactions and Feedbacks in the Turtle Retina

M. G. F. FUORTES

Visual recognition in vertebrates requires the transformation of an optical image into a composite electrical signal generated by a mosaic of retinal photoreceptors. Faithful transcription could be obtained if photoreceptors were independent units, the response of each being controlled only by the light falling upon it. Indeed, independence of photoreceptors appears to be desirable for maximal spatial resolution of the image and it has often been assumed to be a basic principle of retinal organization (see e.g. Naka and Rushton, 1966). It was, therefore, a surprise to find that this is not the design employed in the retina.

If receptors were independent units, the response of one of them would be the same for illumination covering that cell only or extending over other receptors. Experiments show instead that the response of a cone changes not only with the intensity of the light falling upon it but also with the distribution of light on other parts of the retina (Baylor et al., 1971; O'Bryan, 1973). These results indicate that cones are subject to interactions. In the retina of the turtle, two types of interaction have been demonstrated: enhancement from neighboring cones and inhibition from horizontal cells. More recently, interactions have been found between rods and from cones to rods (Schwartz, 1975a and b).

O'Bryan (unpubl.) and Baylor and Hodgkin (1973) have investigated the cone-to-cone enhancement and have determined that it operates only between cones with the same visual pigment: red cones are connected only to other red cones, green to green and blue to blue. Hence, these interactions do not impair discrimination of color, even though they may decrease spatial resolution and the detection of contrasts.

There is an exception to this rule, however, which has recently been described by Richter and Simon (1974) and by Baylor and Fettiplace (1975). In the turtle, as in other animals, a sizeable proportion of photoreceptors are double cones - closely attached receptor pairs with green-absorbing pigment in one cell and red-absorbing pigment in the other. It has been found that the two receptors are closely coupled and therefore the activity recorded from one of them reflects the properties of both green and red pigments. In the experiment of Fig. 1 the red member of a double cone was impaled. When stimuli were applied from darkness the response to a green or to a red flash were similar; but when the flashes were delivered over a monochromatic background the responses to red or green flashes became remarkably different. These results can be readily explained assuming close electrical coupling of the two cells. The red background light accelerates and decreases the responses of the red member but, not being absorbed by the green pigment, does not change the response of the green member of the pair. Therefore, a red flash on a red background will evoke a small rapid response in the red cone and negligible activation of the green. A green flash evokes only a small response in the red member but a large slow response in the green and, due to electrical coupling this

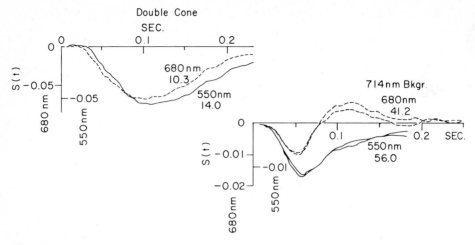

Double Cone
SEC.

Fig. 1. Responses of a double cone to green and red flashes applied from darkness
and over a 714 nm background (Bkgr). Each tracing is the sum of 40 responses. The
quantity S(t) in the ordinates is the voltage V(t) of the summed response, divided
by the total number of photons delivered over 1 μm². The unlabeled numbers near
each trace, when multiplied by 10³ give the number of photons delivered by one
flash over 1 μm². Note how the responses to green or red become different when the
double cone is adapted by red light. (From Fuortes and Simon, 1974)

response can be picked up from the red member. Consistent with these
results and interpretations it is found that spectral sensitivity has
a broad band when the retina is dark adapted and becomes more restricted under chromatic adaptation (Fig. 2).

The absorption properties of the visual pigments explain also some
surprising features of the interactions between horizontal cells and
cones. The action spectra obtained by Baylor and Hodgkin (1973) indicate that there are in the turtle three types of cones which are maximally sensitive to red, green and blue light respectively. Reflecting
the property of pigment absorption, sensitivity of each cone type
drops steeply for wave lengths larger than λ_{max} but falls by only
about one log unit for shorter wavelengths. Thus, the red cones can
be activated by lights of any wavelength within the visible spectrum,
even if light intensity is moderate. Green cones instead absorb red
light very poorly, and blue cones absorb well only blue light.

Evidence collected in our Laboratory (Baylor et al., 1971; Simon, 1973;
Fuortes et al., 1973) indicates that the feedback upon the cones is
initiated by a horizontal cell (called L-cell) with a spectral sensitivity similar to that of red cones. Thus, a deep red light may evoke
a hyperpolarizing response in red cones and L-cells without directly
exciting green or blue cones. In these circumstances the cones develop
a purely depolarizing response.

These effects are shown in Fig. 3, which illustrates that response of
a green cone may reverse with the color of the stimulus. It has not
been possible so far to demonstrate that blue cones also reverse polarity with wavelength. By analogy with the green cones it would be
expected, however, that they are hyperpolarized by blue light and depolarized by green and red. Considering these results together with
those quoted before, it seems safe to conclude that the responses of

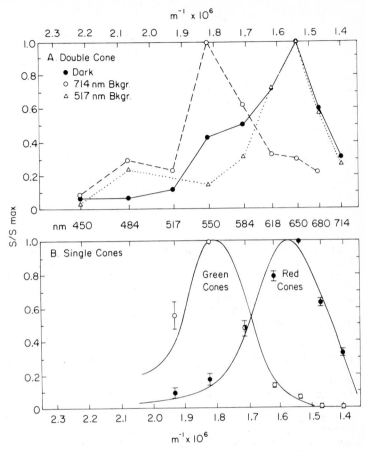

Fig. 2A and B. (A) Relative sensitivity of a double cone to flashes of the wave-lengths indicated in abscissa (<u>nm</u>) applied from darkness or over red or green back-grounds. The points in (B) measure the relative sensitivity of red and green single cones. The continuous curves are similar measurements replotted from Baylor and Hodgkin (1973). (From Richter and Simon, 1974)

retinal cones are controlled not only by the light they absorb, as would happen in the absence of interactions, but also by the distribution and color of the light on other parts of the retina.

Since these features of receptor responses must be reflected in the second-order neurons, Dr. Simon and I have recently reexamined the properties of horizontal cells in the turtle. These cells have been classified in two types: luminosity cells (L-cells) which are hyper-polarized by lights of any wavelength and chromaticity cells (C-cells) which are hyperpolarized by some wavelengths and are depolarized by others. According to Simon (1973) horizontal cells divide in two mor-phological types: type I with thick processes and no well-defined cell body, and type II with thin, less extensive arborizations and a clearly-outlined soma. Simon (1973) recorded luminosity responses from both type I and type II cells and found that the summation area is appreciably larger in cells of type I. Saito et al. (1974) confirmed Simon's results and determined in addition that chromacity responses can be recorded from cells similar to Simon's type II, but never from type I cells.

124

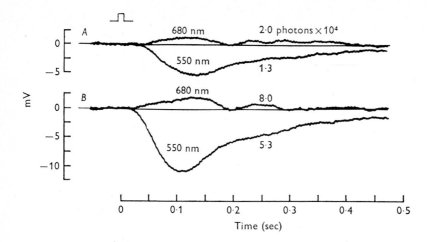

Fig. 3. Comparison of the responses of a green cone to flashes 550 nm and 680 nm covering a circle of 1250 μm radius. The numbers near each trace give the wavelength and the number of photons delivered by the stimulus over an area of 50 μm². (From Fuortes et al., 1973)

Evidence has recently been obtained which suggests that the morphological classification outlined above may have to be modified. Intracellular injections of a rapidly diffusible and strongly fluorescent dye synthesized by W. Stewart at the NIH have shown that a single injection may result in the staining of two separate structures: one type I and one type II horizontal cell. In these cases, careful examination shows that the two structures are connected by a thin axon.

Similar evidence has recently been obtained by Lasansky and Vallerga (1975) in the salamander. It seems likely, therefore, that the structure of horizontal cells in reptiles and amphibia is similar to that discovered by Boycott (unpubl.) and described by Kolb (1974) in the cat. Using the Golgi staining method, Boycott found that some horizontal cells which he called type B consist of a small cell body surrounded by dendritic processes, a thin axon, and an extensive arborization of thick terminal processes. Boycott type A horizontal cells also have a small round soma surrounded by processes but do not have the long thin axon and the terminal arborization of type B cells.

The suggestion which arises from these observations is that Simon type I cells are in fact the terminal arborizations of Boycott type B cells. Simon type II cells could be either a cell without axon (Boycott type A) or the soma-dendritic area of a type B cell.

As already stated above both soma and terminals can give luminosity type responses while chromaticity type responses have been recorded so far only from the somatic part of horizontal cells. It is not yet known, however, if chromatic cells are Boycott type B cells.

Whatever the case may be, the results described above on cone responses suggest a simple interpretation of the different properties of luminosity and chromaticity horizontal cells. There are in the turtle two types of chromaticity cells. One type called R/G cell is depolarized by green or blue; the other type, called G/B is depolarized by red or green and is hyperpolarized by blue. Thus, the responses of R/G cells are similar to those of the green cones and those of

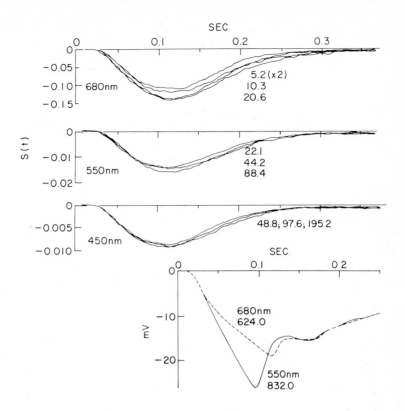

Fig. 4. The tracings are sums of 10, 20 or 40 responses to flashes at intensities varying by factors of 2, so that the total number of photons delivered is the same for each set of tracings. The numbers near each set are as in Fig. 1. The notation (x2) indicates that a control run is included. The inset shows that when light intensity is increased, L-cell responses differ for red or green flashes. (From Fuortes and Simon, 1974)

G/B cells to the responses expected of blue cones. L-cells are hyperpolarized by all wavelengths and thus their responses resemble those of the red cones. The responses of horizontal cells, therefore, might be explained (at least in first approximation) assuming that one cone type is connected exclusively to a corresponding type of horizontal cell: red cones to L-cells; green cones to R/G cells and blue cones to G/B cells.

Horizontal cell responses to bright flashes are exceedingly complicated but, if dim flashes are used, simpler results are obtained and the similarity between cone and horizontal cell responses becomes apparent.

L-cell responses to these dim stimuli are simple hyperpolarizing waves roughly invariant with respect to color (Fig. 4).

The responses of R/G cells to green or blue light also are hyperpolarizing waves, which resemble the conventional cone responses. The responses to red light are depolarizing waves similar to upside-down L-cell responses (Fig. 5). In the G/B cells, only blue light evokes hyperpolarizing responses; the responses to green flashes are simple

126

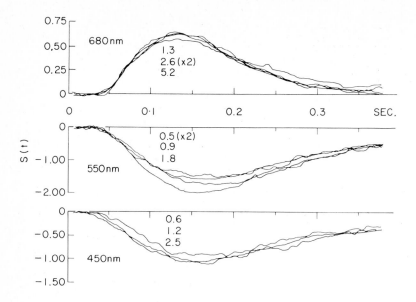

Fig. 5. Responses of an R/G cell to red, green and blue flashes. Hyperpolarizing responses to the shorter wavelengths are appreciably slower than the depolarizing responses to red flashes. (From Fuortes and Simon, 1974)

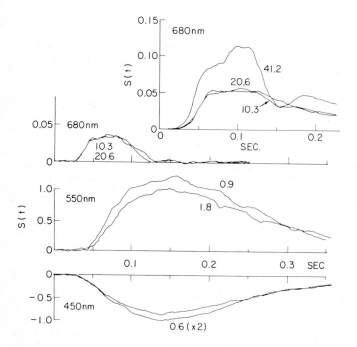

Fig. 6. Responses of a G/B cell. Experiment as in Figs. 4 and 5. Depolarizing responses to red flashes are small and short-lasting. They may grow more than in proportion with light intensity as shown in the cell illustrated in the inset. (From Fuortes and Simon, 1974)

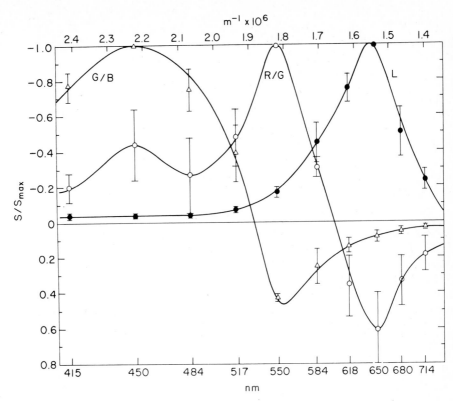

Fig. 7. Relative spectral sensitivities of horizontal cells. Points are the average of measurements taken in 23 L-cells, 16 R/G cells and 4 G/B cells. Hyperpolarizing responses are plotted up and vice-versa. (From Fuortes and Simon, 1974)

depolarizing waves and the responses to red flashes are small, short-lasting depolarizing waves of complicated time course (Fig. 6).

From results such as illustrated above one can construct "linear" spectral sensitivity curves (Fig. 7). It is then seen that the hyper-polarizing responses of each horizontal cell type follow closely the sensitivity curves of a corresponding type of cone, as determined by Baylor and Hodgkin (1973) using localized stimuli which evoke only negligible activation of horizontal cells. In both cases, peak sensi-tivities were at 450 nm, 550 nm and 640 nm. This agreement strongly suggests that the hyperpolarizing response of each type of horizontal cell is caused by impingement from a corresponding type of cone.

The depolarizing response of R/G cells has a peak for red light as is the case for red cones and L-cells. Since it has been shown that green cones are depolarized by red lights, this result is consistent with the view that the depolarizing responses of R/G cells are also the result of green cone impingement.

The interpretation of G/B cell responses, however, is more complicated. If they received impingement only from blue cones their depolarizing responses would be expected to follow the spectral sensitivity of L-cells, and thus to be largest for red light. The results show in-stead the depolarization is greater for green than for red light. It is necessary, therefore, to assume that G/B cells are under the in-

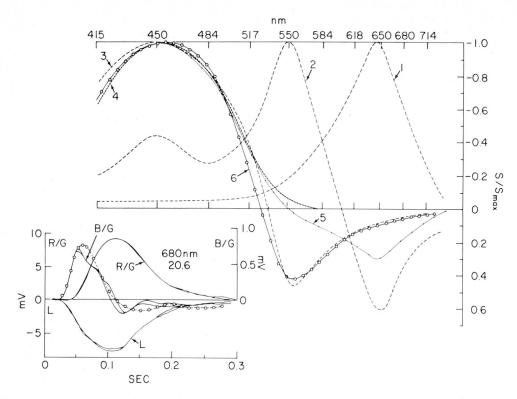

Fig. 8. Reconstruction of G/B cell responses. Curves 1, 2 and 3 are retraced from preceding figure. Curve 4 (<u>small dots</u>) is the spectral sensitivity of blue cones as determined by Baylor and Hodgkin (1973) using localized stimuli. Curve 5 (curve 4 - 0.3 x curve 1) does not follow the spectral curve of the G/B cell but curve 6 (curve 5 - 0.35 x curve 2) gives a good approximation of the experimental data. In the inset, the tracings are responses of an L-cell, an R/G cell and a G/B cell to 680 nm flashes. The dashed curve with circles is the sum of the L-cell and R/G cell responses, and resembles the experimental response of the G/B cell

fluence of some other source in addition to the blue cones. A clue for identifying this source is provided by the curves of Fig. 7.

If G/B cells were activated by blue cones only, their spectral sensitivity curve should be a combination of the direct response of blue cones and of the responses of L-cells. Curve 5 in Fig. 8 is such a combination. Peak sensitivity of the hyperpolarizing responses agrees well with the experimental measurements but the sensitivity predicted by the reconstructed curve is too small for green lights and too great for red lights. This deviation suggests that G/B cells receive impingement not only from blue cones but also from cells which develop responses of opposite polarity for red and for green. Cells with these properties are the R/G cells: if they were to impinge upon G/B cells with reversal of sign, they might provide the additional impingement which is required to explain the results. Curve 6 was constructed based on this assumption and it is seen that it fits the experimental curve satisfactorily. The view that depolarization of G/B cells is brought about by the action of L-cells on blue cones and is modified by impingement from R/G cells seems, therefore, to be tenable. Confirming this interpretation, it is seen in the inset that the time course

Fig. 9. Diagram of connections between cones and horizontal cells. Main connection: solid arrows; additional (and more tentative): dashed arrows. + and - denote transmission without or with inversion of polarity. Both cones and horizontal cells probably have additional output to bipolar cells. Other connection may also exist.
(From Fuortes and Simon, 1974)

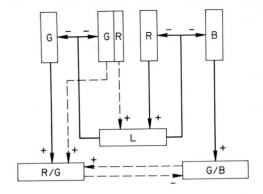

of the response of a G/B cell to a red flash can be roughly reproduced by addition of L-cell and R/G-cell responses to the same stimulus.

In conclusion, it appears that horizontal cell responses can be explained assuming that each cone type is connected to corresponding type of horizontal cell (Fig. 9). The crucial feature of the proposed organization is that the red cones, which absorb fairly well light of all wavelengths, activate a recurrent circuit which causes depolarization of all cones. With this design, red cones and L-cells are hyperpolarized by all wavelengths; green cones and R/G cells are hyperpolarized by blue and green but are depolarized by red, while blue cones are hyperpolarized by blue and are depolarized by green and red. This simple scheme, therefore, is sufficient for explaining the reversal of polarity which is observed in the responses of the so-called chromatic cells. In order to explain certain important details of horizontal cell responses, however, additional interactions must be assumed. An example is the inhibitory impingement of R/G cells on G/B cells which appears to be required to explain why green is more effective than red in eliciting depolarization of G/B cells. Other modifying interactions which seem to be needed are impingement of double cones upon L-cells, resulting in color-dependence of L-cell responses, and excitatory connections from G/B to R/G cells, increasing the sensitivity of R/G cells to blue light.

References

Baylor, D.A., Fettiplace, R.: Light peak and photocapture in turtle photoreceptors. J. Physiol. 248, 433-464 (1975)
Baylor, D.A., Fuortes, M.G.F., O'Bryan, P.M.: Receptive fields of cones in the retina of the turtle. J. Physiol. 214, 265-294 (1971)
Baylor, D.A., Hodgkin, A.L.: Detection and resolution of visual stimuli by turtle photoreceptors. J. Physiol. 234, 163-198 (1973)
Fuortes, M.G.F., Schwartz, E.A., Simon, E.J.: Colour-dependence of cone responses in the turtle retina. J. Physiol. 234, 199-216 (1973)
Fuortes, M.G.F., Simon, E.J.: Interactions leading to horizontal cell responses in the turtle retina. J. Physiol. 240, 177-198 (1974)
Kolb, H.: The connections between horizontal cells and photoreceptors in the retina of the cat: Electron microscopy of Golgi preparations. J. Comp. Neurol. 155, 1-14 (1974)
Lasansky, A., Vallerga, S.: Horizontal cell responses in the retina of the larval tiger salamander. J. Physiol. 251, 145-165 (1975)
Naka, K.I., Rushton, W.A.H.: S-potentials from colour units in the retina of fish (Cyprinidae). J. Physiol. 185, 536-555 (1966)

O'Bryan, P.M.: Properties of the depolarizing synaptic potential evoked by peripheral illumination in cones of the turtle retina. J. Physiol. 235, 207-223 (1973)

Richter, A., Simon, E.J.: Electrical responses of double cones in the turtle retina. J. Physiol. 242, 673-683 (1974)

Saito, T., Miller, W.H., Tomita, T.: C- and L-type horizontal cells in the turtle retina. Vis. Res. 14, 119-123 (1974)

Schwartz, E.A.: Rod-rod interaction in the retina of the turtle. J. Physiol. 246, 617-638 (1975a)

Schwartz, E.A.: Cones excite rods in the retina of the turtle. J. Physiol. 246, 639-651 (1975b)

Simon, E.J.: Two types of luminosity horizontal cells in the retina of the turtle. J. Physiol. 230, 199-211 (1973).

1.7 Interactions between Cones and Second-Order Neurons in the Turtle Retina

L. Cervetto

1.7.1 Introduction

The vertebrate photoreceptors detect light by absorbing it within the mass of pigment molecules located in specialized membranes of their outer segments. This process sets off a sequence of intracellular events which eventually lead to an electrical change in the plasma membrane of the visual cell. The electrical change generated by light consists of a graded increase in the transmembrane potential which is transmitted to adjoining neurons for further processing.

The outer plexiform layer is the first synaptic relay between photoreceptors and other retinal cells. Here, cone pedicles engage a variety of junctional contacts (Lasansky, 1971) which could provide the morphological substrate, for the mutual interactions occurring between photoreceptors and second-order elements (Baylor et al., 1971; Fuortes and Simon, 1974; 1.6 of this vol.).

1.7.2 Mechanisms of Excitation in the Vertebrate Photoreceptors

In the photoreceptors of different species of vertebrate the hyperpolarizing responses to light are associated with a resistance increase of the cell membrane (Bortoff and Norton, 1967; Toyoda et al., 1969; Baylor and Fuortes, 1970; Cervetto et al., in prep.). This result, as well as the observation that the inward-directed current flowing through the membrane of rod outer segments is reduced during illumination (Hagins et al., 1970) has suggested that the light responses of both rods and cones are generated by a decrease of membrane permeability to sodium ions.

The effect of changing ions in the extracellular medium has been studied in retinas of frog and pigeon by recording extracellular responses attributed to the photoreceptors. In both cases the amplitude of the responses was linearly related to the log of the external sodium ion concentration (Sillman et al., 1969; Arden and Ernst, 1970).

More recently the effects of ionic substitution on the intracellular activity of photoreceptors have been studied in turtle cones (Cervetto, 1973) and in toad rods (Brown and Pinto, 1974).

The results of an experiment in which intracellular responses from a cone were recorded during perfusion of the retina with a sodium free solution are illustrated in Fig. 1 (for the Methods see Cervetto, 1973). The superimposed records a, b and c shown in (A) were taken at the times indicated by the arrows in the diagram in (B). It is shown that as the sodium-free solution substitutes the normal medium, the membrane potential in darkness (circles) increases, while the potential at the peak of photoresponse is little changed. The effects

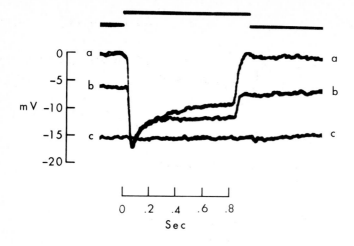

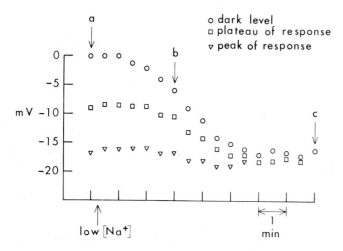

Fig. 1. Effects of low external sodium concentration on the intracellular responses of turtle cone. The records a, b and c in the top figure were taken at the
times indicated respectively by the arrows a, b and c in the diagram below. At
the time indicated by the upward directed arrow a sodium free solution made iso-
osmolar with appropriate amounts of choline chloride starts substituting the control Ringer. The change of membrane potential is indicated relative to the transmembrane potential in darkness during the control period at the beginning of the
experiment. The light intensity used to elicit the responses is 2.4 log units attenuated with respect to the total available energy. (o) dark level; (□) plateau
of response; (∇) peak of response. (From Cervetto, 1973)

of low sodium concentration were reversed when the normal concentration of sodium was restored.

Fig. 2 illustrates the behavior of the intracellular activity of a
turtle cone during perfusion of the retina with a solution containing
an excess of potassium ions. Records in (A) were obtained during perfusion with the control solution a and at the end of a 5 min exposure
to a solution containing 50 mM of potassium ions (20 times higher than

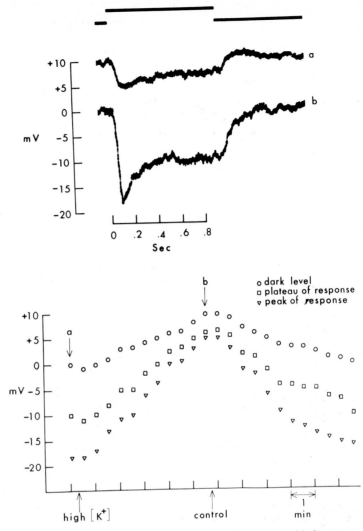

Fig. 2. Effects of high external potassium concentration on the intracellular responses of turtle cones. The records in a and b at the top of the figure were taken at the times indicated respectively by the arrows in a and b in the diagram below. The zero level of membrane potential indicates the level of membrane potential in darkness. At the time indicated by the upward directed arrow a solution containing 50 mM of potassium starts substituting the control Ringer. The stimuli parameters are the same as reported in the legend of Fig. 1. (o) dark level; (□) plateau of response; (∇) peak of response. (From Cervetto, 1973)

in normal Ringer) b. (B) is a plot of the membrane potential in darkness (circles) and at the peak of response (triangles).

Fig. 3 shows that both membrane potential and size of the response to light of a turtle cone are not affected by the absence of chloride ions.

The results indicate that the membrane potential of turtle cones is controlled predominantly by the concentration of sodium ions in dark-

134

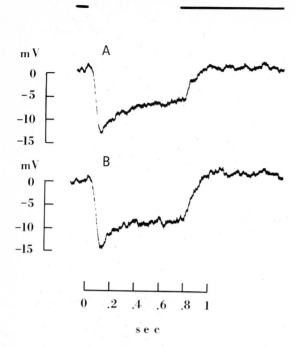

Fig. 3A and B. Intracellular
responses of a turtle cone in
normal control conditions (A)
and during perfusion with a
free chloride solution (B)

ness and by the concentration of potassium ions at the peak of the
photoresponse. This is consistent with the idea that the visual cells
are highly permeable to sodium in darkness and that light reduces
sodium permeability and thereby increases the ratio of potassium to
sodium permeability. The external potassium concentration, however,
also consistently affects the membrane potential in darkness. This
suggests that the ratio of potassium to sodium permeability is not
negligible even in darkness. Thus one cannot rule out the possibility
that light slightly modifies the permeability of the cell to potassium.

To explain how light controls the observed conductance changes of the
photoreceptors membrane, it has been supposed that light, interacting
with the visual pigment, activates a substance which diffuses intracellularly to the cell membrane and modulates the ionic conductance
(Baylor and Fuortes, 1970). This idea has recently received much attention and Ca^{2+} has been indicated as the possible internal transmitter.
The effect of increasing the extracellular calcium on the intracellular
activity of a turtle cone is shown in Fig. 4. The superimposed records
were obtained during perfusion of the retina with the control Ringer
a and with a solution containing 10 mM of Ca^{2+} b and c. The increased
concentration of extracellular calcium hyperpolarizes the membrane and
suppresses the response to light. This is consistent with the observation that the current flowing in darkness along the rat rods is reduced when external Ca^{2+} is increased (Yoshikami and Hagins, 1973).
Although the evidences in favor of the internal transmitter hypothesis
are far from being conclusive, a number of quantitative implications
are satisfied by this model (Cone, 1973).

Fig. 4. Effect of high external
calcium concentration on the
intracellular responses of a
turtle cone. Record a is the
control, records b and c were
taken at different times during
perfusion of the retina with a
solution containing 10 mM of
calcium (10 times more concentrated than the normal)

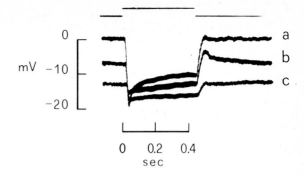

1.7.3 Transmission of signals between Photoreceptors and Second-Order Neurons

The mechanism by which receptor hyperpolarization affects second-order
neurons and causes a hyperpolarization in the luminosity-type horizontal cells has been investigated in retinas of different animals. On
the basis of results obtained from horizontal cells by stimulating
the retina with pulses of radial current, Trifonov (1968) formulated
the hypothesis that transmission of signals from photoreceptors may
be mediated by a depolarizing transmitter continuously released in
darkness. Accordingly, the hyperpolarization of the receptor membrane
would reduce the liberation of transmitter and hyperpolarize the horizontal cells.

Fig. 5 illustrates the results of an experiment in which pulses of
radial current were applied to the retina of turtle while recording
the intracellular activity from a horizontal cell. Pulses of current
flowing radially from the scleral to the vitreal surface of the retina
produce transient depolarizations graded by the intensity of the stimulating current. The size of the responses to radial currents is increased during illumination. It is supposed that the depolarizing potentials induced in horizontal cells by radial currents reflect transient releases of the transmitter from the pedicles of photoreceptors.
Pulses of current flowing in the opposite direction (i.e. from the
vitreal to the scleral surface) should produce hyperpolarizations:
Fig. 5B, however, shows that short pulses of this polarity are ineffective on the horizontal cell in both darkness and light, while long
steps of either polarity still produce measurable potential changes
(C and D). In explaining the different effectiveness of brief pulses
of current of opposite polarities, one should consider that uptake and
inactivation of a transmitting agent at the postsynaptic membrane may
imply processes with different kinetics and thus different time scales.

More recently the mechanism of synaptic transmission between photoreceptors and horizontal cells has been analyzed by using procedures
which specifically affect the release of transmitter (Dowling and
Ripps, 1973; Cervetto and Piccolino, 1974; Kaneko and Shimazaki, 1975).

Fig. 6 shows the continuous record of the activity of a horizontal
cell in the turtle retina during perfusion with a solution containing
2 mM of cobalt. Shortly after the test solution was substituted for
the control Ringer, the membrane became hyperpolarized and the response
to light was gradually reduced. After the retina was returned to the

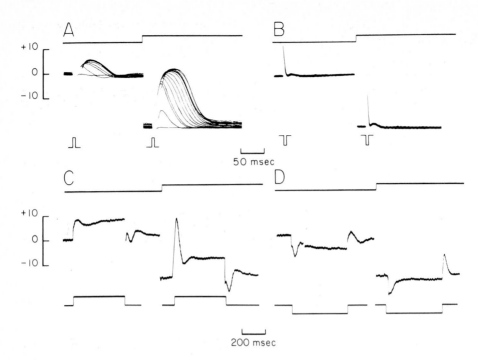

Fig. 5A – D. Responses of horizontal cells to radial polarization of the retina. (A) responses to 5 ms pulses of intensity from 60 to 300 μA in darkness and during illumination. Current flowed from the scleral to the vitreal surface. (B) as in (A) with current flowing in the opposite direction. (C) responses to steps of current 200 ms duration in darkness and during illumination. Polarity as in (A). (D) as in C for steps of opposite polarity. Displacements between each pair of records reflects the actual increase in membrane potential caused by light. Raised bars at the top: light events. The duration of current is monitored at the bottom of each record

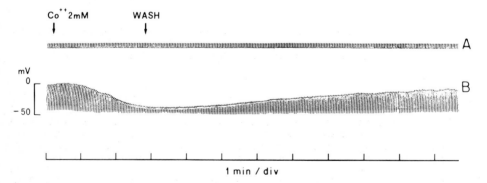

Fig. 6A and B. Effect of 2mM of cobalt chloride on the intracellular activity of a turtle horizontal cell (B). Light events are indicated by the record in (A). (From Cervetto and Piccolino, 1975)

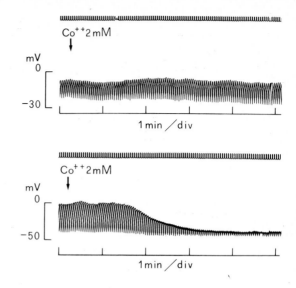

Fig. 7. Chart records showing the intracellular activity of a turtle cone (second record from the top) and horizontal cell (record at the bottom) during the exposure of the retina to a solution containing 2 mM of cobalt. Top and middle traces indicate light periods. (From Cervetto and Piccolino, 1974)

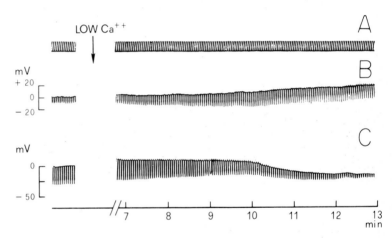

Fig. 8A - C. Chart records showing the intracellular activity of a cone (B) and of a horizontal cell (C) in the turtle retina during the effects of low external calcium. Trace (A) monitors the light stimulus. (From Cervetto and Piccolino, 1975)

control solution, both membrane potential and the response to light recovered. Fig. 7 illustrates the results from an experiment in which the effects of cobalt were studied on both cones and horizontal cells. It is seen that the photoreceptor responds to light in conditions in which the horizontal cell is unresponsive. Similar effects were also obtained when magnesium was added and/or calcium removed.

The effect of calcium deprivation on both cones and horizontal cells is shown in Fig. 8. As in the experiment of adding cobalt, calcium removal produced hyperpolarization of the horizontal cell associated with a decrease of the response to light. By contrast the membrane of the cone slightly depolarized and the size of the response to light eventually increased. If one assumes that the presence of cobalt and/ or the absence of calcium all prevent the transmission of presynaptic

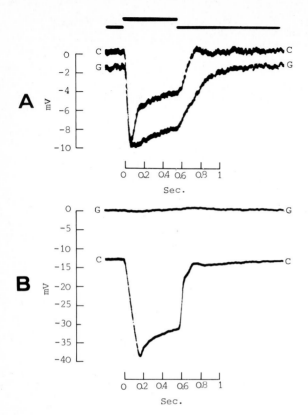

Fig. 9A and B. Effects of sodium glutamate on the activity recorded from a turtle cone (A) and from a horizontal cell (B). Responses to light were obtained during perfusion with control Ringer C and during the subsequent perfusion with 50 mM of sodium glutamate G. (From Cervetto and MacNichol, 1972)

signals from photoreceptors to horizontal cells, these results are consistent with the hypothesis that a transmitter is continuously released by cones in darkness and maintains the membrane of horizontal cell depolarized. Recent evidences indicate that a similar mechanism can explain also the transmission of signals from photoreceptors to bipolar cells (1.8 of this vol.).

Trifonov et al. (1974) have used extracellular current to analyze the electrical properties of subsynaptic and nonsynaptic membranes of horizontal cells in the fish retina. They concluded that the light responses of horizontal cells follow from the interruption of a sustained depolarization maintained in darkness by a high conductance mechanism with the reversal potential close to zero membrane potential for the cell.

The chemical nature of the transmitter released by photoreceptors has not been identified. Different substances have been indicated as possible candidates. One class of agents having a potent depolarizing activity on the horizontal cells are the short chain acidic amino-acids, aspartate and glutamate. These substances, when applied to the retina, depolarize promptly and reversibly the horizontal cells with little effect on the photoreceptor activity (Cervetto and MacNichol, 1972).

Fig. 9 illustrates the results from an experiment where the intracellular activity of a turtle cone and a horizontal cell was recorded during the exposure to a solution containing glutamate. The receptor can produce responses to light when the horizontal cell is depolarized

Fig. 10A - C. Intracellular responses of a cone (A) to a small spot of light 100 μm diameter; (B) to a 100 μm spot followed by an annulus 500 μm internal diameter, (C) to a large spot 1000 μm diameter.
(From Cervetto and Piccolino, 1975)

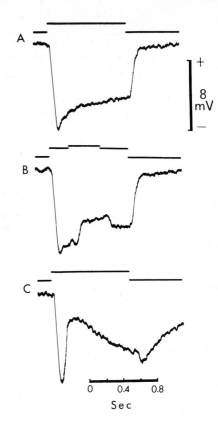

and unresponsive. Also, the drug produces hyperpolarization of the receptor membrane and alters the response to light by decreasing the ratio between the peak and the plateau. At present, however, the qualifications of these substances as transmitter agents of cones are yet unconclusive.

1.7.4 Effects of Blocking the Horizontal Cell Activity on the Cone Responses

The hyperpolarizing responses of the cones of the turtle were shown to be affected by the size of the stimulating spot (Baylor et al., 1971). The facilitatory effect observed when the diameter of a circle of light is increased up to about 100 μm has been attributed to the influences coming from the adjacent cones. When the diameter of the stimulating spot is further increased up to 1 mm or more, a typical delayed depolarization follows the peak of the hyperpolarizing response to light. The correlation of this effect with the responses of horizontal cells, and the observation that injection of hyperpolarizing currents into horizontal cells produces depolarization in cones indicate that horizontal cells have an output on cones (Baylor et al., 1971).

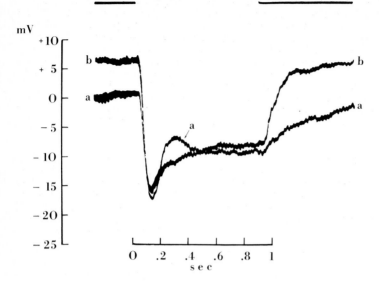

Fig. 11. Superimposed records obtained from a single cone in the turtle retina during perfusion with normal Ringer a and during subsequent perfusion with 3 mM of cobalt b

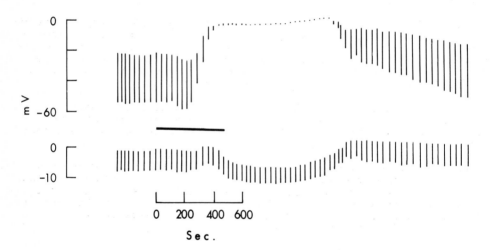

Fig. 12. Effects of sodium glutamate on the activity recorded from a horizontal cell (top) and a cone (bottom) in the turtle retina. Vertical bars trace the amplitude of light responses

Fig. 10 shows the responses of a turtle cone to different patterns of illumination. (A) is the response to a spot of light 100 μm in diameter; when during the illumination of the same spot a concentric annulus is presented, a sustained depolarization superimposes on the hyperpolarizing response to the spot (B). Depolarizations produced by the annulus showed spatial dependent properties as the responses of horizontal cells. Simultaneous illumination of both spot and annulus gives the response shown in (C). Here a depolarization with a time course similar to the responses of horizontal cells follows the initial hyperpolarization. Measurements of the resistance of the cone membrane performed during the depolarization have revealed conductance changes consistent with the activity of a chemical transmitter (O'Bryan, 1973).

On the basis of the present knowledge on synaptic transmission, one cannot explain how a hyperpolarizing potential may start the release of transmitter. This difficulty, however, can be overcome by assuming that the depolarization produced in cones by horizontal cells is caused by a decrease of the release of an agent which keeps the membrane hyperpolarized. This hypothesis is supported by the observation of the behavior of the cone membrane when synaptic transmission is blocked.

Fig. 11 shows two records obtained from a turtle cone in normal conditions and during the action of cobalt. The peak amplitude of the light responses is little affected by the cobalt, while consistent changes are observed in the shape and in the level of the membrane potential. During the exposure to cobalt the delayed depolarization disappears and the membrane depolarizes. Both effects are consistent with the existence of a feedback loop and with the possibility that a hyperpolarizing transmitter is continuously released by horizontal cells in darkness.

As already shown (see Fig. 9), application of glutamate suppresses the light responses of horizontal cells and keeps their membrane depolarized. In these conditions the cell may be particularly liable to release a transmitter whose action should be reflected on cones. The effect of glutamate on a horizontal cell and a cone is shown for comparison in Fig. 12. The plotted data are from records obtained in a single experiment. It is seen that, when the horizontal cell is depolarized by the effect of glutamate, the cone is concurrently hyperpolarized. The result is consistent with the hypothesis that upon depolarization the horizontal cells release a transmitter which hyperpolarizes cones.

The experiments of blocking the release of transmitter and those which show the action of glutamate, both could be best interpreted by assuming that horizontal cells during darkness continuously release a transmitter which keeps cones hyperpolarized. Accordingly, the horizontal cell hyperpolarized by light would reduce the release of transmitter and cause depolarization in cones.

Acknowledgments. Some of the results reported in this paper were obtained in collaboration with Drs. A.L. Byzov, E.F. MacNichol and M. Piccolino. I thank Professor R. Wittmer for generous support and Drs. G. Niemeyer and E. Pasino for helpful discussion.

1.7.5 References

Arden, G.B., Ernst, W.: The effect of ions on the photoresponses of pigeon cones
J. Physiol. 211, 311-339 (1970)

Baylor, D.A., Fuortes, M.G.F.: Electrical responses of single cones in the retina
of the turtle. J. Physiol. 207, 77-92 (1970)

Baylor, D.A., Fuortes, M.G.F., O'Bryan, P.M.: Receptive fields of cones in the
retina of the turtle. J. Physiol. 214, 265-294 (1971)

Bortoff, A., Norton, A.L.: An electrical model of the vertebrate photoreceptor cell.
Vis. Res. 7, 253-263 (1967)

Brown, J.E., Pinto, L.H.: Ionic mechanism for the photoreceptor potential of the
retina of Bufo Marinus. J. Physiol. 236, 575-591 (1974)

Cervetto, L.: Influence of sodium, potassium and chloride ions on the intracellular
responses of turtle photoreceptors. Nature (London) 241, 401-403 (1973)

Cervetto, L., MacNichol, E.F. Jr.: Inactivation of horizontal cells in turtle
retina by glutamate and aspartate. Science 178, 767-768 (1972)

Cervetto, L., Piccolino, M.: Synaptic transmission between photoreceptors and horizontal cells in the turtle retina. Science 183, 417-419 (1974)

Cervetto, L., Piccolino, M.: Mechanism of synaptic transmission in the vertebrate
retina. In: Golgi Centennial Symposium: Perspectives in Neurobiology.
Santini, M. (ed.). New York: Raven 1975, pp. 577-581

Cone, R.A.: The internal transmitter model for visual excitation: some quantitative
implications. In: Biochemistry and Physiology of Visual Pigments. Langer, H. (ed.).
Berlin: Springer 1973, pp. 275-282

Dowling, J.E., Ripps, H.: Effects of magnesium on horizontal cell activity in the
skate retina. Nature (London) 242, 101-103 (1973)

Fuortes, M.G.F., Simon, E.J.: Interactions leading to horizontal cell responses in
the turtle retina. J. Physiol. 240, 177-199 (1974)

Hagins, W.A., Penn, R.D., Yoshikami, S.: Dark current and photocurrent in retinal
rods. Biophys. J. 10, 380-412 (1970)

Kaneko, A., Shimazaki, H.: Effects of external ions on the synaptic transmission
from photoreceptors to horizontal cells in the carp retina. J. Physiol. 252,
509-522 (1975)

Lasansky, A.: Synaptic organization of cone cells in the turtle retina.
Phil. Trans. R. Soc. London 262B, 365-381 (1971)

O'Bryan, P.M.: Properties of the depolarizing synaptic potential evoked by peripheral illumination in cones of the turtle retina. J. Physiol. 235, 207-223 (1973)

Sillman, A.J., Ito, H., Tomita, T.: Studies on the mass receptor potential of the
isolated frog retina II. On the basis of the ionic mechanism.
Vis. Res. 9, 1443-1451 (1969)

Toyoda, J., Nosaki, H., Tomita, T.: Light-induced resistance change in single
photoreceptors of Necturus and Gekko. Vis. Res. 9, 453-463 (1969)

Trifonov, Yu. A.: Study of synaptic transmission between photoreceptors and
horizontal cells by means of electrical stimulation of the retina.
Biophysica 13, 809-817 (1968)

Trifonov, Yu. A., Byzov, A.L., Chailahian, L.M.: Electrical properties of subsynaptic and nonsynaptic membranes of horizontal cells in fish retina.
Vis. Res. 14, 229-241 (1974)

Yoshikami, S., Hagins, W.A.: Control of the dark current in vertebrate rods and
cones. In: Biochemistry and Physiology of Visual Pigments. Langer, H. (ed.).
Berlin: Springer 1973, pp. 245-255.

1.8 Synaptic Transmission from Photoreceptors to the Second-Order Neurons in the Carp Retina

A. Kaneko and H. Shimazaki

1.8.1 Introduction

Visual information received by the receptor mosaic in the vertebrate retina is extensively processed within the subsequent neural network, and the ganglion cells, at the output end of the retina, behave in highly specific ways. This processing is performed at the two successive synaptic layers. The first stage is the outer plexiform layer where receptors interact with second-order neurons, i.e. bipolar and horizontal cells. The second stage of neural interaction takes place at the inner plexiform layer in which bipolar, amacrine and ganglion cells are connected. With intracellular recordings and a reliable identification of recorded cells by intracellular staining, ample observations have been made on response types of each retinal neuron and we now understand many of the essential connections from receptors to ganglion cells.

The photoreceptors respond to light with hyperpolarization. It has been puzzling how the hyperpolarizing signal of photoreceptors is transmitted to bipolar and horizontal cells, since at common synapses, the depolarizing presynaptic potential is directly related to the transmitter release. From the discovery that a transretinally applied current pulse, which depolarized receptor terminals, produced transient depolarization in horizontal cells, Trifonov (1968) proposed a hypothesis that the type of signal transmission at receptor-horizontal cell synapse is <u>disfacilitation</u>: tonic release of an excitatory transmitter in the dark, and its reduction with illumination. This idea has been supported by Toyoda et al. (1969) and by Werblin (1975), who demonstrated that the membrane resistance of horizontal cells was low in the dark and increased by illumination. Furthermore, by blocking the chemical transmission by application of Ca^{2+} antagonists, a number of investigators demonstrated that the horizontal cells are hyperpolarized and light responses suppressed (Dowling and Ripps, 1973; Cervetto and Piccolino, 1974).

Experiments along similar lines have been made in the carp retina providing evidence which supports the above hypothesis (Kaneko and Shimazaki, 1975a). It has also been found that the same transmitter acted on the on-center (depolarizing) bipolar cells with a quite different mechanism: the endogenous transmitter is believed to hyperpolarize bipolar cells by decreasing membrane permeability (Kaneko and Shimazaki, 1975b). The transmitter substance has not yet been identified, but among several substances tested pharmacologically, glutamate showed effects which can be interpreted with the present hypothesis (Murakami et al., 1972, 1975).

This chapter aims to summarize our recent studies on the carp retina toward understanding the mechanisms of synaptic transmission from photoreceptors to the second-order neurons. Intracellular recordings were made from cones and horizontal and bipolar cells in the isolated retina of the carp. The retinas were superfused with various artificial saline solutions saturated with oxygen. By perfusing with standard Ringer solution, retinas were maintained in good condition for up to several hours, and the cells showed good responses to light flashes. Identification of recorded cells were made mainly from their response patterns (Kaneko, 1970). Those cells from which no light responses were seen were identified morphologically by intracellular staining (Stretton and Kravitz, 1968).

1.8.2 Transmission from Receptors to Horizontal Cells

Horizontal cells in the carp retina respond to light with a graded, sustained polarization and show a characteristic large spatial summation. No spike discharges are seen. Horizontal cells are classified into several types, according to response types to monochromatic light of different wavelengths. In this chapter, however, only those horizontal cells showing hyperpolarizing responses to all wavelengths (luminosity or L-type responses) and receiving cone inputs will be considered.

1.8.2.1 Blocking of Chemical Transmission

Since the horizontal cell response is hyperpolarizing, the first question is whether this is a sustained IPSP, produced by the continuous release of inhibitory transmitter from the photoreceptors during illumination, or disfacilitation in which a tonically released excitatory transmitter is reduced by illumination? A clear answer to this was obtained by blocking the chemical transmission by changing the external medium to a test solution containing Co^{2+} or high concentration of Mg^{2+}. These divalent cations are known to antagonize the action of Ca^{2+}, thus blocking the transmitter release presynaptically (for Co^{2+} see Weakly, 1973). Fig. 1 shows that the addition of 1 mM Co^{2+} to the perfusing solution quickly hyperpolarized the horizontal cell and its light responses disappeared, while cones maintained the dark membrane potential and continued to respond to light flashes. This observation immediately suggests that the light-evoked hyperpolarizing response in horizontal cells is the result of above-mentioned disfacilitation.

In the turtle, it has been shown that cones are under the continuous inhibitory influence from horizontal cells (Baylor and Fuortes, 1970). When the chemical transmission was interrupted, responses of turtle cones were enhanced due to the dropout of the feedback from horizontal cells (Cervetto and Piccolino, 1974). On the contrary, feedback inhibition was not obvious in the carp cones, and also in the present experiment no enhancement of cone responses were seen after blocking the synaptic transmission.

After the synaptic inputs have been removed, the carp horizontal cells showed a membrane potential close to the equilibrium potential of K^+. The membrane potential changed by almost 58 mV to a ten-fold change in K^+ concentration, indicating that the membrane potential of horizontal cells was determined by E_K.

1.8.2.2 Effect of Synaptic Facilitation

In contrast to the effect of synaptic blocking, an external ionic environment which facilitates transmitter release from the presynaptic terminals produced a strong depolarization of horizontal cells. Fig. 2

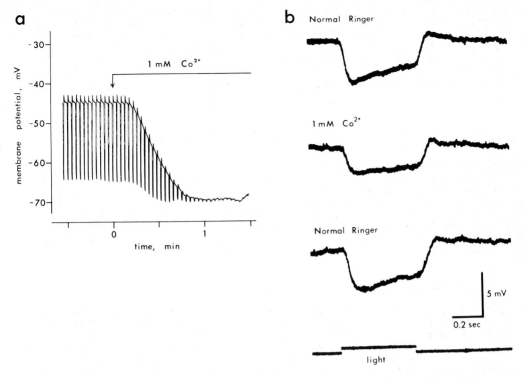

Fig. 1a and b. Effect of blocking chemical transmission on a luminosity (L-) type horizontal cell and a red-sensitive cone. (a) intracellular recording from a horizontal cell. Test solution containing 1 mM Co^{2+} was applied at 0 min. (b) cone response in normal Ringer (<u>top</u>), 2 min after application of Co^{2+}-Ringer (<u>middle</u>), and 2 min after washing with standard Ringer (<u>bottom</u>). Note that the horizontal cell was hyperpolarized and light responses suppressed already 1 min after Co^{2+} application. Diffuse illumination of 620 nm, 3×10^{12} photon/mm^2 s was given to the retina. (From Kaneko and Shimazaki, 1975b)

shows that the addition of 2 mM La^{3+} to the perfusing Ringer solution produced a sustained depolarization in horizontal cells, after an initial transient hyperpolarization. Lanthanum ions have been shown to facilitate the transmitter release from the presynaptic terminals at the neuromuscular junction as revealed by the increased frequency of miniature end plate potentials (Heuser and Miledi, 1971). Contrary to horizontal cells, La^{3+} hyperpolarized cones.

The effect of La^{3+} was very persistent and the reversibility was poor. Even by washing the retina with normal Ringer solution, horizontal cells remained depolarized. However, it seems unlikely that this prolonged depolarization is an indication of cell deterioration, because by application of Na-free solution, this horizontal cell was repolarized to about -70 mV level. This observation may rather indicate that the effect of La^{3+} remained even after washing and the endogenous transmitter depolarized horizontal cells by increasing membrane permeability to Na^+.

146

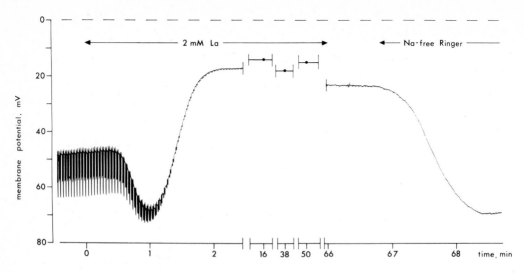

Fig. 2. Effect of facilitation of transmitter release on the membrane potential of a horizontal cell. Successive records from 5 different cells. Test solution containing 2mM La^{3+} was applied for 66 min from 0 min. One min after washing, Na^+-free Ringer was applied. Retina was illuminated with diffuse red (620 nm) and green (520 nm) lights of equal intensity (3 x 10^{12} photon/mm^2/s). Responses to red were larger in amplitude than those to green. (From Kaneko and Shimazaki, 1975a)

1.8.2.3 Responses of Horizontal Cells to Transretinal Current Pulses

Electrical current flowing through the retina has been shown to be an effective stimulus to the retina, evoking light sensation (electric phosphene) and electrical responses in retinal cells (Trifonov, 1968; Knighton, 1975). Light sensation is evoked by a current flowing from the cornea to the sclera. Current flowing in the opposite direction evokes a 'darkening' sensation and also a transient depolarization of horizontal cells (Trifonov, 1968).

1.8.2.3.1 Responses to Current Pulses of Short Duration

Figs. 3a and b show responses in a horizontal cell to transretinal current pulses flowing from the receptor side to the vitreous side. The current pulse evoked in the horizontal cell a transient depolarization with a delay of about 0.7 ms. Current pulses of the opposite polarity given either in the dark or during illumination was ineffective (Figs. 3c and d). In Fig. 3a current pulses of constant intensity were given superimposed on two kinds of illumination; the first was a light spot and the second an annulus. Although these two kinds of illumination hyperpolarized the horizontal cell by almost an equal amount, current pulse of the equal intensity applied during annular illumination was more effective and evoked a larger depolarization than that applied during the spot illumination. This observation indicates that current pulse acted not directly on horizontal cells. Alternative interpretation seems more reasonable that the current pulse polarized receptor terminals which was synaptically transmitted to the horizontal cells. This idea is supported by the finding in Fig. 4 that after blocking chemical transmission by a test medium containing Co^{2+}, the current pulse failed to evoke the transient depolarization.

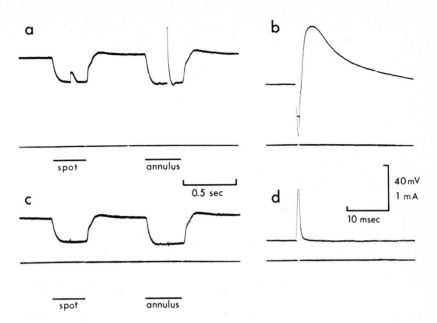

a

spot   annulus

0.5 sec

b

c

d

40 mV
1 mA

10 msec

spot   annulus

Fig. 3a – d. Horizontal cell responses to transretinal current pulses of 0.5 ms duration. (a) and (b) terminal-depolarizing current (current from receptor side to vitreous side). (c) and (d) terminal-hyperpolarizing current (from vitreous to receptor side). In (a) and (c) current pulses were applied during two kinds of white light illumination, 700 μm spot (maximal intensity) and an annulus (900 μm inner and 10 mm outer diameter, attenuated with a 0.75 log unit ND filter). (b) is the same record as in (a) during annulus illumination on an expanded time scale. (d) is the same record as in (c) showing the current artifact. The upper trace in each pair indicates the membrane potential and the second trace the applied current

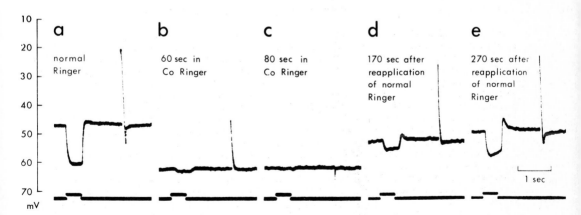

10
20
30
40
50
60
70
mV

a
normal
Ringer

b
60 sec in
Co Ringer

c
80 sec in
Co Ringer

d
170 sec after
reapplication
of normal
Ringer

e
270 sec after
reapplication
of normal
Ringer

1 sec

Fig. 4. Effect of blocking chemical transmission on the horizontal cell response evoked by a transretinal current pulse

Since it now seems clear that the current works on the receptor terminals and modulates transmitter release, the current flowing from the receptor side to the vitreous side will be referred to as 'terminal-depolarizing current', while the current in the opposite direction as 'terminal-hyperpolarizing current'. The short delay of about 0.7-1 ms observed with the transient depolarization evoked by the terminal-depolarizing current seems to be the synaptic delay for the chemical transmission between receptors and horizontal cells. The difference in effectiveness between the terminal-depolarizing and hyperpolarizing currents may be due to a nonlinear current-voltage relationship of the terminal membrane. Probably the effect of the terminal-depolarizing current is very much enhanced. This nonlinear characteristic of the terminal membrane seems to be time-dependent, since the transient depolarization in horizontal cells showed a refractory, and a similar transient depolarizing response was evoked at the cessation of long terminal-hyperpolarizing current. The transient response in horizontal cells remained after application of 10^{-7} g/ml TTX, suggesting that the nonlinear property of the terminal membrane is not due to sodium-dependent action potentials. The nature of this nonlinearity needs further investigation.

1.8.2.3.2 Responses to Current Pulses of Long Duration

The responses of horizontal cells to current pulses having a relatively long duration differed from those to short pulses. Fig. 5 shows such an experiment. The terminal-depolarizing current depolarized the horizontal cell, more during illumination and less during darkness. With long pulses the terminal-hyperpolarizing current was also effective, and hyperpolarized the horizontal cell, this time more during darkness than during illumination.

This observation can be interpreted, based on the present hypothesis, as follows. During illumination, receptor terminals are hyperpolarized and no transmitter is released. Therefore, further hyperpolarization by current cannot make any more change. On the other hand, during darkness receptor terminals have been depolarized and, therefore, hyperpolarization of the receptor terminals by current would effectively reduce the amount of transmitter. Under similar conditions, an additional depolarization by current would soon saturate the release mechanism and thus could not significantly increase the amount of transmitter.

It is worth noting that in an aged or anoxic retina, in which horizontal cells showed a large membrane potential and small light responses, little difference was seen in the effectiveness of terminal hyperpolarizing current between darkness and illumination. It seems likely that deterioration of the retina is accompanied by the malfunction of the transmitter release mechanism from the receptor terminals, and therefore hyperpolarization of the receptor terminal cannot alter the amount of transmitter released.

The experiment in Fig. 5 also gives a satisfactory interpretation of the light sensation evoked by transretinal current. Current flowing from the cornea to the sclera (terminal-hyperpolarizing current) hyperpolarizes the receptor terminal as light illumination. Its effect on the horizontal cell is again equivalent to the effect of illumination. Current flowing in the opposite direction (terminal-depolarizing current) evokes a 'darkening' sensation, which is reasonably understood since the receptor terminals are more depolarized. Response polarity of the horizontal cell (and bipolar cells, as will be shown below) to

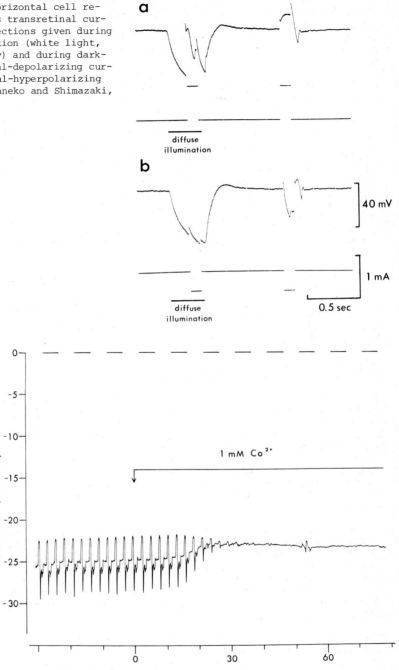

Fig. 5a and b. Horizontal cell responses to 100 ms transretinal current of both directions given during diffuse illumination (white light, maximal intensity) and during darkness. (a) terminal-depolarizing current. (b) terminal-hyperpolarizing current. (From Kaneko and Shimazaki, 1975b)

Fig. 6. Effect of blocking chemical transmission on an on-center bipolar cell. Spot (700 μm, maximal intensity) and annular (900 μm inner and 10 mm outer diameter, -1 log unit) light flashes were given alternately. Test solution containing 1 mM Co^{2+} was applied at 0 min. (From Kaneko and Shimazaki, 1975b)

terminal-depolarizing current also agrees to what might be evoked by 'more darkening' the light illumination.

1.8.3 Transmission from Receptors to Bipolar Cells

Bipolar cells respond to light also with graded responses. They have concentrically organized, antagonistic receptive fields. Two types of bipolar cells have been reported. In one type, illumination of the receptive field center with a light spot evoked a depolarizing response and the illumination of the receptive field surround by an annulus produced a hyperpolarizing response. Cells of this type are called on-center type (or depolarizing) bipolar cells. The other type is called off-center type (hyperpolarizing), which responded just in the opposite way: hyperpolarization to spot and depolarization to annulus. In both of them, the size of the receptive field center was found to be close to the dendritic field, thus suggesting that the direct input from receptors to the bipolar cell produced center-type responses (Kaneko, 1973). The size of the receptive field surround exceeded that of the dendritic field and it is supposed that the surround-type responses is mediated by horizontal cells (Kaneko, 1973; Richter and Simon, 1975). Although antagonistic, the input to the receptive field center is always dominant, and therefore in the present work only the direct transmission between receptors to bipolar cells will be considered.

1.8.3.1 On-Center Bipolar Cells

1.8.3.1.1 Effect of Blocking the Chemical Transmission

The first question deals with the type of synaptic transmission between receptors and on-center bipolar cells. The main question is whether the depolarizing response to light in the on-center bipolar cells is a prolonged EPSP or it is the manifestation of disinhibition. A similar approach as used in the study of horizontal cells is used here, again. As in Fig. 6, when the synaptic transmission was blocked by adding 1 mM Co^{2+} into the perfusate, the on-center bipolar cell was depolarized by about 5 mV, reaching the potential level close to that during spot illumination. This result is in favor of the second hypothesis that during darkness the endogenous transmitter maintained the on-center bipolar cells in hyperpolarization, and by removal of transmitter during illumination the bipolar cell is released from the continuous inhibition.

The ionic mechanism underlying the responses of on-center bipolar cells has not been systematically studied yet, but a slight hint was found in an experiment in which the retina was perfused with Na^+-free Ringer solution. Removal of Na^+ from the external medium also abolished responses, but the on-center bipolar cell was hyperpolarized after a transient depolarization, reaching finally the level of maximal hyperpolarization during surround illumination. This observation suggests that the bipolar cell membrane has a relatively high permeability to Na^+ when the endogenous transmitter is absent. The transmitter reduces the Na^+ permeability.

The present conclusion is in accord with the previous observation by Toyoda (1973) that the input resistance of the on-center bipolar cell is higher in the dark than during illumination, which suggests that the membrane conductance is decreased by the action of the endogenous transmitter.

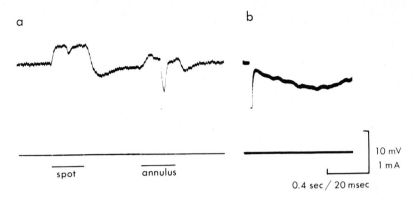

a   b

10 mV
1 mA

spot   annulus

0.4 sec / 20 msec

Fig. 7a and b. Responses of an on-center bipolar cell to a 0.5 ms terminal-depolarizing current pulses given during spot (700 µm) and annular (900 µm inner and 10 mm outer diameter) illumination. (b) is the same record as in (a) on an expanded time scale. (From Kaneko and Shimazaki, 1975b)

1.8.3.1.2 Effects of Transretinal Current

As observed in horizontal cells, the terminal-depolarizing current produced a transient response in on-center bipolar cells (Fig. 7). In this type of cell, this transient response was hyperpolarizing with a delay of about 1 ms. The terminal hyperpolarizing current was again ineffective. These observations indicate that the transmitter, released by the terminal depolarization, acted on the bipolar cell membrane to hyperpolarize.

In the molluscan central neurons a new type of transmitter action has been suggested (Gerschenfeld and Paupardin-Tritsch, 1974). In those synapses the transmitter substance reduced the membrane permeability, resulting in a potential change. The postsynaptic membrane of the on-center bipolar cell seems to have a similar property to this type of synapse.

1.8.3.2 Off-Center Bipolar Cells

Because off-center (hyperpolarizing) bipolar cells were hard to penetrate and to hold, only one observation was made on the effect of synaptic blocking and to a few cells transretinal current was applied. Fundamentally, the behavior of off-center bipolar cells was similar to that of horizontal cells.

In response to a short current pulse of terminal-depolarizing direction, the off-center bipolar cell showed a biphasic response, the initial transient depolarization with a short delay and then a more delayed hyperpolarization (Fig. 8). The delayed hyperpolarization seems to be the secondary effect coming through neighboring horizontal cells since it appeared with an opposite polarity to the initial response and with a longer delay.
Although this experiment is only preliminary, it is tempting to think that the transmitter from the photoreceptors depolarized the off-center bipolar cell with a similar mechanism as in horizontal cells. The measurement in the input resistance of off-center bipolar cells by Toyoda (1973) is also in accord. In those cells, he observed a resistance increase accompanying the hyperpolarizing response to spot illu-

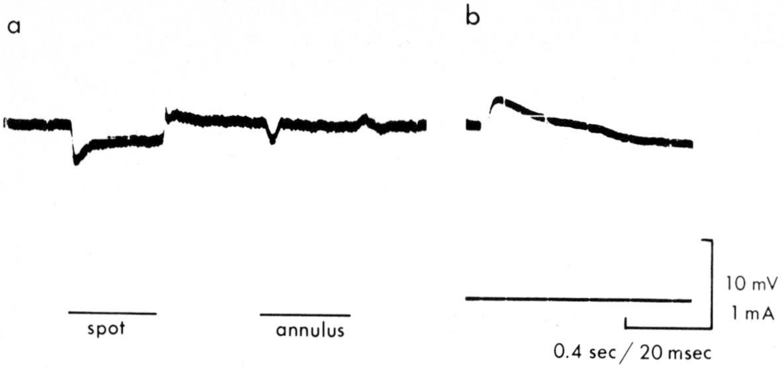

a   b

			10 mV
spot	annulus		1 mA

0.4 sec / 20 msec

Fig. 8. Responses of an off-center bipolar cell under a similar experimental condition as in Fig. 7. (From Kaneko and Shimazaki, 1975b)

mination. However, we have an impression that the ionic mechanism underlying the membrane potential of the off-center bipolar cell does not
seem to be all the same as for the horizontal cell, since, for example,
the response amplitude of the off-center bipolar cell is not so large
and the membrane potential reached after blocking the synapse was only
-40 mV.

1.8.4 Pharmacological Studies of Transmission from Photoreceptors to Second-Order Neurons

From the above results it is concluded that the receptors release the
transmitter substance in the dark and that this substance depolarizes
horizontal and off-center bipolar cells by increasing the membrane
permeability chiefly to Na^+, while it hyperpolarizes the on-center
bipolar cells by decreasing membrane permeability to Na^+. Chemical
compounds which produces the above-mentioned effects on horizontal
and bipolar cells can be a candidate for the transmitter substance.
So far identification of the transmitter substance has not been determined.

A few examples of pharmacological studies on horizontal and bipolar
cells will be presented. It has been shown that glutamate and aspartate produce a strong depolarization in horizontal cells (Dowling and
Ripps, 1972; Murakami et al., 1972; Cervetto and Piccolino, 1974)
and hyperpolarization in on-center bipolar cells (Murakami et al.,
1975). Our results also agreed with these previous investigators.

1.8.4.1 Glutamate

1.8.4.1.1 Effects of L-Glutamate on Horizontal Cells

Fig. 9 illustrates membrane potential changes of horizontal cells produced by the application of glutamate. Addition of 5 mM L-glutamate
in the perfusing Ringer solution depolarized the horizontal cell to
a membrane potential level between -15 and -20 mV and suppressed light
responses. The depolarization was maintained as long as glutamate was
in the perfusate. By washing with the normal Ringer the membrane potential returned to the control level and responses to light flashes

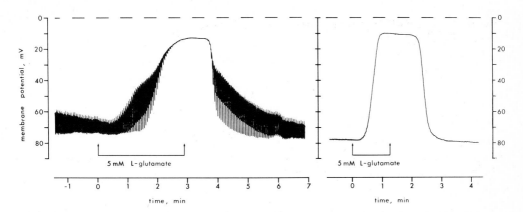

A: in normal Ringer   B: in high Mg Ringer

Fig. 9a and b. Effect of L-glutamate on a luminosity type horizontal cell. (a) test solution containing 5 mM L-glutamate has been applied between two arrows. Diffuse illumination with red (620 nm) and green (520 nm) light of equal intensity was given alternately. (b) response of the same cell as in (a) after the chemical transmission has been blocked by high Mg^{2+}-Ringer

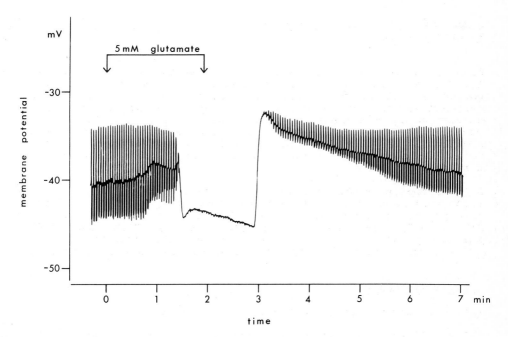

Fig. 10. Effect of L-glutamate on an on-center bipolar cell. Depolarizing responses are to the 700 μm spot and hyperpolarizing responses are to the annulus of 900 μm inner and 10 mm outer diameter. These two stimuli were of white light and given alternately

154

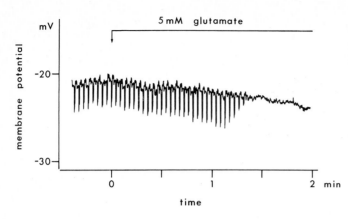

Fig. 11. Effect of L-glutamate on an off-center bipolar cell. Similar experimental conditions as in Fig. 10

reappeared. The least effective concentration of glutamate was about 1 mM. L-glutamate was still effective after the chemical transmission had been blocked by the application of high Mg^{2+} solution (Fig. 9b). This indicates that L-glutamate is acting directly on the horizontal cells. The effect of L-aspartate and its effective concentration was very similar to that of L-glutamate.

1.8.4.1.2 Effects of L-Glutamate on On-Center Bipolar Cells

Fig. 10 shows the effect of L-glutamate on an on-center bipolar cell. Application of the test solution containing 5 mM L-glutamate suppressed responses to light flashes and slightly hyperpolarized the bipolar cell to a level close to the maximum hyperpolarization produces by an annular illumination. The change is in the opposite polarity to that observed when the synaptic transmission has been blocked. L-glutamate is therefore acting as has been expected from the present hypothesis.

1.8.4.1.3 Effects of L-Glutamate on Off-Center Bipolar Cells

Fig. 10 shows the effect of L-glutamate on an off-center bipolar cell. L-glutamate suppressed the light responses as in the on-center bipolar cells, but did not produce a detectable membrane potential change. Unfortunately the cell shown in Fig. 11 was not held long and recovery was not seen.

1.8.4.2 ACh

L-glutamate was not the only chemical substance which depolarized horizontal cells. Acetylcholine was more potent to depolarize horizontal cells when it is applied in the normal Ringer. Fig. 12a illustrates depolarization of a horizontal cell and response enhancement by the application of $10^{-5}M$ ACh. However, ACh became totally ineffective when applied after the chemical transmission had been blocked by high Mg^{2+} Ringer. It seems that the ACh is acting on other cells, the effect of which is further transmitted to horizontal cells synaptically.

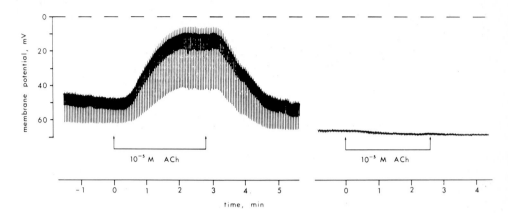

A: in normal Ringer   B: in high Mg Ringer

Fig. 12 a and b. Effect of ACh on a luminosity type horizontal cell. (a) test solution containing 10^{-5}M ACh has been applied between the two arrows. (b) recording from the same cell as in (a), after the chemical transmission has been interrupted by high Mg^{2+}-Ringer. Other experimental conditions are the same as in Fig. 9

1.8.5 General Discussion and Conclusion

The observations described above give one an idea how the signal received at the photoreceptors is forwarded to the second order neurons. It has been established that the vertebrate photoreceptors and second-order neurons are connected with chemical synapses. The receptors release the transmitter in the dark and release reduces with illumination. This transmitter depolarizes horizontal and off-center bipolar cells by increasing membrane permeability chiefly to Na^+, while hyperpolarizes the on-center bipolar cell by decreasing membrane permeability to Na^+.

It is known that three types of cones exist in the carp retina; red-sensitive, green-sensitive and blue-sensitive cones (Tomita et al., 1967). It is necessary to find out the synaptic mechanisms of each type of cone separately. For technical reasons, the present experiments were made on the assumption that the three types of cones have a common synaptic mechanism. Nevertheless, the experimental results can be understood without contradiction.

It is particularly interesting that one transmitter produces very different effects according to the type of postsynaptic cells. The difference of postsynaptic responses must be due to the different characteristics of the postsynaptic membrane. Different types of membrane potential change produced by L-glutamate seem to support this idea. It would be very interesting if one could find morphological and biochemical characters of the individual postsynaptic membrane.

Identification of the transmitter substance is also an important question. Although glutamate shows a similar behavior to what has been expected of the endogenous transmitter, one is puzzled why such a high dose of glutamate (more than 1 mM in concentration) is necessary to evoke responses in horizontal and bipolar cells. Diffusion barriers or poor accessibility to the effective sites might be one reason, but this may not give a full account for such low sensitivity. The effective concentration is more than three orders of magnitude higher compared to other transmitter substances at various synapses. For the identification of transmitter, other kinds of approach, such as neurochemical analysis, are strongly required.

The present discussion has been focused only to the transmission between photoreceptors and second-order neurons. As has been mentioned before, horizontal cells are believed to be mediating lateral information to bipolar cells. In the present study this synaptic transmission was not considered, but this mechanism is playing an important role on the construction of receptive field surround in the bipolar cells.

The vertebrate retina has a very high complexity in the outer plexiform layer. As a result, information concerning space, time and color has already been processed in the bipolar cell (Kaneko, 1973; Richter and Simon, 1975). Further studies of the synaptic mechanisms in the outer layer of the vertebrate retina must give us very important information for understanding the visual system.

Acknowledgments. This work was supported in part by grants from the Education Ministry of Japan Nos. 811014, 911114 and 057297, by the Bionics Research Grant from the Japanese Ministry of International Trade and Industry and by U.S. Public Health Service Grant No. EYOOO17 (T. Tomita, Principal Investigator).

1.8.6 References

Baylor, D.A., Fuortes, M.G.F.: Electrical responses of single cones in the retina of the turtle. J. Physiol. 207, 77-92 (1970)

Cervetto, L., Piccolino, M.: Synaptic transmission between photoreceptors and horizontal cells in the turtle retina. Science 183, 417-419 (1974)

Dowling, J.E., Ripps, H.: Adaptation in skate photoreceptors. J. Gen. Physiol. 60, 698-719 (1972)

Dowling, J.E., Ripps, H.: Effect of magnesium on horizontal cell activity in the skate retina. Nature 242, 101-103 (1973)

Gerschenfeld, H.M., Paupardin-Tritsch, D.: Ionic mechanisms and receptor properties underlying the responses of molluscan neurones to 5-hydroxytryptamine. J. Physiol. 243, 427-456 (1974)

Heuser, J., Miledi, R.: Effects of lanthanum ions on function and structure of frog neuromuscular junctions. Proc. R. Soc. London B 179, 247-260 (1971)

Kaneko, A.: Physiological and morphological identification of horizontal, bipolar and amacrine cells in goldfish retina. J. Physiol. 207, 623-633 (1970)

Kaneko, A.: Receptive field organization of bipolar and amacrine cells in the goldfish retina. J. Physiol. 235, 133-154 (1973)

Kaneko, A., Shimazaki, H.: Effects of external ions on the synaptic transmission from photoreceptors to horizontal cells in the carp retina. J. Physiol. 252, 509-522 (1975a)

Kaneko, A., Shimazaki, H.: Synaptic transmission from photoreceptors to bipolar and horizontal cells in the carp retina. Cold Spring Harb. Symp. Quant. Biol. 40, 537-546 (1975b)

Knighton, R.W.: An electrically evoked slow potential of the frog's retina. I. Properties of response. J. Neurophysiol. 38, 185-197 (1975)

Murakami, M., Ohtsu, K., Ohtsuka, T.: Effects of chemicals on receptors and horizontal cells in the retina. J. Physiol. 227, 899-913 (1972)

Murakami, M., Ohtsuka, T., Shimazaki, H.: Effects of aspartate and glutamate on the bipolar cells in the carp retina. Vis. Res. 15, 456-458 (1975)

Richter, A., Simon, E.J.: Properties of center-hyperpolarizing, red-sensitive bipolar cells in the turtle retina. J. Physiol. 248, 317-334 (1975)

Stretton, A.O.W., Kravitz, E.A.: Neuronal geometry: determination with a technique of intracellular dye injection. Science 162, 132-134 (1968)

Tomita, T., Kaneko, A., Murakami, M., Pautler, E.L.: Spectral response curves of single cones in the carp. Vis. Res. 7, 519-531 (1967)

Toyoda, J.-I.: Membrane resistance changes underlying the bipolar cell response in the carp retina. Vis. Res. 13, 283-294 (1973)

Toyoda, J.I., Nosaki, H., Tomita, T.: Light-induced resistance changes in single photoreceptors of Necturus and Gecko. Vis. Res. 9, 453-463 (1969)

Trifonov, Yu. A.: Study of synaptic transmission between photoreceptors and horizontal cells by means of electric stimulation of the retina. Biophysica 13, 809-817 (1968, in Russian)

Weakly, J.N.: The action of coblat ions on neuromuscular transmission in the frog. J. Physiol. 234, 597-612 (1973)

Werblin, F.S.: Anomalous rectification in horizontal cells. J. Physiol. 244, 639-657 (1975).

1.9 Retinal Physiology in the Perfused Eye of the Cat

G. NIEMEYER

1.9.1 Introduction

The physiology of the vertebrate retina is frequently studied in vitro. The major advantages of this approach are mechanical stability on the one hand and the opportunity to completely control the biochemical environment of the retina on the other. The mechanical stability, as accomplished mainly by the elimination of cardiovascular pulsations and respiratory movements, facilitates intracellular recording from and iontophoretic staining of small neurons. The effects of controlled changes in the composition of the perfusate as well as the addition of synaptically active compounds, can be tested at various levels in the retinal network. Effects of chemical manipulations upon functional properties of the retina independent of systemic changes can be followed electrophysiologically.

The major techniques for studying the retina in vitro include incubation of isolated eyes (Küchler et al., 1956), superfusion of isolated retinas (Sickel et al., 1960) or of isolated eye cups (Cervetto and MacNichol, 1972), and arterial perfusion of isolated eyes (Gouras and Hoff, 1970; Niemeyer, 1973a) or eye cups (Nelson et al., 1975).

The advantages of an in vitro cat retina preparation are (1), much is known about the circuitry, (2) the physiology of its ganglion cells has been studied extensively and (3) the vascular supply of the cat eye lends itself to easy cannulation.

This report will (1) outline the arterial perfusion of the isolated eye of the cat, (2) demonstrate the viability of the retina in vitro by morphological and electrophysiological investigations and (3) show results obtained recently on the effects of some compounds that are potentially active at the synapses.

1.9.2 Methods

For details of the technique of the perfusion employed here the reader is referred to previous publications (Gouras and Hoff, 1970; Niemeyer, 1973a; Niemeyer, 1975). Adult cats were anesthetized (ketamine hydrochloride, 40 mg/kg i.m. and thiamyal, 8 mg/kg i.v.) and anticoagulated (heparin, 5000U-USP/kg i.v.) before surgery. An eye was enucleated and transferred to a thermostatically heated (37.5°C) plastic chamber (Fig. 1). The ophthalmociliary artery was connected to a hydrostatically driven perfusion system. The perfusate was tissue culture medium (Medium 199) which contained 20% calf serum. This hemoglobin-free perfusate was oxygenated (pO$_2$ 430-500 mm Hg) and buffered to pH 7.4.

Stimulation by light was carried out by means of a xenon arc source in combination with neutral and narrow band interference filters, an elec-

Fig. 1. Schematic drawing of the perfusion apparatus for isolated mammalian eyes. C: contact-lens-electrode; E: electrical stimulation; F: narrow band interference- and neutral filters; H: heating system; I: input of chemicals, fixative or dye; O: modified ophthalmoscope; P: perfusate under hydrostatic pressure; R: reduction valve for regulation of flow rate; REC: recording equipment; S: xenon arc source

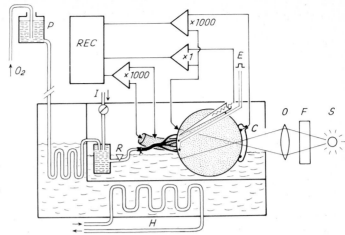

tromagnetic shutter and a modified ophthalmoscope. This system provided monochromatic pulses of light of various durations in Maxwellian view and for observation or photography of the retinal region being stimulated.

Ag-AgCl or Pt-electrodes were placed on the cornea, on the sclera at the posterior pole, and on the optic nerve stump (Fig. 1). For electrical stimulation of the optic nerve a bipolar electrode was passed through the vitreous and positioned at the optic disc. The intraretinal electroretinogram (ERG) and single cell responses were recorded with glass-micropipettes (impedance 20-60 M ohms), introduced through a trephine hole in the pars plana and guided to the retina under visual control.

Light-evoked electrical signals were amplified in either the ac- or dc-mode and recorded by means of conventional electronic equipment. All results were obtained in dark-adaptation. In some experiments, oxygenation, flow rate, pH and temperature were changed for periods of 5 to 12 min. Chemicals were pump-injected into the perfusate through a small chamber and at a slow rate (0.1-0.5 ml/min) compared to the flow of perfusate (2.0-4.0 ml/min). Fixatives or dyes for evaluation of the perfusion were introduced through the ophthalmociliary artery.

1.9.3 Viability of the Retina in vitro

Adequate and homeostatic perfusion maintained integrity of the fine structure and electrophysiological responsiveness of the retina to light for 4 to 10 h. Anatomical investigations, particularly of the inner nuclear layer and functional data from various levels of signal processing in the retina will be presented in order to support the above statements.

1.9.3.1 Morphological Observations[1]

Light- and electron miscroscopic investigations of retinas of perfused cat eyes were carried out after 2 h of perfusion under various controlled conditions (Remé and Niemeyer, 1975a,b). The results revealed, in summary:

1. Low flow rates (0.5-1.0 ml/min) induced drastic changes in about 20% of the cells of the inner nuclear layer as revealed in both semithin and thin sections. The ERGs of such preparations consisted of a-waves only.

2. Perfusion at medium flow rates (2.0-4.0 ml/min) did not induce detectable histological changes at the light microscopic level; electron microscopy of the inner nuclear layer showed clumping of nuclear chromatin in about 3% of the cells but little cytoplasmic changes. The ERGs of these preparations were comparable to in vivo data.

3. High flow rate (5.1 ml/min) of the perfusate further reduced the percentage of damaged cells in the inner nuclear layer; under these conditions the b-waves of the ERG reached higher than normal amplitudes.

4. Horizontal cells appeared to be less vulnerable than other cells of the inner nuclear layer. Fig. 2 shows cells of the inner nuclear layer from an eye which had been perfused at a medium flow rate (3.0 ml/min) for 2 h.

This relation between flow rate and extent of cellular damage parellels previously published data showing that the amplitudes of the b-wave and optic nerve response are grossly related to flow rate (Niemeyer, 1973b). The particular role of oxygen in this context will be discussed in Sect. 4.1 of this chapter.

1.9.3.2 Electrophysiology

Recordings of the ERG and optic nerve responses were routinely obtained, reflecting respectively both the initial and intermediate stages of signal processing on the one hand and the output of the retina on the other. These signals served to monitor the responsiveness and adaptational state of the retina in general. Signals from single cells of various layers of the retina have also been studied.

1.9.3.2.1 ERG

Superimposed ERG responses to a series of increasing intensities are shown in Fig. 3. The b-wave increased in amplitude, decreased in latency and developed oscillatory components parallel with increasing stimulus intensity. The a-waves appeared at high intensities. DC-recordings of the ERG, as shown in Fig. 4, revealed the c-wave. This late and slow positive component of the ERG increased in amplitude and also in its peak time as the duration of the stimulus was increased. Similar changes in the c-wave were obtained by increasing merely the intensity of the stimulus (Niemeyer, 1974a). The c-wave

---

[1]These data were obtained jointly with Dr. Ch. Remé, to whom I am grateful for providing Fig. 2

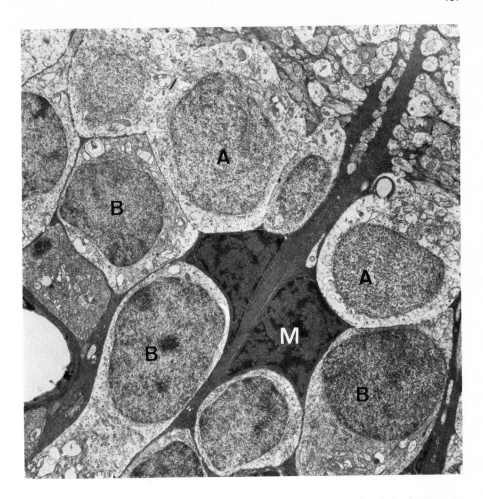

Fig. 2. Electron micrograph of the inner nuclear layer after two h of perfusion at adequate flow rate. A Müller cell M is surrounded by amacrine A and bipolar cells B. All cells reveal regular cytoplasmic organelles and normal appearance of the nuclear chromatin. Uranyl-acetate and lead citrate (x 4250)

appears to be generated in the pigment epithelium but is dependent on light-evoked activity of the photoreceptors (Steinberg et al., 1970). Thus the ERG provides a means of assessing the functional stability of several stages in the retina, i.e. the pigment epithelium, the photoreceptors and the inner nuclear layer (Brown, 1968; Tomita, 1972).

1.9.3.2.2 Recordings from Single Cells

Stabile recordings from single cells of the retina are obtainable in the intact perfused eye. Removal of the anterior segment and the vitreous may further facilitate intracellular recording from this retina (Nelson et al., 1975). Intracellular recording is especially valuable for studying the responses of neurons which do not generate action potentials, and this appears to be a common property of many retinal cells.

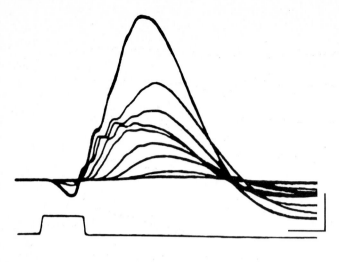

Fig. 3. Electroretinograms from a perfused eye elicited in dark adaptation by stimuli of increasing intensity. The photocell signal (lowermost trace) indicates pulses of 620 nm light of 20 ms in duration. The relative intensity (log units) of the stimuli, from the smallest to the largest amplitude, were 0.4; 1.0; 1.4; 2.0; 2.4; 3.0; 3.4; 4.0 and 4.4, respectively. AC recordings; calibrations: 20 ms and 200 μV

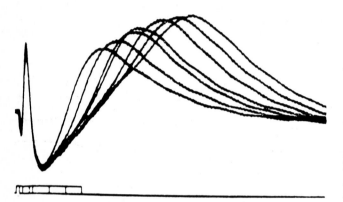

Fig. 4. ERGs, recorded in the DC mode, elicited by stimuli of increasing duration at constant intensity. Note the augmentation and increasing delay of the c-waves with increasing stimulus duration (20-400 ms). The maximum amplitude of the c-wave was 960 μV. The stimuli were unattenuated monochromatic light (λ_{max} 400 nm)

Pigment Epithelium Cells. Glass-micropipettes, advanced slowly towards Bruch's membrane, occasionally detected sudden and large potential changes of up to -65 mV in the region of the pigment epithelium layer. Responses to light were superimposed upon this resting potential and consisted of slow hyperpolarizations that resembled in every respect the c-wave. Fig. 5 shows such responses to monochromatic lights. Removing the electrode from this region produced a sharp shift of the potential back to the former level (Fig. 5, inset). These responses are considered to be obtained from within single pigment epithelium cells. The action spectra of these responses follow rod activity almost exclusively, which appears to be a characteristic of the c-wave.

Horizontal Cells. In mammalian retinas horizontal cells appear to respond to increasing illumination exclusively with hyperpolarization (Gouras, 1972). This negative shift of the membrane potential (S-potential of the luminosity type) is graded with the intensity and more or less sustained for the duration of the stimulus. Most S-potentials appear to show input from both the rods and the cones. The rod component characteristically outlasts the stimulus duration. In the light-adapted state the S-potentials repolarize after the termination of the stimulus (Steinberg, 1969; Niemeyer and Gouras, 1973).

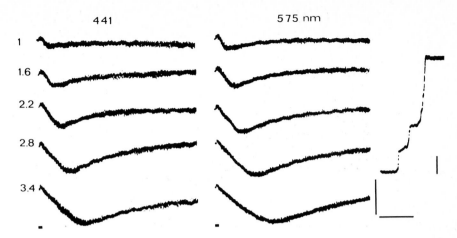

Fig. 5. Intracellular responses from a pigment epithelium cell to incremental stimuli at two wavelengths. Inset on the right side: potential changes recorded during removal of the electrode from the cell. Calibrations: 1 s and 10 mV. (From Niemeyer, 1975, with permission of Dr. W. Junk, B.V., the Hague)

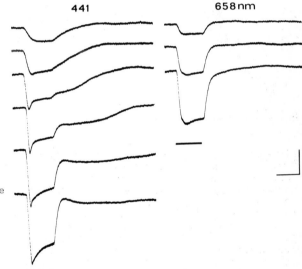

Fig. 6. Intracellular recordings from a horizontal cell. The intensity of 200 ms-pulses of monochromatic light (441 and 658 nm) was increased from top to bottom in 0.6 log steps of relative intensity. Calibrations: 200 ms and 10 vM

Fig. 6 shows examples of S-potentials in response to incremental stimuli from both ends of the visible spectrum. These signals were shown to be characteristic of the "mixed rod-cone" input variety, which was the type most frequently encountered in a previous study (Niemeyer and Gouras, 1973). An analysis of the spectral sensitivity of these horizontal cell responses is illustrated in Fig. 7. The plotted data were derived from response amplitude/intensity functions. Criteria of low amplitude (open circles) and criteria of high amplitude (filled circles) were used to obtain the correponding intensities at different wavelengths. This illustrates that these cells received input from

164

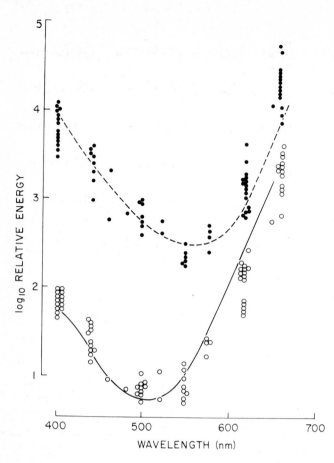

Fig. 7. Threshold versus wavelength curves from S-potentials from 14 horizontal cells at low and at high response criteria. <u>Empty circles</u>: data derived from a low (close to threshold) response amplitude-criterion; solid curve: action spectrum of the rod system of the cat (Dodt and Walther, 1958a). <u>Filled circles</u>: data derived from high (close to saturation) response criteria; <u>Dashed line</u>: action spectrum of the cone-system of the cat (Dodt and Walther, 1958b). Note the decrease of the sensitivity in the range of short wavelengths with increasing intensity of the stimulus. (From Niemeyer (1975) with permission of Dr. W. Junk V.V., The Hague)

rods at threshold (<u>lower curve</u>) and from cones at higher intensities (<u>upper curve</u>). This implies that signals arising from rods and cones converge in most horizontal cells in the cat. For information on putative anatomical pathways for convergence of rod and cone signals within or even before horizontal cells in the cat the reader is referred to three recent publications: Kolb (1974); Raviola and Gilula (1975); Nelson et al. (1975). Two other types of S-potentials were also found, but much less frequently: one with only rod input and the other with only cone input (Niemeyer, 1975). The origin of these S-potentials is not yet established (Nelson et al., 1975).

The first identification of horizontal cells as generators of S-potentials in the cat was carried out by Steinberg and Schmidt (1970); recent work has managed to stain the fine dendrites of these cells and permitted identification of their cell body and axon terminal.

Ganglion Cells. Responses of a retinal ganglion cell to incremental stimuli at various wavelengths are shown in Fig. 8. Spontaneous activity present in the dark can be seen in the upper records. Maintained pulses of light (200 ms) revealed both phasic and tonic components in the discharge patterns of this cell and decreased in latency as stimulus intensity grew. Postexcitatory inhibition and prolonged discharge trains occured at high intensities. Differences in rod and cone input to this ganglion cell are apparent from the responses to different wavelengths.

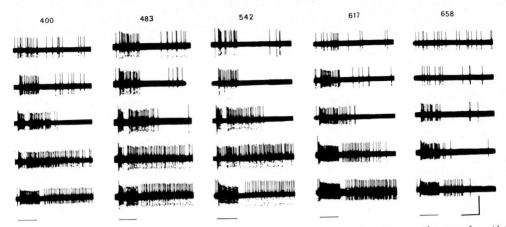

Fig. 8. On-responses from a retinal ganglion cell of a perfused eye. The wavelengths of the stimuli are indicated on top, the intensity was increased, from top to bottom, in 1.2 log steps of relative optical density. The recordings reveal spontaneous activity in the dark and tonic on-responses with inhibition after the termination of the stimulus at high intensities. The cell's responses also exhibit characteristic differences depending on wavelength. The tonic on-responses are interrupted by some inhibition short after the onset of the stimulus in the short, but not in the long wavelength-range. Calibrations: 200 ms and 10 mV

1.9.3.2.3 Optic Nerve Responses

Population responses of the axons of a large number of retinal ganglion cells can be recorded from the optic nerve stump. In this study, the active electrode was placed on the surface of the nerve and the reference point was the cut end (Fig. 1). The responses obtained with this electrode arrangement consisted of negative, temporally dispersed potentials with characteristic oscillations. Fig. 9 shows such responses under two different conditions. In (A), a 20 ms pulse of monochromatic light (λ_{max} 620 nm) was delivered at various intensities. As the intensity increased, the response became faster and its amplitude grew. The oscillations changed only slightly with the intensity. In (B), responses to 10 ms duration white flashes, delivered by a ganzfeld-stimulator, were superimposed. The oscillatory character of these on-responses was a consistent feature and appears to be a characteristic property of intact cat optic nerve (Doty and Kimura, 1963; Steinberg, 1966).

Fig. 10 demonstrates on-responses as described above, tonic components and negative off-effects (arrows). The latter became more and more separated as the stimulus duration increased. It should be noted that the polarity and configuration of the off-effect varied considerably with both stimulus duration and electrode position on the nerve.

Adequate electrical stimulation of the optic disc elicited classic compound action potentials in the optic nerve, revealing components of at least 2 fiber groups with different conduction velocities (Fig. 11). In order to separate more clearly these fiber groups the temperature of the eye was reduced to 34°C.

A

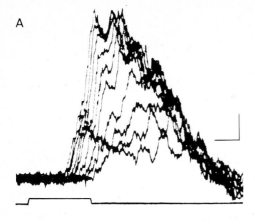

B

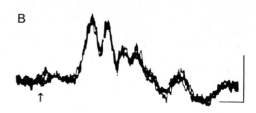

Fig. 9A and B. Light-evoked responses from
the optic nerve. (A) monophasic oscillatory
responses to monochromatic (λ_{max} 620 nm)
pulses of 20 ms in duration at increasing
intensity (from bottom to top 0.4-5.0 log
units). (B) optic nerve responses to xenon
flashes (arrow) of white light; superimpo-
sition of 4 traces. Calibrations: 10 ms
and 50 µV. Positivity is displayed downwards
in all nerve responses

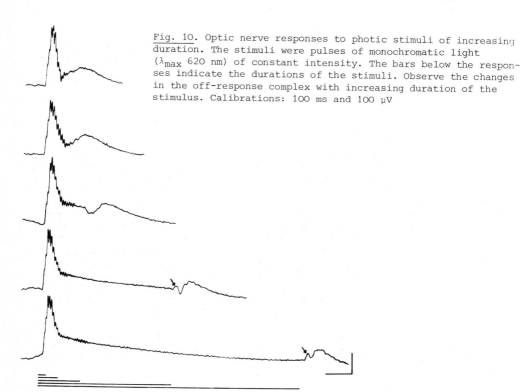

Fig. 10. Optic nerve responses to photic stimuli of increasing
duration. The stimuli were pulses of monochromatic light
(λ_{max} 620 nm) of constant intensity. The bars below the responses indicate the durations of the stimuli. Observe the changes
in the off-response complex with increasing duration of the
stimulus. Calibrations: 100 ms and 100 µV

Fig. 11. Optic nerve responses to electrical shocks applied to the optic disc. The strength of the pulses is indicated on the left margin, and the stimulus duration was kept at 0.05 ms. The dots on the time base appeared at the onset of the electrical pulse and then in intervals of 0.1 ms. Calibration: 1 mV. Several traces were superimposed at each voltage

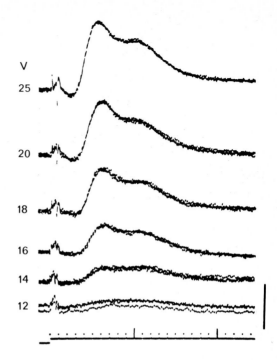

1.9.4 Results

The in vitro situation allows one to study certain aspects of the function of the retina which are more difficult to investigate in the intact animal. Two projects are considered in the following section: (1) the effects of changes in biophysical parameters, and (2) the effects of certain synaptically active agents of the retina.

1.9.4.1 Biophysical Parameters

Previous studies have shown that variations in the flow rate induced parallel changes in the b-wave of the ERG and in the amplitude of the optic nerve responses (Niemeyer, 1973b). With very slow flow rates the b-wave and optic nerve response disappeared, although receptor function, as reflected in the a-wave of the ERG, remained. Fig. 12 shows that keeping the flow rate constant while lowering the concentration of oxygen also resulted in rapid reduction of the amplitude of the b-wave. These changes were reversible. However, the longer and the more severe hypoxia was maintained, the longer was the period required for full recovery. It is interesting in this context, that Hoff and Gouras (1973) reported full recovery of both ERG and optic nerve response after up to two h of circulatory arrest in the perfused cat eye. Increase of the flow rate of the perfusate can compensate for hypoxia within a certain range (Niemeyer, 1974).

Oxygen appears to be more important than glucose or removal of metabolic products for the maintenance of retinal function. Careful control of the pH and temperature appear to be equally crucial for stabile responsiveness of this retina.

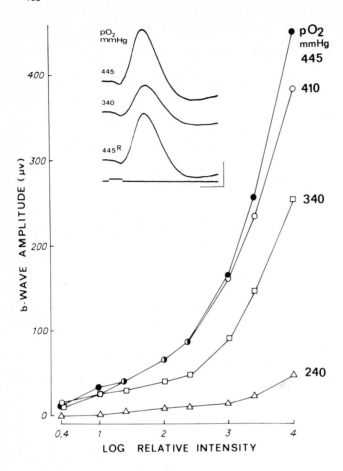

Fig. 12. Effects of transient hypoxia on the amplitude of the b-wave. The data summarize results from several experiments. The control and recovery data are shown in filled symbols of the response-intensity functions, and empty symbols represent data obtained during hypoxia. The inset shows ERGs under control (pO_2 445), hypoxia (pO_2 340) and recovered (pO_2 445R) conditions. Calibrations: 40 ms and 200 µV

1.9.4.2 Effects of Flaxedil, Aspartate and Atropine (Preliminary Results)

Synaptic transmission across several layers of retinal neurons appears to account for the transformation of primary effects of light on photoreceptors into the complex patterns of signals in the optic nerve fibers. In this preliminary attempt only a limited number of substances have been used, namely a putative amino acid transmitter, aspartate, and cholinergic blocking agents, Flaxedil (nicotinic) and atropine (muscarinic blocker). Drugs can be added at various rates and in various concentrations to a continuous flow of perfusate, and their effects can be assessed by electrophysiological measurements as mentioned in Sect. 3 of this chapter. These signals or single components of them reflect the activity of specific retinal layers and can be used to localize effects of chemical compounds.

1.9.4.2.1 Aspartate

Certain amino acids, such as aspartate and glutamate are known to isolate the P_{III} component of the ERG from the remaining components (Sillman et al., 1969). This effect has been used to study photoreceptor activity in isolation. Cervetto and MacNichol (1972) were able to show by intracellular recording that glutamate and aspartate block responses of horizontal cells to light with little effect on photoreceptors in the turtle retina. In the perfused cat eye, aspartate, when transiently applied in concentrations from 7 to 100 mM to the perfusate, induced dose-dependent and reversible changes in the ERG and optic nerve response. The amplitude of both the b-wave and optic nerve response decreased rapidly and were completely abolished under high concentrations of aspartate. The a-wave remained unchanged or showed a moderate increase in its amplitude. This implies that in the cat aspartate blocks synapses from the photoreceptors to both horizontal as well as bipolar cells (see also Niemeyer, Experientia 32, 759, 1976).

1.9.4.2.2 Flaxedil (Gallamine)[2]

Flaxedil is a nicotinic blocking agent which does not cross the blood-brain barrier. Injection of Flaxedil into the systemic circulation has no effect on retinal ganglion cells, as one would expect (Enroth-Cugell and Pinto, 1970). Flaxedil, when added to the perfusate in concentrations up to 100-fold higher than used for complete relaxation of skeletal muscles, affected neither the ERG nor the optic nerve response in the isolated cat eye. Small quantities of Flaxedil, however, when injected into the vitreous close to the retina, drastically reduced the light-evoked responses of the optic nerve. These preliminary results suggest that the blood-retinal barrier in the perfused cat eye is intact and that Flaxedil may have significant effects on synaptic transmission in this retina. These data are in agreement with Vogel et al. (1974) who identified nicotinic acetylcholine receptors in the developing chicken retina by α-bungarotoxin binding. Vogel (pers. comm.) also found evidence for nicotinic receptors in rabbit, cat, and monkey retina.

1.9.4.2.3 Atropine[2]

Atropine, in contrast to Flaxedil, when injected in low concentrations into the perfusion system, induced dose-dependent reversible changes in the optic nerve response, and, to a lesser extent also in the ERG. At low concentrations, atropine has its most impressive effect in decreasing the amplitude of the on-response of the optic nerve. At these concentrations there is a slight increase in the implicit time and a slight augmentation of the b-wave of the ERG. There is a curious inversion of the off-responses in the optic nerve, suggesting that off-responses may actually be enhanced. These effects were completely reversible after 15 min.

Depressant effects of atropine on the b-wave of the ERG were first observed by Val'tsev (1966). The present data add effects of atropine on the optic nerve response and suggest that some of the retinal synapses may be muscarinic (Niemeyer and Cervetto, XIVth Symposium I.S.C.E.R.G., 1976, Docum. Ophthal., in press).

1.9.5 Summary

Arterial perfusion of the cat eye maintains physiologic responsiveness of the retina to light in vitro as revealed by electroretinographic

---

[2]The results on Flaxedil and atropine were obtained jointly with Dr. L. Cervetto during his stay as a visiting scientist in Zürich, 1975

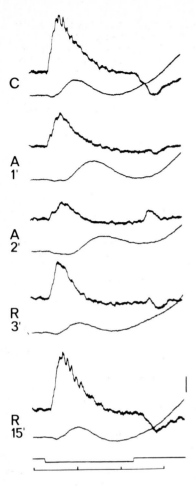

Fig. 13. Effects of atropine on ERG and optic
nerve response. In each pair of responses the
ERG is displayed below the optic nerve response.
The stimulus was a pulse of monochromatic light
(λ_{max} 620 nm) of 200 ms in duration (photocell
response on the bottom). C: control. A 1': 1 min
after onset of injection of atropine in a concentration of 1% at a rate of 0.4 ml/min.
A 2': 2 min after onset of the atropine injection. After this recording the injection was
stopped. R 3': 3 min after stop of injection.
R 15': 15 min after stop of the injection of
atropine, considered full recovery of the response. Calibrations: lowermost trace, 100 ms
divisions vertical bar, 500 µV for ERG and
50 µV for optic nerve

and optic nerve responses. These data are supported by recordings from
the pigment epithelium, horizontal cells and ganglion cells. All these
findings are comparable to in vivo data. Morphological investigations
of the retina indicate only minor cellular damage after 2 h of perfusion at adequate flow rates.

Constant supply of sufficient oxygen, balanced pH and stabile temperature represent important factors for homeostatic perfusion.

The system permits complete control of the biochemical environment including the addition of synaptically active agents. The amino acid
aspartate, when added in low concentrations to the perfusate, abolished
the b-wave of the ERG as well as the optic nerve response reversibly.
The nicotinic blocking agent Flaxedil did not affect retinal function
when administered via the perfusion, but depressed light-evoked responses of the optic nerve when applied directly to the retina from
the vitreal side. Atropine induced dose-dependent and reversible de-
pressions of the on-responses and oscillatory components in the optic
nerve; off-responses appeared to be enhanced at low doses. The ERG
was affected to a lesser extent. These data provide further evidences
for the presence of nicotinic as well as muscarinic mechanisms within
the retina of the cat.

Acknowledgments. I wish to thank Dr. P. Gouras for important suggestions during this study as well as during the preparation of the manuscript. Thanks are due to Prof. R. Witmer for his continuous support of this work. Mrs. R. Fessler provided excellent assistance both in the experiments and in the preparation of the photographs.

This study was supported by the Swiss National Science Foundation (grant No. 3.0630.73) and by the Hartmann Müller-Foundation, Zürich.

1.9.6 References

Brown, K.T.: The electroretinogram: its components and their origins.
  Vis. Res. $\underline{8}$, 633-677 (1968)
Cervetto, L., MacNichol, E.F.: Inactivation of Horizontal cells in Turtle Retina
  by Glutamate and Aspartate. Science $\underline{178}$, 767-768 (1972)
Dodt, E., Walther, J.B.: Netzhautsensitivität, Linsenabsorbtion and physikalische
  Lichtstreuung. Der skotopische Dominator der Katze in sichtbaren und ultravioletten Spektralbereich. Pflügers Arch. Ges. Physiol. $\underline{266}$, 167-174 (1958a)
Dodt, E., Walther, J.B.: Der photopische Dominator im Flimmer-ERG der Katze.
  Pflügers Arch. Ges. Physiol. $\underline{266}$, 175-186 (1958b)
Doty, R.W., Kimura, D.S.: Oscillatory potentials in the visual system of cats
  and monkeys. J. Physiol. $\underline{168}$, 205-218 (1963)
Enroth-Cugell, C., Pinto, L.H.: Gallamine triethiodide (Flaxedil) and cat retinal
  ganglion cell responses. J. Physiol. $\underline{208}$, 677-689 (1970)
Gouras, P.: S-potentials. In: Handbook of Sensory Physiology. Berlin: Springer 1972,
  Vol. VII/2, pp. 513-530
Gouras, P., Hoff, M.: Retinal function in a isolated, perfused mammalian eye.
  Invest. Ophthal. $\underline{9}$, 388-399 (1970)
Hoff, M., Gouras, P.: Tolerance of mammalian retina to circulatory arrest.
  Xth I.S.C.E.R.G. Symp., Los Angeles 1972. Doc. Ophthal. Proc. Ser. Vol. 2
  Dr. W. Junk, The Hague, 1973, pp. 261-268
Kolb, H.: The connections between horizontal cells and photoreceptors in the retina
  of the cat: electron microscopy of Golgi preparations.
  J. Comp. Neurol. $\underline{155}$, 1-14 (1974)
Kuechler, G., Pilz, A., Sickel, W., Bauereisen, E.: Sauerstoffdruck und b-Welle des
  Elektroretinogramms vom isolierten Froschauge. Pflügers Arch. $\underline{263}$, 566-576 (1956)
Nelson, R., v. Luetzow, A., Kolb, H., Gouras, P.: Horizontal cells in the cat retina
  with independent dendritic systems. Science $\underline{189}$, 137-139 (1975)
Niemeyer, G.: Intracellular recording from the isolated perfused mammalian eye.
  Vis. Res. $\underline{13}$, 1613-1618 (1973a)
Niemeyer, G.: ERG dependence on flow rate in the isolated and perfused mammalian
  eye. Brain. Res. $\underline{57}$, 203-207 (1973b)
Niemeyer, G.: C-waves and intracellular responses from the pigment epithelium in
  the cat. Symposium on electrooculography, Augenärztetagung Freiburg/Germany
  (1974a, in press)
Niemeyer, G.: O_2-dependence of the b-wave in the isolated perfused mammalian eye.
  XIth I.S.C.E.R.G. Symp., Bad Nauheim 1973. Doc. Ophthal. Proc. Ser. Vol. 4.
  Dr. W. Junk, The Hague (1974b)
Niemeyer, G.: The function of the retina in the perfused eye.
  Documenta Ophthalmologica $\underline{39}$, 53-116 (1975)
Niemeyer, G., Gouras, P.: Rod and cone signals in S-potentials of the isolated
  perfused cat eye. Vis. Res. $\underline{13}$, 1603-1612 (1973)
Raviola, E., Gilula, B.: Intramembrane organization of specialized contacts in the
  outer plexiform layer of the retina. J. Cell Biol. $\underline{65}$, 192-222 (1975)
Remé, Ch., Niemeyer, G.: Studies on the ultrastructure of the retina in the
  isolated and perfused feline eye. Vis. Res. $\underline{15}$, 809-812 (1975a)
Remé, Ch., Niemeyer, G.: Netzhautfeinstruktur und Funktion des isolierten und
  arteriell perfundierten Säugetierauges. 73. Versamml. Deut. Ophthal. Ges. 1973.
  München: Bergmann 1975b, pp. 411-417

Sickel, W., Lippmann, H.-G., Haschke, W., Baumann, Ch.: Elektrogramm der umströmten menschlichen Retina. Deut. Ophthal. Ges. 63, 316-318 (1960)

Sillman, A.J., Ito, H., Tomita, T.: Studies on the mass receptor potential of the isolated frog retina. Vis. Res. 9, 1435-1442 (1969)

Steinberg, R.H.: Oscillatory activity in the optic tract of cat and light adaptation. J. Neurophysiol. 29, 139-156 (1966)

Steinberg, R.H.: Rod and cone contribution to S-potentials from the cat retina. Vis. Res. 9, 1319-1329 (1969)

Steinberg, R.H., Schmidt, R.: Identification of horizontal cells as S-potential generators in the cat retina by intracellular dye injection. Vis. Res. 10, 817-820 (1970)

Steinberg, R.H., Schmidt, R., Brown, K.T.: Intracellular responses to light from cat pigment epithelium: origin of the electroretinogramm c-wave. Nature 227, 728-730 (1970)

Tomita, T.: The electroretinogram, as analyzed by microelectrode studies. Handbook of Sensory Physiology. Berlin: Springer 1972, Vol. VII/2, pp. 635-666

Val'tsev, V.B.: Role of cholinergic structures in outer plexiform layer in the electrical activity of frog retina. Federation Proc. 25 II, T 765 - T 766 (1966)

Vogel, Z., Daniels, M.P., Nirenberg, M.: Localization of acetylcholine receptors during synaptogenesis in retina. Federation Proc. 33, 1426 (1974).

2 ARTHROPODS

2.1 Adaptations of the Dragonfly Retina for Contrast Detection and the Elucidation of Neural Principles in the Peripheral Visual System

S. B. LAUGHLIN

2.1.1 Neural Principles and Their Analysis

The organisers of this conference chose an excellent title and I will endeavour to show in this chapter by a priori and comparative argument based upon my own experimental analysis of the dragonfly visual system and the excellent work now being carried out upon the outer plexiform layer of the vertebrate retina, that such things as neural principles really do operate in visual systems.

We might suggest that neural principles exist because the visual systems of many animals share common functional properties and have to cope with identical problems during operation. They all sample the same visual world of light intensity and attempt to extract information on shape, colour, movement and position from the visual input parameters of intensity, its spatial and temporal distribution and its spectral content. Thus, although there are many diverse forms of visual system throughout the animal kingdom, many show remarkable analogies and these can be thought of as design principles, related to shared biological principles.

Consider, for example, some of the similarities that exist between the receptors of arthropod and vertebrate visual systems. Both compound eyes and vertebrate "single lens" eyes use a dioptric apparatus to transfer the angular distribution of incident radiation onto a two-dimensional array of photoreceptors and it is this spatial distribution of receptor potential that forms the basis for all subsequent neural processing (e.g. Rodieck, 1973; Laughlin, 1974c). At a molecular level the photopigments of arthropods and vertebrates are rhodopsins and the photoreceptors of cows, squids and diverse insects share 11-cis retinal as a common chromophore (Hamdorf et al., 1973). In the majority of photoreceptors the photopigment is packed into narrow cylinders, which act as waveguides and point in the direction of incident light (Menzel and Snyder, 1975) and, at least in frogs, crayfish and flies the photopigment dipoles are carefully arranged to give the whole photoreceptor a maximum quantum capture efficiency to unpolarized light (Snyder and Laughlin, 1975). These common photoreceptor characteristics have self-evident functional advantages and might, therefore, be termed principles of photoreceptor design and function.

Similarly one might expect that the neurons of the visual system share neural principles. The delineation of neural principles requires that we first describe the neural processes that take place and for several reasons (Kirschfeld, 1972; Laughlin, 1974c) the retina and lamina of the insect eye are particularly favourable material for the application of electrophysiological, optical, behavioural and anatomical techniques. The experiments and results described in this paper are derived from an electrophysiological analysis of neural integration at the level of the first synapse in the dragonfly visual system. This work is made possible by the suitability of dragonfly receptors and interneurons for making stable intracellular recording and the ordered anatomical arrangement of neurons (Laughlin, 1974c). It is possible to perform identical experiments on both receptor and post-synaptic second order interneuron and examine the processes of neural integration by comparing the responses of the two cell types and their sensitivities.

When these studies were begun it was naively expected that the compound eye should show radically different properties from the vertebrate retina. In fact, the converse is the case; neural principles, seen so clearly in the dragonfly retina are seen in the vertebrate retina too, although, for technical reasons they are sometimes obscure and on rare occasions can only be found by someone who already knows what it is he is looking for!

2.1.2 The Dragonfly and Its Lamina

The Australian corduliid dragonfly, Hemicordulia tau, is a fast flying aerial predator. On summer days it is constantly on the wing, catching small flying insects, engaged in sexual encounters and egg-laying or patrolling and defending its territory. Field observations on this and other dragonflies (Corbet, 1962) show that vision plays a major role in almost all these activities. Moreover these animals often feed at low light intensities, just after sunset, when myriads of small insects appear close to the lakes and streams that dragon-flies inhabit.

Clearly, this aerial predator has an effective visual system and it is hardly surprising that the large compound eyes dominate the head-structure (Fig. 1). The visual system follows the standard plan for an apposition compound eye. The outer surface of the eye consists of an array of lenslets and beneath each small lens lies a single optical set of receptors and their accessory cells, termed the ommatidium. The complex arrangement of retinula cells within the ommatidium need not concern us here because we need only consider the receptor in terms of its input properties. Each ommatidium is optically isolated from those around it and its receptors share a common light path and have the same angular sensitivity. Consequently each ommatidium is a single spatial element in the receptor mosaic and within this element the receptors have their own spectral, and polarization sensitivities.

Neuroanatomy tells us that this receptor array maps point for point onto an array of neural elements in the first optic neuropile, the lamina (Horridge and Meinertzhagen, 1970). The lamina is in fact a sheet of these neural elements, lamina cartridges, and there is one cartridge for each ommatidium. Most importantly, the receptor axons from one receptor spatial element, the ommatidium, project to a single lamina cartridge. This convenient anatomical fact considerably simplifies the description of lateral (spatially dependent) interactions between neurons of the lamina and allows neuroanatomists to draw up comprehensive wiring diagrams (e.g. Strausfeld and Campos-Ortega, 1972).

Preliminary studies of the dragonfly neuroanatomy made by Dr. Willi Ribi (pers. comm.) suggest that six retinula cell axons, the short visual fibres, terminate in the lamina cartridge and synapse with large second order neurons, the large monopolar cells or LMCs. These post-synaptic neurons are homologous to the monopolar cells of fly, L1 and L2. In the fly each short visual fibre makes a large number of parallel synapses with each LMC and it is in this respect, extremely likely

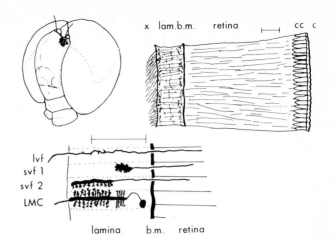

Fig. 1. The peripheral visual system of the dragonfly, Hemicordulia tau. A sketch of the head shows the large pair of compound eyes which wrap around the head capsule. A diagrammatic section through the peripheral visual system shows the relative positions of the dioptric apparatus; cc: crystalline cone; the retina separated from the first optic neuropile, the lamina lam. by the basement membrane b.m. and the fibres leaving the lamina for higher centres via the first chiasma x. The lower diagram shows the morphology of the lamina neural elements examined in this paper; the receptor cell axons, classified as long visual fibres lvf and two types of short visual fibre svf 1 and 2, and the second-order large monopolar cell LMC. Each neural element is shown in a single lamina cartridge and has been redrawn from material kindly provided by Dr. Willi Ribi. The calibration bars show 100 µm

that dragonfly retinula cells and LMCs share the same synaptology and possess similar parallel synaptic arrays.

Needless to say, the lamina is a complex piece or neural machinery (e.g. Strausfeld and Campos-Ortega, 1972) containing a large number of neural elements. The dragonfly lamina is no exception to this (Cajal and Sánchez, 1915; Strausfeld, pers. comm.; Ribi, pers. comm.) but, because we are going to consider some rather simple, but hopefully fundamental, characteristics of signal transfer from retinula cell to LMC this very basic account of dragonfly lamina structure will suffice. However, it will become clear that there are many complex neural interactions taking place in the dragonfly lamina. The elucidation of phenomena such as the generation of response transients, lateral inhibition and neural sensitivity control will depend upon a more complete understanding of the neural wiring diagram.

2.1.3 Neural Integration in the Dragonfly Lamina

2.1.3.1 The Experimental Approach

A detailed description of the experimental method used to derive the results presented in this account can be found in the original papers from which they are taken (Laughlin, 1973, 1974a,b, 1975b). However, two important points relating to the choice of stimulus and the type of measurements made with it are worthy of discussion here. For technical reasons it is impossible, at present, to investigate the voltage input-output relationships of receptors and LMCs directly by recording simultaneously from pre- and post-synaptic sites. Instead one must measure the individual cell's responses separately and then compare different cells under identical stimulus conditions. Fortunately this can be done with precision because stable single unit recordings can be made for up to 3 h with little change in resting potential.

To obtain identical and strictly comparable stimuli a single distant point source was used, positioned at the point of maximum sensitivity in the visual field of the unit under examination, defined as the axis. We can expect that the axis of the LMCs corresponds to the optical axis of the retinula cells because axons from one ommatidium project onto one lamina cartridge. For this reason small axial stimuli are identical for both cell types. Moreover, the use of a small axial point source has two additional advantages. First, it minimises errors resulting from comparison between units with different receptive field properties, and secondly, it standardises and, therefore, controls against chromatic aberrations in the receptors (Eguchi, 1971) and lateral inhibitory effects in the lamina.

Sensitivity functions were derived to describe and quantify the dependence of receptor and interneuron response on changes in stimulus parameters such as wavelength and angle of incidence. Again the techniques for obtaining accurate and valid sensitivity measures are described elsewhere (Laughlin, 1974a,b) and only the basic principles and advantages of this measurement method will be discussed here. Sensitivity measurements are based upon a neural element's intensity/response function and the fact that a change of stimulus in one domain, such as wavelength, can be compensated for by increase or decrease in stimulus intensity. Sensitivity is often defined as the reciprocal of the number of photons required to give a certain criterion response. For example, as the angle of incidence of light is changed and the point source moves away from axis, the response of retinula cells decreases, more incident photons are required to restore the response amplitude to the criterion value, and sensitivity is reduced.

Sensitivity measurements have two tremendous advantages over measurements which simply relate response amplitude alone to a change in wavelength or angle. Both advantages derive from the fact that sensitivity tells us the proportion of incident photons that 'score a hit' and contribute to the response, or apparently score a hit. In receptors the number of photon hits relates directly to the linear optical properties of absorption and transmission (Laughlin, 1975a) whereas in interneurons the number of apparent photon hits describes neural processing in the stimulus intensity domain and can be directly related to behavioral measures such as increment threshold.

The second advantage of sensitivity measurements is based upon the fact that they refer changes in voltage signal in receptor and interneuron back to a common parameter, intensity. The whole basis of neural integration is that intensity signals are changed as they pass from neuron to neuron in a way that is itself a function of other stimulus parameters. This can easily be seen if one considers a chain of repeater stations consisting of identical unity gain amplifiers. This is similar to an action potential in as much as it transmits a signal but does not integrate it. Thus, if we are to understand integration we must, among other things, obtain the voltage transfer properties between neurons under a variety of stimulus conditions. Sensitivity data pinpoints integration very easily. By relating both receptor and interneuron voltage responses to the common intensity parameter the changes of voltage transfer properties between two cell types are seen as differences between their sensitivity functions. In other words, when their sensitivity functions are different we can say that integration

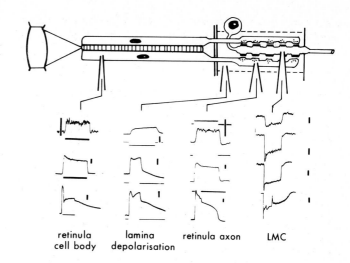

retinula   lamina   retinula axon   LMC
cell body depolarisation

Fig. 2. A diagrammatic representation of two retinula cells with short visual fibres
and a post-synaptic large monopolar cell LMC together with the receptor response recorded intracellularly in the retina and the 3 responses recorded in the lamina.
These lamina responses have been assigned to two intracellular and an extracellular
compartment on the criteria discussed in the text. Each response is to a 500 ms
flash and intensity increases from top to bottom. The flash intensities are not
equal between records and the calibration bars show 10 mV

is at work. Let us now consider the voltage response properties and
sensitivities of both receptor and interneuron and commence the search
for neural principles.

2.1.3.2 The Intensity Signal in Retina and Lamina

2.1.3.2.1 The Receptor Response at the Site of Transduction

Intracellular recordings made from the retinula cell bodies show that
these cells respond to light with a typical invertebrate depolarising
receptor potential, similar to that described in <u>Limulus</u> (Hartline
et al., 1952). At low intensities the response is monophasic and noisy
(Fig. 2). This noise depends upon the random nature of photon absorptions (Kirschfeld, 1966; Laughlin, 1975a). Increasing intensity increases response amplitude and at higher intensities the receptor shows
adaptation. The initial peak response decays to a plateau, probably
through a calcium ion mediated membrane effect (Baumann, 1974). The
maximum saturated receptor potential amplitude, V_{max}, is generally in
in the region of 60 mV.

2.1.3.2.2 The Receptor Response in the Lamina

Positive potentials with receptor-like waveforms are commonly recorded
in the lamina and have been positively identified as emanating from
retinula cell axons (Järvilehto and Zettler, 1973). In dragonfly, dye-
injection experiments have not been carried out on these potentials
but, on the basis of response dependent criteria, positive potentials
fall into two overlapping classes (Laughlin, 1974a): axon responses
and lamina depolarizations. The axon responses have properties iden-

tical to retinula cell bodies and are presumed to represent intracellular recordings from retinula axon terminals. Obviously, as shown in
fly, water bug and barnacle (Ioannides and Walcott, 1971; Shaw, 1972;
Zettler and Järvilehto, 1973) the receptor potential is conducted
electrotonically to the first synapse and action potential propagation
is not used for information transfer.

2.1.3.2.3 Lamina Depolarizations

These potentials appear to be badly smoothed retinula responses
(Fig. 2). They have depressed absolute sensitivity (Fig. 3), broad
angular sensitivity (Fig. 4), show no shot noise (Fig. 2), and no PS.
On the basis of this evidence it is suggested that these potentials
are dominated by an extracellular field potential in the lamina which
results from the lamina having a high input resistance and acting as
a current sink for the receptors. It has been suggested that this extracellular depolarisation could act as an electrotonic negative feedback in the lamina by inhibiting the synaptic terminals of the axons
(Laughlin, 1974a,c).

2.1.3.2.4 The Post-Synaptic LMC Response

Hyperpolarising responses were discovered in the locust lamina by
Shaw (1968) and identified by dye injection as emanating from large
monopolar cells in fly (Autrum et al., 1970) and dragonfly (Laughlin,
1973). In fly lamina another fibre type, the centrifugal basket fibre,
produces a similar hyperpolarising response but penetrations of this
fibre are rare (Järvilehto and Zettler, 1973). In dragonfly only a
few dye injection experiments have been performed and all units examined in this study were assumed to be LMCs on the basis of response
waveform alone. It is possible, therefore, that a proportion of responses used in this study emanate from other fibre types. If this is
so then these other fibres are at present functionally indistinguishable from the majority and conform to the same neural principles.

There are obvious and striking differences between the voltage response properties of dark-adapted LMCs and the presynaptic receptors.
These are:

1. LMCs hyperpolarise in response to light, rather than depolarise.
   Hyperpolarisation increases with stimulus intensity (Fig. 2).

2. The response waveform is very phasic, showing a rapid initial 'on'
   transient lasting 50 ms, a constantly decaying phase of plateau
   potential and often a brief depolarising 'off' transient (Fig. 2).
   These transients are generated neurally in the lamina because over
   the entire dark adapted LMC dynamic range the receptor response
   is monophasic.

3. At low intensities the lamina hyperpolarisation is much larger in
   amplitude than the receptor potential. A 50 mV LMC hyperpolarisation
   is generated by a 3.6 mV receptor depolarisation (Fig. 3)

4. The LMCs show a high level of noise, both when responding and in
   darkness. This dark noise is very prominent and can be up to 7 mV
   peak to peak. It is probably synaptic in origin (Laughlin, 1973)
   because its amplitude decreases and its frequency increases with
   higher level of stimulation (Fig. 2).

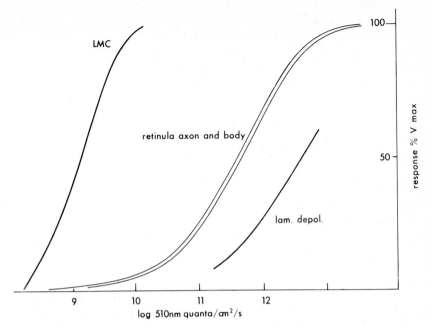

Fig. 3. The dark-adapted intensity/response functions of the classes of responses given in Fig. 2. The curves are averaged from a number of experiments performed with a calibrated axial point source whose intensity is given as corneal quantal irradiance (data from Laughlin, 1973, 1974a, and unpublished work)

2.1.3.3 Intensity and Contrast Coding

2.1.3.3.1 Receptor Body and Axon

Among the most important questions to ask of elements in the peripheral visual system is: "How is intensity information carried?" In the receptors absorption increases with intensity, each photon absorbed opens more membrane conductance channels and the amplitude of depolarising receptor potential rises (Laughlin, 1975a). When the receptor remains in the same adaptation state this simple property gives rise to a fixed relationship between receptor potential amplitude and intensity, the intensity/response function; on a semi-log plot this is seen as the familiar sigmoidal V/log I curve (Fig. 3). The sigmoidal shape is accounted for by the self-shunting effect of the simplified transducer described above (Lipetz, 1971).

The sigmoidal V/log I curve of the self-shunting receptor has the following properties:

1. A broad dark-adapted dynamic range of 5 log units from the toe of the V/log I curve to saturation at V_{max}.

2. At the low and high ends the slope of the V/log I curve (contrast efficiency) is intensity dependent.

3. In the mid-region of the curve the response voltage is approximately proportional to log intensity.

These three properties of the intensity/response function allow us to draw two conclusions about intensity/coding. Firstly, the visual signal can be represented by the receptors as a voltage modulation which is equivalent to a modulation of intensity. In this way intensity information can be encoded as membrane potential and transmitted via the axons, which exhibit identical V/log I curves (Fig. 3), to the lamina. Secondly, over the median part of the dynamic range, which represents about 60% of the total voltage bandwidth, the receptor performs a log transform on the intensity signal so that voltage output is proportional to log intensity. This means that, over this region, equal relative changes in intensity (equal log increments) cause equal voltage increments, regardless of absolute intensity. This is ideal for carrying information on contrast and the log transform corresponds to the well known Weber-Fechner law.

2.1.3.3.2 LMCs

All available evidence suggests that dark-adapted dragonfly LMCs code intensity changes as voltage modulations of hyperpolarising potential. As in the receptors there is an intensity/response function governing the relationship between intensity and response amplitude. However, when the LMC V/log I curve is compared with the retinula cell the following differences in intensity and contrast coding characteristics can be seen (Fig. 3).

1. The dynamic range of the dark-adapted LMC is compressed to 2 log units. This corresponds rather well to the normal range of environmental contrasts at any one ambient intensity because few objects have reflectances of less than 1% or greater than 99%. Thus the dynamic range of the LMC is matched to the environmental signal for this one adaptation state. Obviously this matching is achieved by neural interactions in the lamina.

2. The contrast efficiency of the dark-adapted LMC is high and relatively constant over its dynamic range (about 80% V_{max}/log unit). This is obviously a consequence restricting the dynamic range to 2.0 log units. This high contrast efficiency should be compared with the retinula cells at the same intensity where contrast efficiency is low, around 7% V_{max}/log unit, and varies constantly over the LMCs dark-adapted dynamic range. Neural integration in the dark-adapted lamina has performed two functions, it has increased contrast efficiency and has made it constant over the dynamic range. Thus at low intensities the lamina has compensated for the imperfect log transform made by the dark-adapted photoreceptor and can respond to equal contrast increments with equal voltage output differences. Evidently the dark-adapted LMCs can perform very effectively as a high sensitivity contrast coding pathway and in the next section we will examine the neural mechanisms that are responsible.

2.1.3.4 Voltage Transfer at the First Synapse

2.1.3.4.1 Amplification

The increases in voltage signal and contrast efficiency during transfer of the visual signal across the first synapse lead to the conclusion that the most important integrative mechanism in the dark-adapted lamina is signal amplification. By comparing intensity/response functions of LMCs and retinula cells under identical dark-adapted conditions it is possible to derive a good quantitative measure of the voltage transfer gain at the first synapse. The gain is rather high, x 14, and it has been suggested (Laughlin, 1973) that three general classes of mechanism could be responsible.

The first is an amplification of voltage signal in the receptor terminals by direct interaction between receptor axons but this is ruled out in dragonfly because the axons have dark-adapted intensity/response functions identical to those of their cell bodies (Fig. 3). The second is amplification resulting from the convergence of six receptors onto one interneuron and this post-synaptic summation of receptor signal seems very probable indeed. The third mechanism is the array of parallel synapses that exist between each individual axon and its LMC. Both amplification mechanisms, receptor convergence and parallel synapses, have the added advantage of suppressing noise.

2.1.3.4.2 Transferring Signal without Noise

In general a visual system is subject to two types of noise, extrinsic and intrinsic. Extrinsic noise is inherent to the intensity signal received by the photoreceptors and at low intensities it is dominated by photon shot noise, i.e. the random process of photon absorption. The signal to noise ratio of a photon signal depends on the average number of photons contained in the signal and, because photon absorptions are distributed in time according to the Poisson distribution, the signal-to-noise ratio is proportional to the square root of the number of photons received (Barlow, 1964). We can immediately see that the summation of six receptors' signals at the level of the LMCs improves the signal-to-extrinsic-noise ratio by a factor of $\sqrt{6}$ or 2.4.

The second general class is intrinsic noise, that is spurious signals generated within the components of the nervous system. In our simple two-component system we need only consider intrinsic receptor noise and intrinsic LMC noise. The receptor is subject to transducer noise, resulting from stochastic processes occurring during transduction, and transmission noise picked up by the signal as it is transferred from retina to lamina. If both types of noise are generated independently in each photoreceptor, then again, summation decreases the effectiveness of these two noise sources in degrading the signal. In addition the passive conduction of signal (Ioannides and Walcott, 1971; Shaw, 1972; Zettler and Järvilehto, 1973) reduces to a minimum the number of physical processes involved in transmission and, therefore, potential noise sources. Finally, the passive cable properties of the receptor axon suppress high frequency signals and noise together (Shaw, 1972) and this may aid in the detection of low frequency signals.

Two types of LMC noise must be considered, noise generated at the hyperpolarising retinula → LMC synapse and noise generated by other synaptic inputs. Synaptic noise is suppressed by the prominent array of parallel synapses connecting each retinula axon with the post-synaptic monopolar cell. As a result of electrotonic signal conduction in the pre- and post-synaptic cells, pre- and post-synaptic signals will appear simultaneously at each synapse and their effects will sum. The synaptic noise will appear randomly at each synapse and will, therefore, sum as the square root of the number of synapses. Thus, with respect to synaptic noise, parallel synapses improve the signal-to-noise ratio of synaptic transmission in proportion to the square root of the number of synapses. In fly there are at least 40 synapses per retinula axon so the improvement in signal-to-noise is six-fold (Laughlin, 1973). Note too that parallel synapses increase the absolute voltage noise level of the synaptic noise and this explains the large and steady fluctuations in membrane potential seen in the unstimulated cell. Because of amplification this high LMC noise represents a receptor signal of 300-400 µV (Laughlin, 1973) and we can conclude that the retinula-LMC pathway really is a low noise transmission system.

184

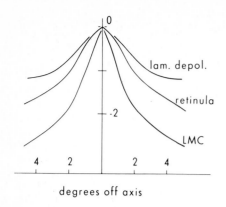

degrees off axis

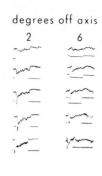

Fig. 4. The average angular sensitivity functions of lamina depolarizations, retinula cell body and axons and LMCs, (Data from Laughlin, 1974a,b) together with the dependence of LMC response waveform on angular position of stimulus. Sensitivity is plotted in log units. Note that lateral inhibition has compressed the LMC angular sensitivity and decreased the sustained hyperpolarization during a 500 ms flash

2.1.3.5 Angular Sensitivity

2.1.3.5.1 Receptors and LMCs

Because each ommatidium of the compound eye points in a unique direction and the receptors of each ommatidium have angular sensitivity, the angular distribution of environmental intensity is mapped out onto the retina as a spatial distribution of receptor potential in a manner exactly analogous to that of the vertebrate eye. In dragonfly retina the receptors have the narrowest measured angular sensitivities of any compound eye. The width of the receptors visual field at the 50% sensitivity level is 1.3-1.4° (Laughlin, 1974a) which means that this is, for an insect, a high acuity system. The orderly projection of retinula axons with the same field of view (from the same ommatidium) to the same lamina cartridge ensures that this acuity is retained in the lamina, and there is no evidence for lateral inhibition between retinula cell axons (Laughlin, 1974a).

It is hardly surprising, therefore, that the LMCs also have narrow angular sensitivity functions (Fig. 4). Nor is it surprising that these are narrowed still further by lateral inhibition (Laughlin, 1974b). We have seen that neural processes in the lamina generate the LMC 'on' transient and cause the plateau phase amplitude to decrease constantly during the response. These two inhibitory effects are potentiated if off-axis light is used (Fig. 4), the maximum attainable 'on' transient amplitude decreases and the plateau phase becomes positive going. Clearly some components of inhibition are favoured by off-axis light and are, therefore, generated in other cartridges. In summary, the LMCs retain spatial acuity and utilise the principle of lateral inhibition (Ratliff, 1965) to emphasise spatial transients.

2.1.3.6 Spectral Sensitivity

2.1.3.6.1 Retinula Cells

The intracellular recordings obtained from retinula cell bodies and axons show that the receptors of the dragonfly retina fall into two classes, the linked pigment cells which are in the majority and have broad spectral sensitivities resulting from inputs from more than one photopigment, and the single pigment cells that have the well tuned rhodopsin spectral sensitivities appropriate for colour vision (Laughlin, 1975a). The spectral sensitivities of receptors in a single class of linked pigment cells are shown in Fig. 5. The vexatious un-

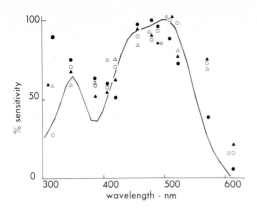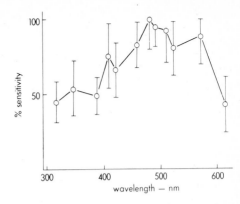

Fig. 5. The spectral sensitivities of 4 individual linked pigment retinula cells (left) and the average sensitivity of 14 LMCs (right). The retinula cell results are fitted with the theoretical absorbance of a mixture of UV, blue and green rhodopsin (Laughlin, 1975a). The LMC spectral sensitivity measurements are very prone to scatter (range bars = ±1 S.D.) but the similar spectral sensitivity functions of several units show that averaging has not broadened the spectral sensitivity function

certainties surrounding pigment linkage and the resulting broad spectral sensitivity functions have been amply discussed elsewhere (Snyder et al., 1973; Wasserman, 1973; Chappell and DeVoe, 1975). At present the available evidence suggests that intracellular recordings do accurately portray the receptor spectral sensitivities (Laughlin, 1975a).

2.1.3.6.2 LMCs

The spectral sensitivities of LMCs are not very different from that of linked pigment retinula cells (Fig. 5) and similarities such as the elevated absorption in both cell types at around 570 nm, suggest that linked pigment cells constitute the LMCs major input. Not only are the LMC spectral sensitivity functions extremely broad, with sensitivity at more than 50% over almost the entire spectral range of 340-570 nm, but there is not a trace of spectral opponency over this wavelength region because the V/log I curves obtained at different wavelengths are parallel (Laughlin, 1975c). Thus the spectral properties of the retinula-LMC pathway are those one would expect of black and white film for use by insects and the LMC spectral sensitivities support the hypothesis that the retinula-LMC pathway is coding contrast. Moreover, it is important to realise that the broad spectral sensitivity increases the number of photons sampled by each spatial channel and gives the system improved sensitivity and a significantly better signal-to-noise at low intensities.

2.1.3.7 Light Adaptation

2.1.3.7.1 Receptors

The dark-adapted receptor responds over a dynamic range of 5 log units, however light-adaptation greatly extends this to at least 7 log units (Fig. 6). A sustained background illumination causes a decrease in

186

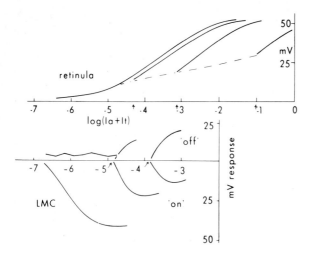

Fig. 6. The light and dark-adapted intensity/response functions of the retinula cell peak response and the LMC 'on' and 'off' transient response. Absolute response amplitude (mV from dark-adapted resting potential) is plotted against the relative total intensity, $\log (I_{adapt} + I_{test})$. Note that there is a sustained background signal in the receptor (---) but none in the LMC and that the LMC 'off' transient response is potentiated by light adaptation. Arrows: appropriate adapting intensities

sensitivity and the maintained stimulus pushes the receptor V/log I curves along the log intensity axis. This shift is small at first but the dependence of sensitivity on background increases with increasing intensity (Fig. 6). The parallel shift of V/log I curves indicate that the receptor is behaving as if light adaptation reduced the proportion of incident photons absorbed or decreased the number of conductance channels opened by each photon (Laughlin, 1975a). Finally it is important to note that the background intensity is always represented in the receptor by a depolarisation or in other words the background intensity signal is always represented in the retina (Laughlin, 1975b).

2.1.3.7.2 LMCs

During identical light-adaptation experiments the LMCs behave in a way that is profoundly different from their receptors. It is true that, as in receptors, the background pushes the LMC V/log I curves along the log intensity axis but here the similarity ends. To begin with the sensitivity change is greater in LMCs (Fig. 6) and in fact these neurons follow the Weber-Fechner relationship rather well. Note that if they did not then their narrow dynamic range would no longer completely cover the range of environmental contrast at the new adaptation state. This brings us to the second difference. In LMCs there is no representation of the background. After 5 s the background signal (the plateau phase) has decayed to dark-adapted resting potential level. Again this is advantageous because any representation of the background would curtail the dynamic range. The third unique property is closely linked, functionally, to the first two. The LMC response waveform becomes far more phasic in nature when the unit is light-adapted and small intensity increments are incapable of sustaining any hyperpolarisation at all, after the initial 'on' transient. In addition the 'off' transient becomes more pronounced and its amplitude becomes intensity dependent, increasing with increasing intensity decrement. In fact the 'off' transient comes to have the same contrast efficiency as the 'on' transient so that the system treats identically both positive and negative deviations from background. Note too that an 'off' transient is a necessity for coding contrasts darker than the average background because the actual background intensity signal is no longer represented at the LMC membrane.

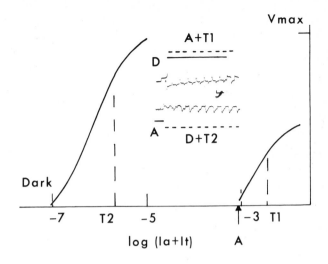

Fig. 7. A simple experiment to demonstrate the speed and effectiveness of neural
sensitivity control in the dragonfly lamina. The curves show the V/log I curves for
a LMC when dark-adapted and when light-adapted by background intensity A. T1 and T2
are the total intensities of the two test flashes used, log $(I_{adapt} + I_{test})$. The
two experimental records show the transition from dark-adapted to light-adapted
state (top) and back to dark-adapted again (bottom). Initially the cell is in the
dark D, the adapting light and the pulsing test-light are then applied A + T1.
Note that 'on' and 'off' transient responses appear within 100-200 ms. In the bottom record the cell is initially light adapted A. The adapting light is then turned
off and only the pulsing test light T2 applied. Again the sensitivity increases
fast enough for the response to appear within 200 ms. Test flash duration is 75 ms,
interstimulus interval is 200 ms and the calibration bars show 20 mV and 200 ms

The final important difference between receptor and interneuron is
that LMC contrast efficiency is normalized so that it is relatively
constant over their entire range of operation despite the changes in
receptor contrast efficiency that occur at low intensities (Fig. 6).
Thus the lamina functions to adjust sensitivity in the pathway, independently from the receptors, to ensure that the voltage bandwidth
of the system is matched to the range of contrasts that are most likely to occur at any one average intensity. Because, at any one adaptation state, the contrast efficiency is high the neural sensitivity
control must act rapidly to ensure that large intensity increments
do not saturate the system for long periods of time. In dragonfly it
is a simple matter to demonstrate the speed with which neural sensitivity control operates (Fig. 7). A 2 log unit saturation is allevi-
ated within 100 ms and the full 2.5 log unit change in sensitivity
is completed within 2 s. Powerful inhibitory forces must be at work
in the lamina to produce such a rapid response.

2.1.3.7.3 Mechanisms of Light-Adaptation

The sensitivity changes that take place during light adaptation can
be thought of as resulting from a series of negative feed-backs or
feed-forwards acting at different vertical levels in the chain of
receptor and subsequent interneurons. To understand light-adaptation
mechanistically we must begin by isolating these inhibitory effects
and elucidating their site of action and their origins in other parts
of the pathway. If we begin with the receptors we can see that present

evidence suggests these cells possess at least two different intrinsic feed-back mechanisms which are driven by receptor absorption and/
or receptor potential and do not require neural input from the lamina.
The first acts quickly and is seen from the rapid (200 ms) decline of
receptor potential from an initial peak to a plateau potential. There
is evidence from drone bee retina to suggest that this is a local membrane effect which depends upon calcium ions (Baumann, 1974). The second is a slower acting inhibition which causes a gradual decrease
in receptor potential over 120 s. The time course of this response
suggests that it is photomechanical in origin and is perhaps similar
to the "longitudinal pupil mechanism" attenuating light in fly photoreceptors (Francescini, 1972; Stavenga, 1975). In the lamina two different functional types of inhibition are apparent. The first annihi-
lates or "backs-off" the background signal and the second is a control
which reduces voltage transfer gain at the first synapse and compen-
sates for the rise in receptor contrast efficiency with intensitiy,
thus normalising the LMC contrast signal. One can draw some general,
but none the less, functionally interesting conclusions on the inputs
driving the neural inhibition of LMCs. As shown in Fig. 8, four possible vertical feed-back and feed-forward loops can act in the lamina
on the retinula cell and the LMC. Inhibition can either be fed forward
via interneurons from receptor to receptor → LMC synapse or to LMCs
themselves. Alternatively inhibition can be feed-back from the LMC
output to these same two sites. Clearly, because there is no perma-
nent LMC background signal an LMC driven feed-back cannot maintain the
lowered sensitivity of the pathway. This must be driven by receptor
input because this is the only component which contains the background
signal. Note that this argument does not deny the existence of LMC
feed-back and this type of inhibition may be extremely important in
sharpening spatial and temporal transients. Finally, with respect to
the spatial distribution of sensitivity control we know that there is
lateral inhibition and there is limited evidence for intra-cartridge
(Laughlin, 1974a) inhibition but their relative roles in light adaptation are at present inexorably entangled.

2.1.4.1 Neural Principles for Contrast Coding

The results obtained from the analysis of neural integration in the
dragonfly lamina leave little doubt that this pathway operates to code
contrast with high sensitivity over a wide range of intensities. In
analysing the intensity/response functions for receptors and interneurons and in elucidating some of the neural interactions that take
place in the lamina one is struck by the number of properties that
appear to be ideally suited for, or in other words are adaptations
for, the task of contrast detection. It is suggested that these properties constitute neural principles and to support this suggestion
examples are taken from the experimental analysis of other complex
visual systems to show that these self-same properties are widely distributed in the animal kingdom. Let us now consider these neural principles in turn.

2.1.4.1.1 The Log Transform

The V/log I functions of LMCs are approximately linear over their entire dynamic range consequently the voltage signal in the pathway is
a log transform of intensity. At higher intensities the receptors show
similar log transform properties but at low intensities the relationship is closer to linear and here, at least, the log transform is a
property resulting from voltage transfer at the first synapse.

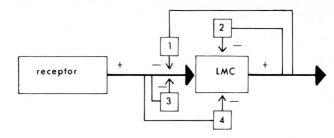

Fig. 8. The vertical distribution of the four possible inhibitory circuits that can mediate neural sensitivity changes at the level of the first synapse in the visual system. These are interneuron feed-back (1 and 2) and receptor feed-forward (3 and 4). The roles of these pathways in accounting for neural sensitivity control in the dragonfly lamina are discussed in the text where it is pointed out that these curcuits have yet to be identified and their lateral distribution in the lamina is unknown

The advantages of the log transform to a system which wishes to code relative intensity differences over a wide range of intensities in self-apparent and its ubiquity in sensory systems is expressed by the widespread application of the Weber-Fechner law. It is interesting, therefore, to note that the receptors and second order interneurons of dragonfly (above), fly (Järvilehto and Zettler, 1971), mudpuppy (Norman and Werblin, 1974; Werblin, 1974) are extremely similar and that, in all three cases the dark-adapted system performs a log transform at the level of the first synapse.

2.1.4.1.2 The Normalisation of the Log Transform

In dragonfly lamina the contrast efficiency of retinula cells is intensity dependent close to their dark-adapted threshold but, over this range and at higher intensities, the slope of the second-order interneuron V/log I curves varies relatively little. Thus, contrast coding is rendered intensity independent in the lamina and the relationship between voltage modulation in the channel and environmental contrast is constant over the entire range of operation (providing of course that the contrast change is not so large that it saturates the channel at any one adaptation state). Again this behaviour is seen in the vertebrate retina and the importance of the normalisation of retinal ganglion cell output has already been stressed (Barlow and Levick, 1969). In mudpuppy retina, normalisation can be seen in the intensity/response functions of bipolar cells (Werblin, 1974).

2.1.4.1.3 Dynamic Range of Contrast Coding Neurons

In dragonflies the 1-2 log unit dynamic range of LMCs is set by neural interaction in the lamina and it is rapidly adjusted by neural sensitivity control to ensure that the channel operates with maximum sensitivity at each new background. This same phenomenon has been better investigated in the mudpuppy retina where powerful lateral surround antagonism controls bipolar cell sensitivity (Werblin, 1974). In both animals neural sensitivity control acts very rapidly and, in order to allow the system to operate with maximum voltage bandwidth and contrast efficiency the background signal is annihilated.

2.1.4.1.4 Wider Receptor Dynamic Range

A large number of retinal receptors in arthropod and vertebrate eyes show similar intensity/response functions and have dynamic ranges of 4 to 5 log units. These similarities are very striking and could result from constraints imposed by many factors such as self-shunting, the resistance of arrays of closely packed membranes, and the events linking photon absorption to membrane conductance change. Even so one might pose the question, "Why do second order interneurons have their sensitivities and dynamic range matched to contrast differences while the receptors do not?" There are at least two additional answers to this question, aside from those very general proposals suggested above. One is that receptors may need to carry information other than contrast such as absolute intensity and the other is that receptors have to have a broader dynamic range than second order interneurons because it is the receptor signal which, directly or indirectly, drives neural sensitivity control. As we have seen in dragonfly, feed-forward inhibition from the receptors is required to suppress the background signal. If suppression were to be carried out by feed-back the gain required would be large and instability of the system could result. Obviously one is highly unlikely to find a chronically unstable system upon which to test this last proposal; nonetheless, it will be interesting and informative to examine the relationship between feed-back and feed-forward in retinal pathways and to analyse in greater depth the manner in which large sensitivity changes can be brought about rapidly without severe overshoot or oscillation.

2.1.4.1.5 Lateral Inhibition

This phenomenon is clearly seen in the lamina of the fly (Zettler and Järvilehto, 1972) and in dragonfly where it compresses the angular sensitivity of second order monopolar cells. Its functional significance in producing contrast enhancing transients in the nervous system has been well documented (Ratliff, 1965) and demonstrated experimentally in the retina of the horseshoe crab Limulus (Hartline and Ratliff, 1972). Finally lateral inhibition is clearly seen in mudpuppy retina where it plays a dominant role in neural sensitivity control (Werblin, 1974).

2.1.4.1.6 Receptor Summation

The summation of receptor signals is a familiar concept in the vertebrate visual system where it plays an important role in retinal sensitivity control (Rushton, 1965) and in matching the receptor sampling pool size to the noise levels inherent in the incident photon flux (Barlow, 1964) to obtain optimal acuity at low luminances. By comparison receptor pooling in dragonfly or fly lamina (Kirschfeld, 1972) is modest; nonetheless, it serves the same function of increasing sensitivity and reducing the effects of both extrinsic and intrinsic receptor noise (Kirschfeld, 1966).

2.1.4.1.7 Parallel Synapses

It is clear that an array of identical parallel synapses connecting two neurons will improve the signal-to-noise ratio for signal transmission across the synapse as the square root of the number of synapses (Laughlin, 1973). When one considers the similarities in neural morphology (Cajal and Sanchez, 1915) between neurons in the lamina of many insects and crustacea one must conclude that parallel synapses are almost universal among monopolar cells. It would be surprising if similar arrays of parallel synapses were not found in the verte-brate retina. In fact recent Golgi-EM studies suggest that a cat cone makes about 12 synapses with a bipolar cell and in monkey retina the number of parallel synapses between a foveal cone and its midget bi-polar is about 25 (Kolb, 1970; Boycott and Kolb, 1973). In the frog retina a similar parallel synaptic array may be created by the elec-trical coupling of rods (Fain, 1975).

2.1.4.1.8 Graded Potentials

Perhaps the most obvious similarity between the responses of retinal receptors and second order interneurons in arthropod and vertebrate visual systems is that they fail to generate action potentials. In-stead the visual signal is transmitted as a modulation of membrane potential. It has been suggested (Laughlin, 1973) that graded poten-tials are an adaptation that aids in transmitting large amounts of information per unit volume of neuropile. Spike coding is required for propagation of the signal over large distances but is superfluous for short neurons. In fact all spike trains are initiated by graded potentials and the encoding and decoding processes must limit the rate of information transfer and introduce additional and potentially noisy membrane reactions. It is quite logical, therefore, to dispense with action potentials in compact retinae.

2.1.5 Conclusions

The analogies between the receptors and interneurons of insect and vertebrate visual systems suggest that these shared properties are neural principles of real functional significance that contribute toward the satisfactory operation of complex visual systems. Of course much of the data is still incomplete and in dragonfly lamina several properties require further analysis before they can be fully endorsed as neural principles. In particular the role of the receptor background signal in driving neural sensitivity control and the spatial distribu-tion of sensitivity control is still poorly understood. Moreover the qualitative advantages of transmission by graded potentials must be transformed to quantitative actualities by a combination of theoreti-cal analysis with experimental measurement. Nonetheless, I think that we can answer the theoretical question proposed by Cajal and Sanchez (1915) who asked, a propos of their analysis of the neural organisa-tion of the insect visual system: "Would it be too ambitious to hope that, at least with reference to certain sensory centres or to the determined mechanisms of nervous reaction, the invertebrates, and in particular the insects, offer some interpretative criteria for the extremely complex nervous system of the higher vertebrates?" No, it is not too ambitious, and these comparisons may be invaluable for see-ing the wood from the trees.

Acknowledgments. I am extremely grateful to many of colleagues, past and present, and in particular to Martin Wilson, Allan Snyder, Adrian Horridge and Randolf Menzel for many exciting discussions; Margaret Blakers and Steve McGinness for technical assistance; and Flora Jack-son for typing this manuscript.

2.1.6 References

Autrum, H., Zettler, F., Jarvilehto, M.: Postsynaptic potentials from a single
monopolar neuron of the ganglion opticum I of the blowfly Calliphora.
Z. Vergl. Physiol. 70, 414-424 (1970)

Barlow, H.B.: The physical limits of visual discrimination. In: Photophysiology.
Giese, A.C. (ed.). New York, Academic Press 1964, Vol. II, pp. 163-202

Barlow, H.B., Levick, W.R.: Three factors limiting the reliable detection of light
by retinal ganglion cells of the cat. J. Physiol. (London) 200, 1-24 (1969)

Baumann, F.: Electrophysiological properties of the honey bee retina.
In: The Compound Eye and Vision of Insects. Horridge, G.A. (ed.).
Oxford: Oxford Univ. Press 1974, pp. 53-74

Boycott, B.B., Kolb, H.: The connections between bipolar cells and photoreceptors
in the retina of the domestic cat. J. Comp. Neurol. 148, 91-114 (1973)

Cajal, S.R., Sânchez, D.: Contribucion al conocimiento de los centros nerviosos
de los insectos. Parte I, retina y centros opticos.
Trab. Lab., Invest. Biol. Univ. Madr. 13, 1-164 (1915)

Chappell, R.L., DeVoe, R.D.: Action spectra and chromatic mechanisms of cells in
the median ocelli of dragonflies. J. Gen. Physiol. 65, 399-419 (1975)

Corbet, P.S.: A Biology of Dragonflies. London: Witherby 1962

Eguchi, E.: Fine structure and spectral sensitivities of retinula cells in the
dorsal sector of compound eyes in the dragonfly Aeschna.
Z. Vergl. Physiol. 71, 201-218 (1971)

Fain, G.L.: Quantum sensitivity of rods in the toad retina.
Science 187, 838-841 (1975)

Franceschini, N.: Pupil and pseudopupil in the compound eye of Drosophila.
In: Information Processing in the Visual System of Arthropods.
Wehner, R. (ed.). Berlin-Heidelberg-New York: Springer 1972

Hamdorf, K., Paulsen, R., Schwemer, J.: Photoregeneration and sensitivity control
of photoreceptors of invertebrates. In: Biochemistry and Physiology of Visual
Pigments. Langer, H. (ed.). Berlin-Heidelberg-New York: Springer 1973

Hartline, H.K., Ratliff, F.: Inhibitory interaction in the retina of Limulus.
In: Handbook of Sensory Physiology. Fuortes, M.G.F. (ed.).
Berlin-Heidelberg-New York: Springer 1972, Vol. VII/2

Hartline, H.K., Wagner, H.G., MacNichol, E.F. Jr.: The peripheral origin of nervous
activity in the visual system. Cold. Spr. Harb. Symp. Qant. Biol. 17, 125-141
(1952)

Horridge, G.A., Meinertzhagen, I.A.: The exact neural projection of the visual
fields upon the first and second ganglia of the insect eye.
Z. Vergl. Physiol. 66, 369-378 (1970)

Ioannides, A.C., Walcott, B.: Graded illumination potentials from retinula cell
axons in the bug Lethocerus. Z. Vergl. Physiol. 71, 315-325 (1971)

Järvilehto, M., Zettler, F.: Localised intracellular potentials from pre- and postsynaptic components in the external plexiform layer of an insect retina.
Z. Vergl. Physiol. 75, 422-440 (1971)

Järvilehto, M., Zettler, F.: Electrophysiological-Histological studies on some
functional properties of visual cells and second order neurons of an insect
retina. Z. Zellforsch. 136, 291-306 (1973)

Kirschfeld, K.: Discrete and graded receptor potentials in the compound eye of
the fly Musca. In: Functional Organization of the Compound Eye.
Bernhard, C.G. (ed.). Oxford: Pergamon Press 1966

Kirschfeld, K.: The visual system of Musca: Studies on optics, structure and
function. In: Information Processing in the Visual Systems of Arthropods.
Wehner, R. (ed.). Berlin-Heidelberg-New York: Springer 1972

Kolb, H.: Organization of the outer plexiform layer of the primate retina: electron
microscopy of Golgi-impregnated cells.
Phil. Trans. R. Soc. London 258B, 261-283 (1970)

Laughlin, S.B.: Neural integration in the first optic neuropile of dragonflies.
I. Signal amplification in dark-adapted second order neurons.
J. Comp. Physiol. 84, 335-355 (1973)

Laughlin, S.B.: Neural integration in the first optic neuropile of dragonflies.
II. Receptor signal interactions in the lamina. J. Comp. Physiol. 92, 357-375
(1974a)

Laughlin, S.B.: Neural integration in the first optic neuropile of dragonflies.
III. The transfer of angular information. J. Comp. Physiol. 92, 377-396 (1974b)

Laughlin, S.B.: The function of the lamina ganglionaris. In: The Compound Eye and
Vision of Insects. Horridge, G.A. (ed.). Oxford: Oxford Univ. Press 1974c,
pp. 341-348

Laughlin, S.B.: Receptor function in the apposition eye: an electrophysiological
approach. In: Photoreceptor Optics. Snyder, A.W., Menzel, R. (eds.).
Berlin-Heidelberg-New York: Springer 1975a, pp. 479-498

Laughlin, S.B.: Receptor and interneuron light-adaptation in the dragonfly visual
system. Z. Naturforschung 30 C, 306-308 (1975b)

Lipetz, L.E.: The relationship of physiological and psychological aspects of
sensory intensity. In: Handbook of Sensory Physiology. Loewenstein, W.R. (ed.).
Berlin-Heidelberg-New York: Springer 1971, Vol. I

Menzel, R., Snyder, A.W.: Introduction to photoreceptor optics - an overview.
In: Photoreceptor Optics. Snyder, A.W., Menzel, R. (eds.). Berlin-Heidelberg-
New York: Springer 1975, pp. 1-13

Normann, R.A., Werblin, F.S.: Control of retinal sensitivity. I. Light and dark-
adaptation of vertebrate rods and cones. J. Gen. Physiol. 63, 37-61 (1974)

Ratliff, F.: Mach Bands. San Francisco: Holden Day 1965

Rodieck, R.W.: The Vertebrate Retina: Principles of Structure and Function.
San Francisco: Freeman 1973

Rushton, W.A.H.: The Ferrier lecture, 1962. Visual adaptation.
Proc. R. Soc. (London) 162B, 20-46 (1965)

Shaw, S.R.: Organization of the locust retina. Symp. Zool. Soc. London 23,
135-163 (1968)

Shaw, S.R.: Decremental conduction of the visual signal in the barnacle lateral
eye. J. Physiol. (London) 220, 145-175 (1972)

Snyder, A.W., Laughlin, S.B.: Dichroism and absorption by photo-receptors.
J. Comp. Physiol. 100, 101-116 (1975)

Snyder, A.W., Menzel, R., Laughlin, S.B.: Structure and function of the fused
rhabdom. J. Comp. Physiol. 87, 99-135 (1973)

Stavenga, D.G.: Optical qualities of the fly eye - an approach from the side of
geometrical, physical and waveguide optics. In: Photoreceptor Optics.
Snyder, A.W., Menzel, R. (eds.). Berlin-Heidelberg-New York: Springer 1975,
pp. 126-144

Strausfeld, N.J., Campos-Ortega, J.A.: Some inter-relationships between the first
and second synaptic regions of the fly's (Musca domestica L.) visual system.
In: Information Processing in the Visual Systems of Arthropods .
Wehner, R. (ed.). Berlin-Heidelberg-New York: Springer 1972, pp. 23-30

Wasserman, G.S.: Invertebrate color vision and the tuned receptor paradigm.
Science 180, 268-275 (1973)

Werblin, F.S.: Control of retinal sensitivity. II. Lateral interactions at the
outer plexiform layer. J. Gen. Physiol. 63, 62-87 (1974)

Zettler, F., Järvilehto, M.: Lateral inhibition in an insect eye.
Z. Vergl. Physiol. 76, 233-244 (1972)

Zettler, F., Järvilehto, M.: Active and passive axonal propagation of non-spike
signals in the retina of Calliphora. J. Comp. Physiol. 85, 89-104 (1973)

2.2 Voltage Noise in Insect Visual Cells

U. Smola

2.2.1 Introduction

2.2.1.1 The Quantum Nature of Light and Its Effect on Visual Cells

Since Planck and Einstein developed the quantum theory at the beginning of the century, we know that the emission and absorption of light
radiation only occurs in discrete energy quanta. Even when all the
conditions are kept constant the emission of individual light quanta
is subject to statistical fluctuations. Hence, the intensity of radiation emitted by a light source fluctuates about a constant mean.
These fluctuations give rise to the shot effect of the light quanta.
To my knowledge Barnes and Czerny (1932) were the first to investigate the possibility of visual perception of this shot effect of the
quanta. They showed that certain fluctuations in brightness perceived
while observing very weak light sources are at least partly the result
of the shot effect. The statistical methods applied by Barnes and
Czerny were used by Hecht et al. (1942) in expanded form to determine
the visual threshold, i.e. the minimum energy required to produce a
visible sensation. In research which followed this classical work of
Hecht et al. (1942) the main themes were the absolute thresholds and
the quantum fluctuations at the absolute threshold (cf. Pirenne, 1956,
1962; Baumgart, 1972). The fundamental findings of Hecht et al. (1942)
that a single light quantum absorbed by a human rod is sufficient to
trigger an intraretinal response were thereby confirmed. Yeandle (1958)
made intracellular measurements on individual visual cells of Limulus
and observed irregular discrete waves. Using the same statistical
methods which Hecht et al. (1942) applied, he was able to show that
the absorption of a single quantum is sufficient for generation of a
discrete wave. This result was confirmed in the work of Adolph (1964),
Fuortes and Yeandle (1964), Dowling (1968) and Borsellino and Fuortes
(1968). Discrete waves, often called "quantum bumps" were recorded
from the photoreceptors of some other species: Locust (Scholes, 1965),
Leech (Walther, 1966), Fly (Kirschfeld, 1966; Wu and Pak, 1975). Fig. 1
shows some of these quantum bumps. They were measured in a visual cell
of a cockroach eye under very weak illumination. In the top row individual bumps can be seen which rise very clearly from the resting potential of the visual cell. In the second row two bumps following each
other closely form a single discrete occurrence of longer duration. At
higher light intensities the superposition of a series of bumps brings
about strong fluctuations of the potential as can be seen in the bottom row. When the intensity of the light is further increased individual bumps in the response can no longer be distinguished. Fig. 2
shows the generator potential measured in a single visual cell of the
cockroach. This results when the visual cell is subjected to a maintained-step light stimulus of high intensity. The response can be regarded as being composed of a DC-component and random fluctuations of
the voltage (voltage noise). Dodge et al. (1968) analyzed such responses to steps of light in Limulus. They showed that these potentials
are composed of a summation of quantum bumps in which amplitude and
duration of the bumps depended on the light intensity. Barnes' and
Czerny's (1932) initial question can now be answered thus: the shot

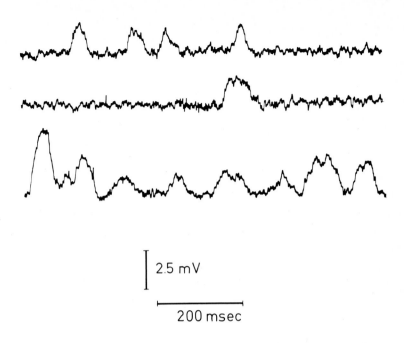

2.5 mV

200 msec

Fig. 1. Quantum bumps recorded in a cockroach retinula cell. Upper row: separate quantum bumps. Middle row: two bumps in close succession make up a longer voltage deflection. Bottom row: the overlapping of bumps in very close succession brings about more marked fluctuations of the potential

effect of light can be measured at the level of the visual cells as fluctuations of the generator potential.

The fact that fluctuations of the generator potential are caused by the shot effect of light is of consequence for the transmission of information by the visual cells. I would like to go into this rather briefly. For the detection of a visual object and the determination of its nature it is necessary that the spatial and temporal distribution of optical characteristics (e.g. reflection characteristics) of different points in space be projected onto the mosaic screen of the visual cells by means of light radiation. Each visual cell has its own spatial angle. The sum of the optical characteristics of the objects contained in this angular element change with time. The usual cause of this change is movement of the environment relative to the cell. By means of light radiations, changes in the optical characteristics of the objects contained in an angular element must be converted into corresponding changes in the generator potential of the visual cell. But as we have already seen the quantum nature of light gives rise to bumps with a characteristic amplitude and duration. This is the reason why the visual cell is not always capable of resolving rapid changes in the environment. The random fluctuations of the voltage also caused by the shot effect of the light set the limit to the resolution of the amplitude of these changes. The importance of time resolution and of the amplitude resolving power for the transmission

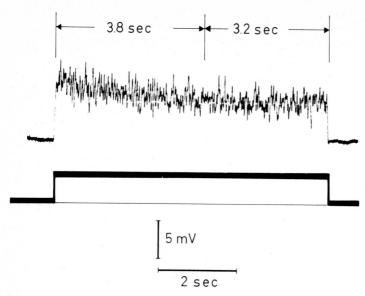

Fig. 2. Generator potential recorded in a cockroach retinula cell in response to a maintained-step light stimulus. Note the strong voltage fluctuations (voltage noise) of the generator potential. To calculate the standard deviation of the voltage fluctuations a time interval of 3.8-7.0 s after the start of the light stimulus was used

of information about the optical environment by visual cells is shown very clearly by the famous Shannon equation

$$C = B \cdot \text{ld} \left(1 + \frac{\bar{U}^2_{sig}}{\bar{U}^2_{noise}} \right) \text{[bit/s]}.$$

C is the maximum channel capacity in bit/s. B is the bandwidth of the visual cell expressed in cycles/s. ld is the binary logarithm. $\bar{U}^2_{sig}$ is the square of the effective value of the signal voltage. $\bar{U}^2_{noise}$ is the square of the effective value of the noise voltage. Numerator and denominator together make up the so called signal-to-noise ratio. It can easily be seen by means of this equation that the larger the bandwidth and the smaller the noise, the better the transmission of information by the visual cell. This paper deals with two important aspects relating to the quantum noise of the voltage in visual cells:

1. Voltage noise in the soma of visual cells in relation to light intensity in cockroach, fly and bee.

2. Two examples of mechanisms to reduce voltage noise in the visual cells themselves.

2.2.1.2 Some Remarks about the Insects Investigated

Experiments were carried out on visual cells of the cockroach _Periplaneta americana_, the blowfly _Calliphora erythrocephala_ and the

worker bee <u>Apis mellifica</u>. The way in which the optical environment
influences these insects' behavior can be briefly described in the
following manner. It seems that the cockroach only uses optical information in order to find dark corners and cracks as fast as possible. Blinded cockroaches find their way about just as well as normal
ones. Daylight has the opposite effect on the behavior of the other
two insects. The worker bee has outstanding color vision and the as-
tounding manoevrability of a blowfly in flight is due to the high
efficiency of its eyes. The differences in the importance of optical
information from the environment to each of these insects makes it
interesting to compare the transmission characteristics of their visual cells. Let us first of all take a look at the arrangement of visual cells inside the ommatidia of the insect in question (Fig. 3).
The three pictures show electron microscope photographs. They were
all taken at the same magnification and each shows a cross section
through one ommatidium from the compound eye of the cockroach
(Fig. 3a), the blowfly (Fig. 3b) and the worker bee (Fig. 3c) respectively. All three ommatidia were sectioned roughly 130 μ underneath
the cristal cone. At this level the fused rhabdom of the cockroach
and the bee is composed of the rhabdomeres of eight visual cells. The
open rhabdom of the fly on this picture consists of seven rhabdomeres.
The areas of the cross sections of the rhabdoma differ from one another remarkably. The ratio of the areas is as follows: cockroach: 16,
fly: 1.7, bee: 1.0. When one takes the number of visual cells in the
cross section of each ommatidium into account this means that the
average cross section of a fly's rhabdomere has roughly twice the
area of that of a bee and that of a cockroach 16 times that of the
bee. These distinct differences point towards differences in the
transmission characteristics of the visual cells in accordance with
the fact that each specific adaptation to the environment has its
counterpart in the structure and function of the sensory organs.

2.2.2 Methods

Preparations of cockroach and fly have already been described in
Smola and Meffert (1975). The preparation of the bee was in accordance
with Autrum and Zwehl (1964). In the case of the fly the direction of
the electrode and the exact position of its tip were histologically
examined, to ensure that only recordings from the retina or from the
lamina ganglionaris be taken into account. Intracellular staining with
Procion yellow made sure that the intracellular recordings came either
from the soma or from the axon of a visual cell as required. The light
source was a 150 W XBO-lamp with quartz condenser. The measurements
of the potentials were accomplished using a point source (3 mm diameter and 70 mm distant) positioned so that a maximal potential could
be elicited. The light intensity was regulated by combining neutral-
density filters. To begin with the characteristic curve of the visual
cell was determined. Then, using monochromatic light stimuli spectral
sensitivity of the visual cell was determined. There followed maintained-step stimuli of white light at various intensities. The spectrum of the white light extends from 300-700 nm. Each of these stimuli lasted 7 s. For fly and bee they were given at intervals from
0.5-2 min according to intensity. For the cockroach the dark gaps
were between 1-5 min long. In all the successful measurements the sensitivity of the visual cell remained constant throughout the experiment. This could be confirmed by recording the resting potential, by
giving test stimuli during the dark gaps and by measuring the characteristic curve at the end of the experiment. The potentials were stored
on magnetic tape using a PCM-device (John and Reilhofer, 3 K 12). The

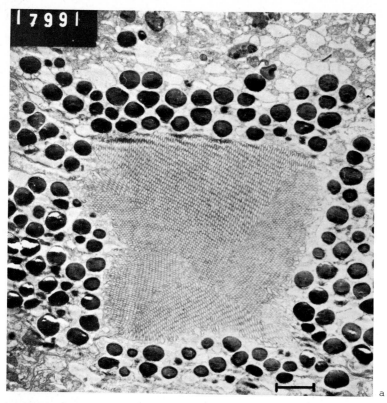

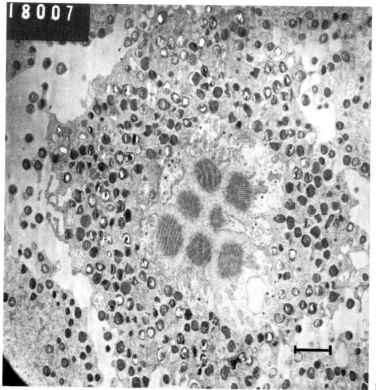

Fig. 3 a and b

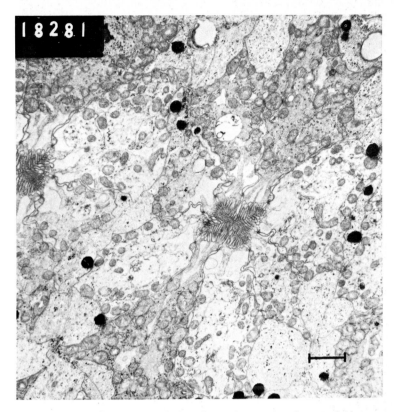

Fig. 3a-c. a) Cross section through one ommatidium from the compound eye of the cockroach. The ommatidium was sectioned roughly 130 μ underneath the cristal cone. The fused rhabdom is composed of the rhabdomeres of 8 visual cells. The scale drawn in the bottom right hand corner represents 1 μ.

b) Cross section through one ommatidium from the compound eye of the blowfly. The ommatidium was sectioned roughly 130 μ underneath the cristal cone. The picture shows the rhabdomeres of 7 visual cells. The scale drawn in the bottom right hand corner represents 1 μ.

c) Cross section through one ommatidium from the compound eye of the worker bee. The ommatidium was sectioned roughly 130 μ underneath the cristal cone. At this level the fused rhabdom of the worker bee is composed of the rhabdomeres of 8 visual cells. The scale drawn in the bottom right hand corner represents 1 μ.

bandwidth of the recording channel was 1 Kcycles/s. The potential measurements taken for evaluation were from intervals beginning 3.8 s after the stimulus. These intervals lasted 3.2 s. The potentials were sampled at intervals of 0.2 ms. The values obtained were analyzed at the Leibniz Rechenzentrum der Bayerischen Akademie der Wissenschaften (TR 440). Using the least square line, the potentials were split up into DC and noise components. The standard deviation of the noise was calculated. Apparatus noise was low and need not be taken into consideration. In the cases of fly and cockroach several measurements from several intervals at the same light intensity were analyzed.

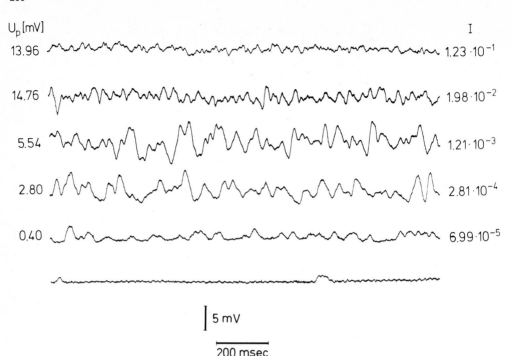

Fig. 4. Voltage noise of the generator potential from the soma of a cockroach visual cell at different intensities of maintained-step light stimuli. The DC-component of the potential is not shown. The value for the DC-components is given to the left of the picture. The dimming factor of the light stimuli is shown to the right

2.2.3 Results

2.2.3.1 Voltage Noise in the Soma of Visual Cells of the Cockroach

Fig. 4 shows voltage fluctuations of the generator potential in the soma of a visual cell of the cockroach. The spectral sensitivity measured for this cell makes it a green receptor as defined by Mote and Goldsmith (1970). The fluctuations shown were taken from the intervals used for analysis (3.8-7.0 s after the start of a step of light). The DC-component of the potential is not shown in this picture. The value of the DC-component of the potential is, however, given to the left of the recordings in mV. Starting at the top of the picture the records were taken at progressively dimmer light conditions - in each case dimmer by a factor which is shown at the right hand side of the picture. The recording at the bottom of the picture shows the noise in the dark. The next recording shows a more concentrated sequence of bumps. In the next one it is already impossible to distinguish single bumps. The fluctuations are considerably larger. They reach their maximum in the recording with the dimming factor of $1.2/10^3$. At higher intensities the frequency of the fluctuations increases and the amplitude becomes smaller. The quantitative relationship is shown in Fig. 5. To enable comparison between the noise and the other potential parameters the maximum response-versus-intensity curve was plotted (curve I). To the left is the initial region and to

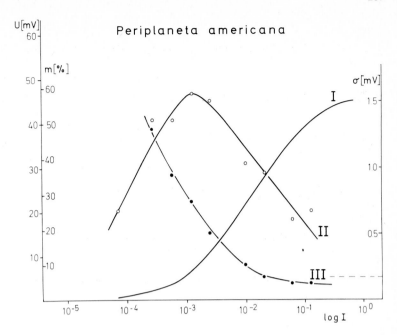

Fig. 5. Relationship between standard deviation of the voltage noise in the soma of a cockroach visual cell and the intensity of light stimuli. Curve I: maximum-response-versus-intensity curve. The scale for this curve is shown on the left (U[mV]). Curve II: standard deviation. The scale for this curve is shown on the right (σ[mV]). Curve III: noise modulation degree. The scale for this curve is shown on the left (m[%]). Dashed line on right: standard deviation for noise in the dark

the right the saturation region of the characteristic curve. The linear region of this curve ranges over 2.5 log units. The scale for the response-versus-intensity curve is given on the left hand side of the picture (U[mV]). Curve II shows the standard deviation of the voltage fluctuations in relation to the light intensity. The scale for this curve is given on the right hand side of the picture (σ[mV]). At low intensities the relatively small number of fluctuations in the generator potential gives only a small standard deviation. The standard deviation increases with the intensity until it reaches a maximum and from then on it declines. The maximum is situated at the beginning of the linear region of the characteristic curve. The dashed line on the right of the picture represents the standard deviation for the noise in the dark. When we examine the relationship between the standard deviation and the DC-component of the generator potential the importance of noise in the transmission of information start to become clear. The relationship

$$m = \frac{\sigma}{U_p} \cdot 100 \ [\%]$$

represents the degree of noise modulation (U_p represents the value of the DC-component of the interval taken for analysis). Curve III shows the dependence of noise modulation on intensity. The scale for noise modulation is drawn on the left hand side of the picture [m(%)]. At very low light intensities it reaches values near 100% because of the

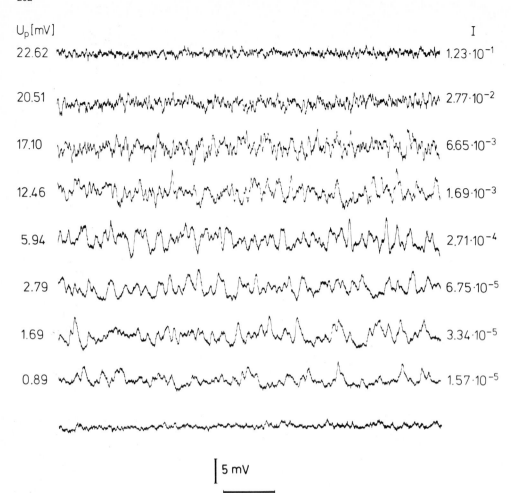

U_p [mV]   I

22.62 ⟨waveform⟩ $1.23 \cdot 10^{-1}$

20.51 ⟨waveform⟩ $2.77 \cdot 10^{-2}$

17.10 ⟨waveform⟩ $6.65 \cdot 10^{-3}$

12.46 ⟨waveform⟩ $1.69 \cdot 10^{-3}$

5.94 ⟨waveform⟩ $2.71 \cdot 10^{-4}$

2.79 ⟨waveform⟩ $6.75 \cdot 10^{-5}$

1.69 ⟨waveform⟩ $3.34 \cdot 10^{-5}$

0.89 ⟨waveform⟩ $1.57 \cdot 10^{-5}$

5 mV

200 msec

Fig. 6. Voltage noise of the generator potential from the soma of a fly visual cell at different intensities of maintained step light stimuli. For further details see legend Fig. 4

small DC-component (not shown in the picture). With increasing light intensities it decreases exponentially. At the saturation region of the response-versus-intensity curve it amounts to about 5%. As we can imagine signals which only give rise to low signal voltage of the visual cell get masked by the noise like music in a badly tuned radio. So there is every reason to suppose that the visual cells of the cockroach are hardly equipped at all to transmit light signals with low intensity variations.

2.2.3.2 Voltage Noise in Visual Cells of the Fly

2.2.3.2.1 Recordings from the Soma

Let us now take a look at the performance of the fly. Fig. 6 shows the relationship between light intensity and noise measured in the

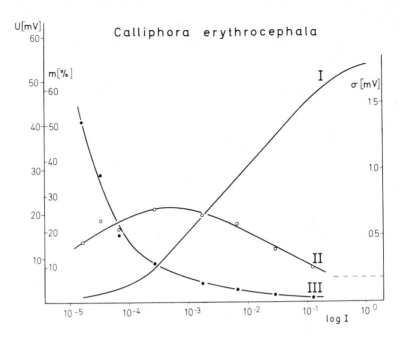

Fig. 7. Relationship between standard deviation of the voltage noise in the soma of a fly visual cell and the intensity of light stimuli. For further details see legend Fig. 5

soma of a fly's visual cell R_{1-6}. The scales are the same as those in Fig. 4. Compared to the noise shown by the cockroach that in the fly is of higher frequency. That is to say, the fluctuations are faster and more numerous. Their amplitudes are, however, on average smaller than those shown at these intensities by the cockroach (Fig. 4). The difference becomes clear when we look at the relationship between the standard deviation of the noise and the light intensity as shown in Fig. 7. Here, as in Fig. 5, the maximum-response-versus-intensity curve (curve I), the standard-deviation-versus-intensity curve (curve II) and the noise-modulation-versus-intensity curve (curve III) are plotted. The scales are the same as those in Fig. 5. The curve for the standard deviation has the same shape as that from the cockroach except that it is flatter. Here the maximum of curve II is only 0.70 mV. It was 1.55 mV in the case of the cockroach. The degree of noise modulation shows up the difference even more. Lower noise and higher DC-component in the generator potential lead to a much lower degree of noise modulation in the case of the fly (curve III). The degree of noise modulation in the case of the fly is roughly 10-15% in the transition phase between the initial region and the linear region of the characteristic curve. In the saturation region of the characteristic curve the degree of noise modulation sinks to about 1%. The cockroach gave noise-modulation values of 25-35% in the transition phase just mentioned. It falls at the beginning of the saturation region of the characteristic curve to about 5%. From this it can be concluded that the visual cells of the fly at the level of the soma possess a greater resolving power for the amplitude of a light signal than those of the cockroach. This is not surprising. One could have guessed at this result since information from the optical environment is of enormous importance for the behavior of the fly.

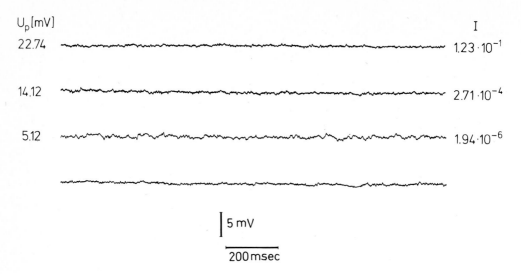

Fig. 8. Voltage noise of the generator potential from the axon of a fly visual cell
at different intensities of maintained step light stimuli. For further details see
legend Fig. 4

2.2.3.2.2 Recordings from the Axon

Compared to measurements from the soma of the visual cells R_{1-6} intracellular measurements from the axons of these cells present a quite
different picture. Fig. 8 shows the voltage fluctuations of 3 responses to maintained step light stimuli at different intensities recorded
from the axon of a visual cell. The scales are the same as those in
Fig. 4 and 6. The quantitative relationships for the recordings shown
in Fig. 8 are drawn up in Table 1. Just as in the case of the soma the
standard deviation is at its greatest in the transition phase between
initial and linear phase in the characteristic curve. And just as in
Fig. 7 it falls at higher light intensities. In addition, the values
of the DC-component for the three regions of the characteristic curve
correspond well with values for the same regions of the characteristic
curve recorded in the soma. The standard deviations of soma and axon
in the transition phase between initial and linear regions of each
characteristic curve differ by the factor 2.3. The surprisingly low
noise in the axons of the fly's visual cells will be discussed more
closely towards the end of this paper.

2.2.3.3 Voltage Noise in the Soma of Visual Cells of the Worker Bee

Let us now examine noise measurements from the soma of worker bee's
visual cell (Fig. 9). These were taken in a cell whose spectral sensitivity is at a maximum in green (Autrum and Zwehl, 1964). The scales
are the same as those in the earlier pictures. It is at first sight
significant that the noise measured in the bee is less than the cockroach's and even less than the fly's. The noise has roughly the same
amplitude at high light intensities as it has in the dark. Fig. 10
shows the standard deviation and the degree of noise modulation in
relationship to the characteristic curve. From these curves it can be
seen that the worker bee of all the three insects in question has the

Table 1. Standard deviation, DC-component and degree of noise modulation of the responses shown in Fig. 8 in relation to the light intensity

Relative intensity	Region of the characteristic curve	Standard deviation of noise σ[mV]	DC-component U_p[mV]	Degree of noise modulation m[%]
1.23/10	Begin of the saturation region	0.193	22.74	0.85
2.71/10^4	Middle of the linear region	0.217	14.12	1.53
1.94/10^6	Transition phase between initial region and linear region	0.310	5.12	6.05
Dark	-	0.207	-	-

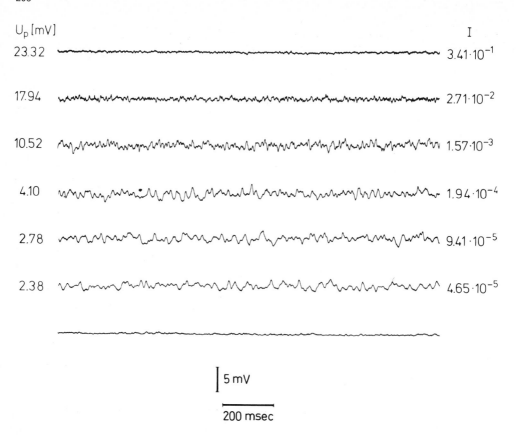

Fig. 9. Voltage noise of the generator potential from the soma of a worker bee visual cell at different intensities of maintained light stimuli. For further details see legend Fig. 4

least noise in the soma of its visual cells. Especially at hight light intensities the degree of noise modulation sinks to markedly lower values.

2.2.4. Noise-Reduction Mechanisms on the Level of Visual Cells

2.2.4.1 Noise-Reduction in Worker Bees' Visual Cells

I shall now attempt to explain why there is such low noise in the worker bee's visual cells. In research based on Shaw's findings (Shaw, 1969), Snyder et al. (1973) assumed that the visual cells of the worker bee in the region of the soma are electrically coupled (see Menzel, 1975; Laughlin, 1975). This coupling is of advantage in that it reduces the polarization sensitivity of the visual cells. It has another important advantage and that is that it also reduces the noise in the visual cells as I shall show. Let us adopt the assumption of Snyder et al. (1973) that two visual cells are electrically coupled and let us assume further that the coupling is independent of light induced current and of the input resistance of the cells. In this case

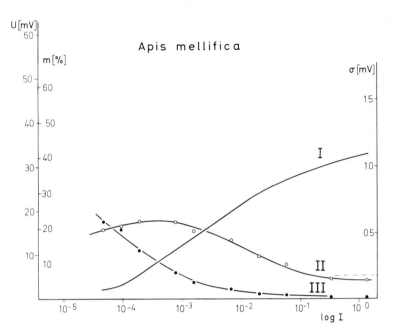

Fig. 10. Relationship between standard deviation of the voltage noise in the soma of a worker bee visual cell and the intensity of light stimuli. For further details see legend Fig. 5

the voltage in each of these cells is expressed by the following equations

$$U_{10} = U_1 + \frac{k}{2} (U_2 - U_1) \tag{1}$$

$$U_{20} = U_2 + \frac{k}{2} (U_1 - U_2) . \tag{2}$$

U_{10} and U_{20} are the measured voltages in the coupled cells. U_1 and U_2 are the measured voltages in the noncoupled cells. k is the coupling factor.

$$0 \leq k \leq 1 \tag{3}$$

If k is zero the cells are completely decoupled and if it is one they are completely coupled. All the rhabdomeres of a rhabdom in the bee's eye receive light from the same point source in the environment. The signal voltage is thus equal in all cells with the same spectral sensitivity (Bernard (1975) has reservations here, but even, then, if rhabdomeres of a single fused rhabdom possess different visual fields the signal voltages in the cells in question are strongly correlated). The noise, too, is equal in these cells. If as we may suppose noise is a stationary random process it can be expressed in each of the coupled cells as the square of the standard deviation (variance). If we now take the square of the effective signal voltage and the square of the standard deviation of noise fluctuations we are in a position to determine the signal-to-noise ratio which we mentioned in the Shannon equation

$$r = \frac{\bar{U}^2_{1,sig}}{\bar{U}^2_{1,noise}} = \frac{\bar{U}^2_{2,sig}}{\bar{U}^2_{2,noise}} = \frac{\bar{U}^2_{sig}}{\sigma^2} \cdot \tag{4}$$

r is the signal-to-noise ratio in each of the uncoupled cells. Now if the two cells are electrically coupled with the coupling factor k we can use Eqs. (1) and (2) to determine the signal-to-noise ratio of the coupled cells r_k.

$$r_k = (\frac{1}{1 - k + \frac{k^2}{2}}) \cdot \frac{\bar{U}^2_{sig}}{2} = (\frac{1}{1 - k + \frac{k^2}{2}}) \cdot r \cdot \tag{5}$$

If the condition of Eq. (3) is fulfilled, r_k is never less than 1. From this we can conclude that every electric coupling between two cells with the same signal voltage (or strongly correlated signal voltages) brings about an improvement in the signal-to-noise ratio. This improvement is expressed in the ratio

$$\frac{r_k}{r} = \frac{1}{1 - k + \frac{k^2}{2}} \cdot \tag{6}$$

The maximum improvement is reached when coupling is complete. In the case of two cells the signal-to-noise ratio has then improved by a factor of 2. In reality we do not know how many cells are coupled. Nor do we know how large the coupling factors are. However, the low noise in the bee's visual cell somata is a strong indication that several cells are coupled together. This improvement of the signal-to-noise ratio is an example of how cooperation between visual cells can reduce noise without impairing the quality of the signals. And this takes place at the level of the cell somata.

2.2.4.2 Noise Reduction in Fly's Visual Cells

As you may remember there was a noticeable difference between the voltage fluctuations of responses from the soma and those of responses from the axon of the fly's visual cells. The assumption that recordings of responses from the laminal region were extracellular seems the most likely explanation for their low noise. However, Järvilehto and Zettler (1970) were able to give a convincing demonstration that low-noise generator potentials from the lamina ganglionaris were in fact intracellular recordings from the axons of the visual cells R_{1-6} (see 2.4 of this vol.). On several occasions we also succeeded in our laboratory in showing that the low-noise potentials from the lamina ganglionaris were recorded intracellularly in the axons. This was ascertained by staining visual cells and by localizing the electrode histologically (Kabiersch, unpubl.). It could also be assumed that the reduction of the voltage noise results from capacitative loss in decremental conduction of the generator potential along the axons (Scholes, 1969). Smola and Gemperlein (1972) were, however, able to show that the frequency responses in soma and axon are essentially

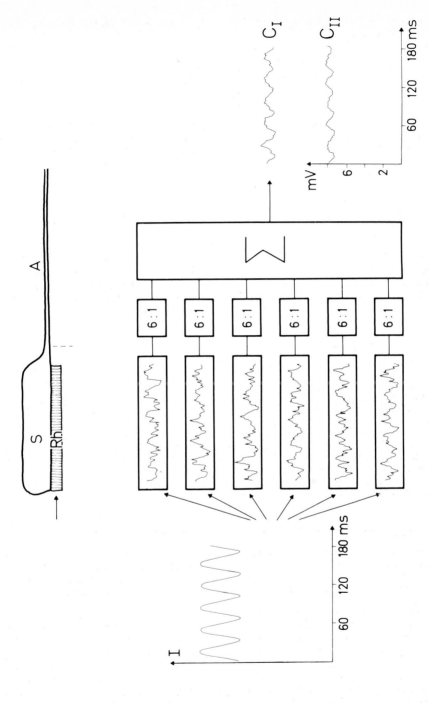

Fig. 11. Improvement of the signal-to-noise ratio by means of summation of diminished generator potentials from 6 visual cells of a fly's compound eye. (From Smola, 1975)

the same. The frequency components in the spectrum of the noise must, therefore, be transmitted as far as the axon. It is now known that light from a given point in the environment is received by six cells of the type R_{1-6} and that these cells are situated in separate adjacent ommatidia (Kirschfeld, 1967; we shall disregard in this context the visual cells R_{7+8}). The axons of these same six visual cells join together in the lamina ganglionaris to form a single cartridge (Braitenberg, 1967). Electrophysiological measurements have shown that the generator potentials of each of these cells are diminished in the course of their conduction along the axons. In the cartridge a summation of the potentials takes place prior to transmission into the neurons (Scholes, 1969). The summation of the diminished potentials does not result in a potential greater than that in each of the six cell somata. Thus the question arises as to what advantage this summation might have. As Gemperlein and Smola (1972) have shown this advantage is the reduction of noise. The noise is random in character and thus at the same point in time the signal-evoked potential in one of the six cells may be more or less increased while in another cell it may be more or less diminished by superposition of the voltage noise. At the summation stage the random noise-fluctuations in the six cells coupled together at the end of the axons average each other out. The potential which remains is the signal component which is equal in all of the six cell somata. The averaging of the six potentials brings a six-fold improvement in the signal-to-noise ratio. Fig. 11 should help make this clear. At the top of this diagram is a simplified drawing of a fly visual cell. To the left the intensity of a sinusoidally modulated light stimulus is plotted against time. The degree of modulation is 16% and the stimulus is of low average intensity. It is radiated from a point source and received by the six visual cells whose axons meet to form a single cartridge in the lamina. The stimulus generates equal signal components in the somata of the six cells. The six generator potentials differ from one another but only because of the noise. In reality the noise recordings shown all come from the same cell and not six different cells. The six different noise specimens taken at different times in one cell serve to illustrate the fluctuations which occur simultaneously in six different cell somata. A sinus wave form is not recognizable in any of the six recording specimens. The generator potential from each cell is then divided by six. The sum of these six reduced potentials gives a potential like one shown to the right (C_I). The random noise fluctuations have averaged each other out. And what is left is a recording in which the sinus form of the stimulus is clearly recognizable. Underneath this computed recording (C_I) there is a real recording (C_{II}) carried out in just the same way as the six on the left of Fig. 11, except that it was measured in the axon of a visual cell. The striking resemblance in quality between the calculated result (C_I) and the actual result (C_{II}) speaks for itself.

Let us now use the Shannon equation for a closer examination of the improvement achieved. For this purpose we can take a light intensity from the middle range of the characteristic curve. At this intensity the noise superimposed upon the DC-component of the generator potential amounts to about 0.65 mV. From earlier measurements (Smola and Gemperlein, 1972) we know which bandwidth of the visual cell corresponds to this intensity. It is about 70 cycles/s. Using these figures we can work out the channel capacity as a function of the signal voltage evoked by variations about the mean light intensity. We assume that the frequency distribution of the signal voltage amplitude has a shape like a normal curve (the same holds good for voltage noise; we shall not elucidate this prerequisite any further). The channel capacity as a function of the signal voltage is shown in Fig. 12. The

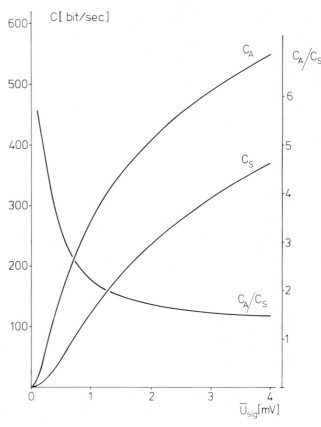

Fig. 12. Channel capacity of a fly's visual cell as a function of the effective signal voltage. The bandwidth used (B = 70 cycles/s) for calculation and the standard deviation of the noise (σ = 0.65 mV) correspond to conditions in the middle of the linear region of the characteristic curve (see Fig. 7). Curve C_S: channel capacity in the soma. Curve C_A: channel capacity in the axon. Curve C_A/C_S: improvement quotient from curves C_A and C_S. The scale for this curve is given on the right

abscissa is the effective signal voltage. The ordinate on the left hand side is the channel capacity in bit/s. The curve C_S shows the channel capacity for the soma under constant noise and bandwidth conditions. Curve C_A is the calculated result of a six-fold signal-to-noise ratio improvement. This curve is for the axon. The improvement quotient calculated from the two curves gives the curve C_A/C_S. The scale for this curve is given on the right hand side of this picture. Curve C_A/C_S clearly shows that the averaging mechanism brings the biggest improvement in channel capacity when the signal voltage is low. The fly possesses about 6000 of these averaging microcomputers which are without doubt all highly important for signal transmission.

Acknowledgments. This work was supported by the Deutsche Forschungs-gemeinschaft (Sm 3/16). I wish to thank Mr. P. Meffert and Mr. E. Enger for their invaluable cooperation on the electrophysiological measurements and the computation of the data. I also wish to thank Mrs. H. Tscharntke for the preparation of the electron micrographs. I am most grateful to Prof. Dr. H. Autrum for his interest in this work. Thanks are also due to P. Norris for translating this paper.

2.2.5 References

Adolph, A.: Spontaneous slow potential fluctuations in the limulus photoreceptors. J. Gen. Physiol. 48, 297-322 (1964)

Autrum, H., v. Zwehl, V.: Die spektrale Empfindlichkeit einzelner Sehzellen des Bienenauges. Z. Vergl. Physiol. 48, 357-384 (1964)

Barnes, R.B., Czerny, M.: Läßt sich ein Schroteffekt der Photonen mit dem Auge beobachten? Z. Physik. 79, 436-439 (1932)

Baumgardt, E.: Threshold quantal problems. In: Handbook of Sensory Physiology. Jameson, D., Hurvich, L.M. (eds.). Berlin-Heidelberg-New York: Springer 1972, Vol. VII/4, pp. 29-55

Bernard, G.D.: Physiological optics of the fused rhabdom. In: Photoreceptor Optics. Snyder, A.W., Menzel, R. (eds.). Berlin-Heidelberg-New York: Springer 1975, pp. 78-97

Borsellino, A., Fuortes, M.G.F.: Responses to single photons in visual cells of Limulus. J. Physiol. (London) 196, 507-539 (1968)

Braitenberg, V.: Patterns of projection in the visual system of the fly. I. Retina-lamina projections. Exp. Brain Res. 3, 271-298 (1967)

Dodge, F.A. Jr., Knight, B.W., Toyoda, J.: Voltage noise in Limulus visual cells. Science 160, 88-90 (1968)

Dowling, J.E.: Discrete potentials in the dark-adapted eye of the crab Limulus. Nature (London) 217, 28-31 (1968)

Fuortes, M.G.F., Yeandle, S.: Probability of occurrence of discrete potential waves in the eye of Limulus. J. Gen. Physiol. 47, 443-463 (1964)

Gemperlein, R., Smola, U.: Übertragungseigenschaften der Sehzelle der Schmeißfliege Calliphora erythrocephala. 3. Verbesserung des Signal-Störungs-Verhältnisses durch präsynaptische Summation in der Lamina ganglionaris. J. Comp. Physiol. 79, 393-409 (1972)

Hecht, S., Shlaer, S., Pirenne, M.H.: Energy, quanta, and vision. J. Gen. Physiol. 25, 819-840 (1942)

Järvilehto, M., Zettler, F.: Micro-localisation of lamina-located visual cell acti-vities in the compound eye of the blowfly Calliphora. Z. Vergl. Physiol. 69, 134-138 (1970)

Kirschfeld, K.: Discrete and graded receptor potentials in the compound eye of the fly (Musca). In: The Functional Organization of the Compound Eye. Bernhard, C.G. (ed.). Oxford: Pergamon 1966, pp. 291-307

Kirschfeld, K.: Die Projektion der optischen Umwelt auf das Raster der Rhabdomere im Komplexauge von Musca. Exp. Brain Res. 3, 248-270 (1967)

Laughlin, S.B.: Receptor function in the apposition eye - an electrophysiological approach. In: Photoreceptor Optics. Snyder, A.W., Menzel, R. (eds.). Berlin-Heidelberg-New York: Springer 1975, pp. 479-498

Menzel, R.: Polarisation sensitivity in insect eyes with fused rhabdoms. In: Photo-receptor Optics. Snyder, A.W., Menzel, R. (eds.). Berlin-Heidelberg-New York: Springer 1975, pp. 372-387

Mote, M.I., Goldsmith, T.H.: Spectral sensitivities of colour receptors in the compound eye of the cockroach Periplaneta. J. Exp. Zool. 173, 137-146 (1970)

Pirenne, M.H.: Physiological mechanisms of vision and the quantum nature of light. Biol. Rev. 31, 194-241 (1956)

Pirenne, M.H.: Quantum fluctuations at the absolute threshold. In: The Eye. Davson, H. (ed.). New York-London: Academic 1962, Vol. II, pp. 141-158

Scholes, J.: Discontinuity in the excitation process in locust visual cells.
Cold Spr. Harb. Symp. Quant. Biol. 30, 517-527 (1965)

Scholes, J.: The electrical responses of the retinal receptors and the lamina in
the visual system of the fly Musca. Kybernetik 6, 149-162 (1969)

Shaw, S.R.: Interreceptor coupling in ommatidia of drone honeybee and locust compound eyes. Vis. Res. 9, 999-1029 (1969)

Smola, U.: Übertragung von optischen Signalen durch Sehzellen.
Naturw. Rdsch. 28, 239-250 (1975)

Smola, U., Gemperlein, R.: Übertragungseigenschaften der Sehzelle der Schmeißfliege
Calliphora erythrocephala. 2. Die Abhängigkeit vom Ableitort: Retina-Lamina
ganglionaris. J. Comp. Physiol. 79, 363-392 (1972)

Smola, U., Meffert, P.: A single-peaked uv-receptor in the eye of Calliphora
erythrocephala. J. Comp. Physiol. 103, 353-357 (1975)

Snyder, A.W., Menzel, R., Laughlin, S.B.: Structure and function of the fused
rhabdom. J. Comp. Physiol. 87, 99-135 (1973)

Walther, J.B.: Single cell responses from the primitive eyes of an annelid.
In: The Functional Organization of the Compound Eye. Bernhard, C.G. (ed.).
Oxford: Pergamon 1966, pp. 329-336

Wu, C.F., Pak, W.L.: Quantal basis of photoreceptor spectral sensitivity of
Drosophila melanogaster. J. Gen. Physiol. 66, 149-168 (1975)

Yeandle, S.: Electrophysiology of the visual system. Discussion.
Am. J. Ophthalmol. 46, 82-87 (1958).

2.3 Spectral and Polarization Sensitivity of Identified Retinal Cells of the Fly

M. Järvilehto and J. Moring

2.3.1 Introduction

Many experiments, both training and optomotoric, can demonstrate how many insects, especially bees, are able to utilize the information of polarized light. Many concrete facts about this can be found in the book by Frisch (1965).

Though only few behavioral studies on Dipteran have been published about the positive response to polarized light and different colors stimuli (Kirschfeld and Reichardt, 1970; Eckert, 1971; Schümperli, 1973) many intracellular measurements still indicate that flies as well as bees might be able to detect the oscillation plane of polarized light (Kuwabara and Naka, 1959; Burkhardt and Wendler, 1960; Scholes, 1969; Järvilehto and Moring, 1974) and to perform wave length discrimination (Autrum and Burkhardt, 1961; McCann and Arnett, 1972). The polarization sensitivity is characterized by the dichroic ratio and has been measured by the microspectrophotometric method for the Calliphora eye (Langer, 1967). It has been found that the intensity of absorbed light when the electromagnetic e-vector is parallel to the microvilli is about twice that when the e-vector is perpendicular to the microvilli. Only in the centrally located tandem rhabdomere is the situation reversed (Kirschfeld, 1969).

Using microspectrophotometric methods it was also possible to measure the absorption spectra in individual rhabdomers of the Calliphora compound eye. The centrally located rhabdomere R_7, R_8 was found to have an absorption spectrum with a maximum at approximately 470 nm, while the surrounding rhabdomeres R_1 to R_6 had an absorption maximum at 515 nm.

Studies on excitation in individual visual cells when using intracellular electrodes appear to agree only partially with the absorption curves. An explanation of the difference between microspectrophotometric and electrophysiological measurements may be found in the light-scattering effects and bleaching of rhodopsin at high measuring intensities in microspectrophotometric experiments.

In this report the intention is to relate the anatomical structure of the eye to the electrophysiological findings on polarization and spectral sensitivity of some cell types in the retina of the fly Calliphora erythrocephala.

2.3.2 Methods

2.3.2.1 Preparation and Recording

The flies were about three to ten days old, male and wild. The intracellular recordings were carried out from the left complex eye of the

a

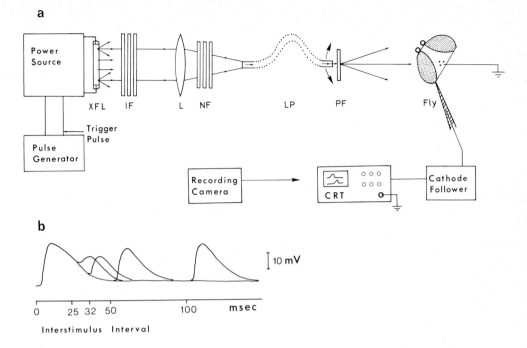

b

0   25 32 50   100   **msec**

Interstimulus Interval

Fig. 1a and b. a) The stimulation and recording apparatus. <u>XFL</u>: xenon flash tube, <u>IF</u>: narrow band interference filter set, <u>L</u>: focusing lens, <u>NF</u>: neutral filter set, <u>LP</u>: light guide, <u>PF</u>: polarization filter. Two different sweep velocities and the optical mean value of about 25 sweeps on the <u>CRT</u>-screen allow a quantitative state- ment of the potential response in one stimulus condition. b) Double stimuli re- sponses with different interstimulus intervals

isolated head. During the experiments the temperature was about 23°C and the relative humidity 70%.

The glass capillary microelectrodes were filled with a 6% Procion yellow M-4RAN aqueous solution using the injection method. The resis- tance of such electrodes was about 70-150 Megohm. The light-induced signals were recorded by a camera from an oscilloscope screen.

2.3.2.2 Light Stimulation

The flash-light stimuli were provided by a Xenon tube and the flash duration was approximately 10 µs.

The details of the recording and stimulation apparatus can be found in Fig. 1a. The flash frequency was programmed by a pulse generator and the focal point of the flash tube (XFL) was projected through a neutral density filter (NF) and a narrow band interference filter (IF) into the end of a quartz glass light guide (LP). The other end of the light guide provided for the eye a punctiform primary light source which was movable, equidistant, and about 70 mm, from the dioptric apparatus of the eye. The polarizing filter (PF), Polarex P-W 44, with a polarization degree of over 99%, was placed when necessary in front of the primary light source on the optic axis of the receptor.

Light energy transmission at ultraviolet wave length was approximately 2%.

The light stimulus was calibrated by measuring the transmitted energy at the output of the light guide by a radiometric flash analyser (8OX OPTOMETER, United Detector Technology, Inc.; USA). The "Optometer" was calibrated by the manufacturer, and also with a pyroelectric radio-meter (Molectron PR 200, Molectron Corp., USA). The energy measured at each wave length was converted to quanta. The intensity-response functions were placed on a calibrated quanta scale, and the amplitude or latency was converted to the "effective" amount of quanta. A response level of 15-30% from the maximum amplitude at 486 nm was used.

2.3.2.3 Experimental Procedure

The dark adapted receptor was constantly stimulated with a flash frequency of about 12.5 pulse/s. Thus the interstimulus interval (ISI) was about 80 ms. As can be seen in Fig. 1b, during this time the potential response returns to the resting potential level and after the next stimulus the potential reaches its full amplitude. Stimulus response function of the cell was determined by changing the intensity of the achromatic light source. The sensitivity to the changes in the oscillation plane of the linear polarized light was examined by turning the polarization filter on the optic axis of the same receptor in 30° steps. Spectral sensitivity of the receptor was tested at 353 nm, 464 nm, 486 nm, 519 nm and 619 nm wave lengths.

This stimulation arrangement allows quantitative recordings from one cell, and in a few minutes it is possible to record 5.000 to 10.000 potentials.

About 25 potentials were always recorded under the same stimulus condition, and the optical mean value was used for calculations, e.g. for one point in the stimulus response function.

2.3.2.4 Cell Identification

After the recording program the cell and the recording site were identified by an intracellular staining technique (Zettler and Järvilehto, 1970, Järvilehto and Zettler, 1973).

The rhabdomeres in a cross section of an ommatidium show an unequivocal picture. It is thus possible to identify the cell type of the stained cell in the cross section of the retinula layer as shown in Fig. 2. The R_5 cell is seen as a bright yellow spot when examined under a fluorescent microscope.

The numbering used here is after Dietrich (1909). The following results are based on measurements from a total of 43 cells.

2.3.3 Results

2.3.3.1 Intensity-Response Characteristics

Light stimulus causes a membrane potential change which depends on the energy of the light stimulus. Potential change is always dynamic, but the potential could also have a static plateau according to the form of the stimulus. If the stimulus has a pulse form, no static components are included in the potential response.

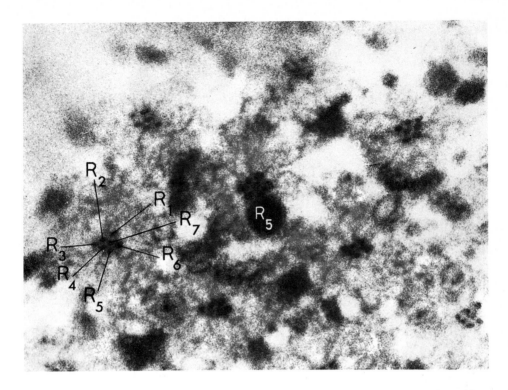

Fig. 2. A cross section of the distal region of the ommatidia. The retinula cells, R_1 to R_7 are recognized in the rhabdomere pattern, the recorded R_5 is visible as a bright yellow spot under the microscope

In this study the light stimulus is very nearly an impulse function if compared to the cell response, which is about seven thousand times slower. This stimulus is naturally responsible only for the activation of the receptors, which are the presynaptic elements for the postsynaptic L-neurons. The L-neurons have as stimulus a quantity of the transmitter substance released by the presynaptic potential of a different character to the postsynaptic one. The presynaptic one (R_{1-6}) is always depolarized and the postsynaptic one, at least L_1 and L_2, hyperpolarized.

An important point when analyzing the information processing in different chains of cells is to find an adequate potential parameter for comparing potential responses of different cells. With arbitrary parameters it is possible to find great differences in the characteristics of different cells.

Fig. 3 shows typical pre- and postsynaptic potential responses to different intensities of achromatic light stimulus. The intensity-response functions with respect to the potential peak, maximum slope and latency are different in pre- and postsynaptic sites.

The receptor group response to the achromatic light pulse is very similar to a depolarized potential wave shown in Fig. 3, but the hyperpolarized potentials can have many other complex wave forms. Most localized $L_{1,2}$-potentials have the same appearance as shown in Fig. 3.

218

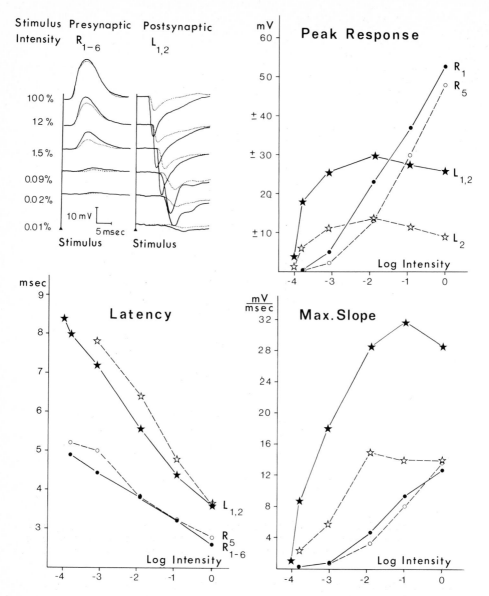

Fig. 3. Pre- and postsynaptic potential responses compared in different achromatic stimulus intensities. The stimulus is a very short (10 μs) flash with the maximal energy of 0.006 μJ. Peak, latency and maximal slope responses from the presynaptic site are compared to those of the postsynaptic site

All intensity functions of the receptors are monotonic throughout the whole range of intensity, but only the latency functions of the $L_{1,2}$-neurons are monotonic and can be compared with the response through-out the whole range of the intensity. Peak response and slope functions are nonmonotonic in the higher intensities of the light stimulus, and so it is hardly possible to transmit unambiguously light information, for example from polarized light, at least within these

Fig. 4. The potential amplitude-latency relationship.
Solid line: relationship at
the presynaptic site; dotted
area: the postsynaptic one.
Every point represents the
mean value of about 25 measurements. Stars shown by
arrows indicate examples
of two different measure-
ments with the same stim-
ulus conditions

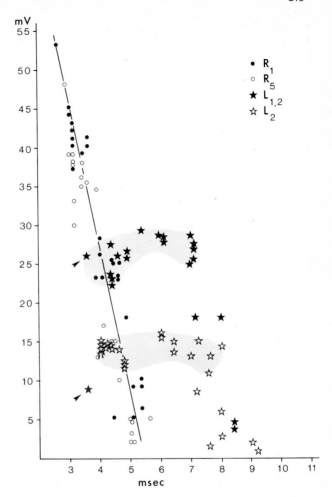

intensities. Differences between the present and the earlier reports
(Järvilehto and Zettler, 1971) can be explained by the higher maximum
stimulus intensity. For example very short latencies recorded indicate
this.

The latency is here defined as the time difference between the time t_O
when the stimulus is given and the time t_L when the maximum tangent
of the potential intersects the resting potential level U_O of the cell.

According to this definition the synaptic delay between the receptor
and the neuron is about 1-3 ms depending on the stimulus intensity.

Differences between cell types in each group R_{1-8} and L_{1-2} are not significant with regard to the potential amplitude, slope and latency. The
potential amplitude-latency relationship (Fig. 4) shows clearly that
it is possible to find relatively monotonic correlation of the potential amplitude and latency in the receptor cells, but not in the L-neurons. Every point in the figure is an optical mean value of about
20 potential responses measured in the same stimulus situation. At
each point the cell was stimulated with a light stimulus of different

220

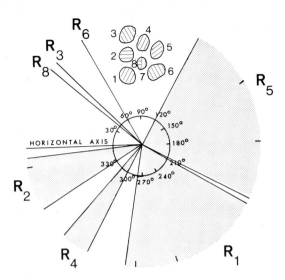

Fig. 5. The different e-vector activation maxima of different retinular cells. The insert shows a rhabdomere cross section of an ommatidium with different anatomical microvilli directions

quality and quantity. The variation of the latency with the same amplitude is much greater in L-neurons than in the receptor cells. Therefore it is possible to get in two different cases with the same stimulus conditions a response of the same latency but different amplitude (arrows in Fig. 4).

2.3.3.2 Polarization Sensitivity

Polarization sensitivity (PS) curves express the reciprocal of the amount of light intensity needed to produce a constant intracellular response at each e-vector. Thus

$$PS = \frac{I_{max}}{I_{min}}$$

where I is the intensity of the light stimulus. This definition of PS expresses nothing about the absorption quantity of the stimulated light.

The microvilli of retinular cells have only one main anatomical direction in each cell. According to their anatomical structure all retinular cells should be capable of detecting the oscillation plane of polarized light, but the axonal convergence of the retinular cells to the large monopolar neurons suggests that the polarization detection ability of the short type retinular cells disappears in the lamina ganglionaris. In the somata at least they should all have approximately the same PS sensitivity (Snyder, 1973).

The previous paper (Järvilehto and Moring, 1974) shows that in spite of theoretical considerations on the activity of R_{1-6} some of them have no measureable polarization activity in the soma. In the present study, the eye was stimulated with green-blue stimulus and it was found that at least R_1, R_3, R_4 and R_5 can either detect the polarized light or not. Thus one cannot predict the ability of the cell to detect the polarized light. All receptor cells except R_7, which have been classified, also have polarization activity with different e-vector activation maxima as shown in Fig. 5.

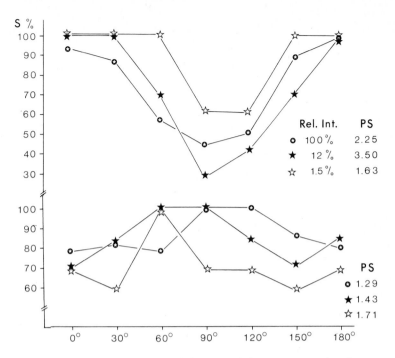

Fig. 6. Two different short type receptor cells with PS activity. Every point is a mean value of about 25 potential measurements

Using the anatomical structure of the direction of microvilli one can construct receptor pairs (Fig. 5 cross section of rhabdomeres). Such pairs were R_1 and R_4, R_2 and R_5, R_3 and R_6, each pair having the same microvilli direction. The pair R_7 and R_8 has other anatomical charac-teristics. Fig. 5 shows the e-vector main activation direction of the receptor cell. Because of the few measurements from each cell type, it is not meaningful to form a mean value. Within the limits of devia-tion it should be possible to construct receptor pairs, but we would suggest two main groups perpendicular to each other: R_1, R_3, R_6 and R_8 forming one group and R_2, R_4, R_5 the other. As far as we know the R_7 does not react to the polarized light (Järvilehto and Moring, 1974). The R_8 (only one localized recording) has nearly the same e-vector direction as the microvilli of the cell.

The PS ratios from localized receptor cells are with 100% relative stimulus intensity between 1.4 and 2.6.

Higher stimulus intensity causes greater amplitude response and also the dependence on the e-vector position seems to become greater, but if the effective intensities are compared, the difference becomes smaller. Fig. 6 shows two different cells with three different stim-ulation intensities. The short type receptor cell in the upper curve group undoubtedly has polarization activity at different stimulus in-tensities, but the lower curve group of another short type receptor shows the situation to be less clear. It is difficult to see if there is a preference direction of activity.

About 20 L-neurons were tested and in all cases no measurable polar-ization sensitivity was found.

222

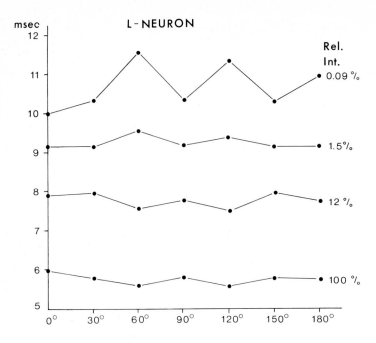

Fig. 7. The L-neuron polarization activity is shown comparing the latencies with different stimulus intensities. The stimulus is an achromatic light pulse and every point represents a mean value of about 25 potential measurements

The L-neurons show no special activity at any stimulus intensity. In Fig. 7 the latencies are compared because of the monotonic character of the latency function.

2.3.3.3 Spectral Sensitivity

Spectral sensitivity S (λ) curves express here the reciprocal of the amount of light (quanta) needed to produce a constant intracellular response at each test wavelength. The results are very inconsistent concerning the cell type and all efforts to compose curve groups seem to be artificial.

One anatomical cell type in different preparations can have great variations of spectral sensitivity, e.g. high ultraviolet and low visible region sensitivity or vice versa.

Phenomenologically it is possible to find three main differences between sensitivity curves. One group has a high UV and lower visible region sensitivity, the second group shows a lower UV, but higher visible region sensitivity. The third group contains two separate observations which were sensitive only to ultraviolet.

One recording from a polarization sensitive R_2 shows a very high UV sensitivity and a much smaller, about 3% of the former, response in the visible region. Another obvious UV cell was recorded in the chiasma, but it was unfortunately possible only to localize a piece of axon running to the medulla. The sensitivity of the cell was very high at UV, and in the visible region it was not at all sensitive, the response was rather hyperpolarized. This cell was also polarization sensitive.

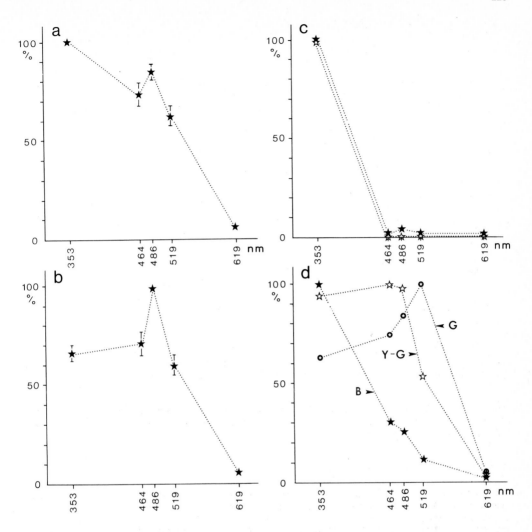

Fig. 8a–d. Three groups of receptor cells on the basis of their UV-visible region sensitivity relationships. The maximum sensitivity in each group is set to 100%, thus the maxima in different curves are not comparable to each other with respect to their absolute value of sensitivity. In (a) and (b) the bars show the standard deviations within each group. (c) The curves of two only UV cells recorded. (d) Spectral sensitivity curves modified after Burkhardt (1962). The maximum in each curve is set to 100% and only those wavelengths used in the present paper are shown. B: blue, G: green, Y-G: yellow-green type. In (a-d) the individual points are connected to each other only to clarify the diagrams, not to show the course of the spectral sensitivity

Fig. 8 shows the spectral sensitivities of the three suggested groups, where the relation between UV and the visible region is used as the criterion. In each curve the maximum sensitivity is set to 100% and does not show the absolute sensitivity, thus the maxima are not comparable to each other.

Receptors

PS

e-vector

direction

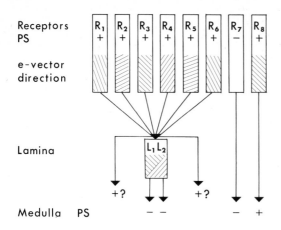

Lamina

Medulla   PS

Fig. 9. The present view of the PS information transmission to higher centres. + means sensitivity, - insensitivity. The e-vector direction at maximum sensitivity is shown. Because of the input arrangement to L_{1-2} neurons the PS-information is summed and thus abolished. Arrows: possible information channels to the medulla

The modified spectral sensitivity curves by Burkhardt (1962) are shown in Fig. 8d; the accordance is obvious. Fig. 8a shows the sensitivity of the green type, 8b yellow-green type and 8c blue type receptor cells, that do not fit with the anatomical R_{1-8} cell types. There must be some other anatomical criteria for classification.

Some L-neurons were also tested, but if the amplitudes in the same intensity range as above are compared the results are very inconsistent. As a result of the inconsistent character of the intensity-response function the latency has been used as measure parameter. The results show many different forms of spectral sensitivity. Some L-neurons have a very high UV sensitivity and some very low. The visible region sensitivity varies, but a typical UV cell was not found.

2.3.4 Discussion

The results presented here are rather more qualitative than quantitative, but results from individual cells are quantitative. More localized studies with different cell types are needed. Attention should be payed especially to localized recordings from different regions in the compound eye of the fly. Most of the present recordings concern the middle equator region, and more particularly the dorsal area of the eye. It might be possible that there are differences in the physiological properties, for example polarization and spectral sensitivity, between two cells both of which are of the same anatomical type, but one in the ventral and the other in the dorsal region. In such a case the coupling in the higher centers would become more complex. In the most difficult case in the information analysis, the anatomical arrangement and the physiological cell type distribution are two completely different things. Especially in the equator region the arrangement of rhabdomeres is often irregular. Such anomalies could possibly explain some of the discrepancies in the present results.

Our view regarding the information transmission of the polarized light is shown in Fig. 9. All short type visual cells react to the polarized light in two main groups which are perpendicular to each other. R_7 does not have PS, but R_8, as supposed in many other theoretical studies, transmits PS information to the medulla (Snyder, 1973; Gribakin, 1973).

A recent study on Calliphora by Horridge and Kimura (1975) has shown that the short type retinular cell can have different maximum activation directions of polarized light depending on the spectral composition of the light stimulus. This fact would make the analysis of polarized light by short type retinular cells complex.

There are many questions open. Is there any equivalent shifting in the R_8? Is the polarization sensitivity of R_{1-6} only a by-product? Anatomical studies show that there are enough connections to the medulla. Some of the thin L-neurons, not so far characterized physiologically, could transmit the PS information of short type retinular cells to the medulla.

2.3.5 Summary

The spectral and polarization sensitivities were recorded intracellularly from different receptors and L-neurons in the left complex eye of the fly C. erythrocephala. All recordings were identified with an intracellular Procion yellow dye injection technique. Retinular cells stimulated with very short flashes and recorded in soma were both polarization sensitive and insensitive. The identified R_8 has PS about 1.8, the other types of R_{1-6} between 1.4 and 2.6. PS of the cell was not constant in different stimulus intensities. The postsynaptic L-neurons did not show any PS activity. A quantitative correlation between preferred planes of polarization and retinal geometry was not found. The preferred planes of polarization varied from one retinular cell to the next. Two groups with perpendicular plane were suggested. Spectral sensitivity was tested at UV and visible region wavelengths. The results in regarding to the anatomical cell type are not uniform. The relation between UV and visible region light as criterion it is possible to classify the sensitivities into three main groups: UV, green and yellow-green. The green and yellow-green types have double peak maxima at 353 nm and 486 nm. The UV type has the maximum at 353 nm. The UV cells are polarization sensitive. The recorded R_8 is an yellow-green type cell having the maximum at 486 nm and lower UV sensitivity. No spikes are obtained.

Acknowledgments. We are indebted to Miss Ann Jennison for revising the English text. We also would like to thank Miss Anneli Moilanen for technical assistance.

2.3.6 References

Autrum, H., Burkhardt, D.: Spectral sensitivity of single visual cells. Nature 190, 639 (1961)

Burkhardt, B.: Spectral sensitivity and other response characteristics of single visual cells in the arthropod eye. Symp. Soc. Exp. Biol. 16, 86-109 (1962)

Burkhardt, D., Wendler, L.: Ein direkter Beweis für die Fähigkeit einzelner Sehzellen des Insektenauges, die Schwingungsrichtung polarisierten Lichtes zu analysieren. Z. Vergl. Physiol. 43, 687-692 (1960)

Dietrich, W.: Die Facettenaugen der Dipteren. Z. Wiss. Zool. 92, 465-539 (1909)

Eckert, H.: Die spektrale Empfindlichkeit des Komplexauges von Musca. Kybernetik 9, 145-156 (1971)

Frisch v., K.: Die Tanzsprache der Bienen. Berlin: Springer 1965

Gribakin, F.G.: Perception of polarised light in insects by filter mechanism. Nature 246, 357-357 (1973)

Horridge,, G.A., Mimura, K.: Fly photoreceptors. I. Physical separation of two visual pigment in Calliphora retinula cells 1-6. Proc. R. Soc. London 190B, 211-224 (1975)

Järvilehto, M., Zettler, F.: Localized intracellar potentials from pre- and post-synaptic components in the external plexiform layer of an insect retina. Z. Vergl. Physiol. 75, 422-440 (1971)

Järvilehto, M., Zettler, F.: Electrophysiological-histological studies on some functional properties of visual cells and second order neurons of an insect retina. Z. Zellforsch. 136, 291-306 (1973)

Järvilehto, M., Moring, J.: Polarization sensitivity of individual retinula cells and neurons of the fly Calliphora. J. Comp. Physiol. 91, 387-397 (1974)

Kirschfeld, K., Reichardt, W.: Optomotorische Versuche an Musca mit linear polarisiertem Licht. Z. Naturforsch. 25B, 228 (1970)

Kirschfeld, K.: Absorption properties of photopigment in single rods, cones and rhabdomeres. Reichardt, W. (ed.). New York-London: Academic 1969, pp. 116-136

Kuwabara, M., Naka, K.: Response of a single retinula cell to polarized light. Nature 184, 455-456 (1959)

Langer, H.: Grundlagen der Wahrnehmung von Wellenlänge und Schwingungsebene des Lichtes. Verh. Dt. Zool. Ges. Göttingen 60, 195-233 (1967)

McCann, G.D., Arnett, D.W.: Spectral and polarization sensitivity of the dipteran visual system. J. Gen. Physiol. 59, 534-558 (1972)

Scholes, J.: The electrical responses of the retinal receptors and the lamina in the visual system of the fly Musca. Kybernetik 6, 149-162 (1969)

Schümperli, R.A.: Evidence for colour vision in Drosophila melanogaster through spontaneous phototactic choice behaviour. J. Comp. Physiol. 86, 77-94 (1973)

Snyder, A.W.: Polarization sensitivity of individual retinula cells. J. Comp. Physiol. 83, 331-360 (1973)

Zettler, F., Järvilehto, M.: Histologische Lokalisation der Ableitelektrode. Belichtungspotentiale aus Retina und Lamina bei Calliphora. Z. Vergl. Physiol. 68, 202-210 (1970).

2.4 Neuronal Processing in the First Optic Neuropile of the Compound Eye of the Fly

F. Zettler and R. Weiler

2.4.1 Introduction

The first optic neuropile of the dipteran eye, the lamina ganglionaris, is structurally very complex, and the function underlying this intricate spatial organization is not at all well understood. One of the chief features of the cellular arrangement in this neuropile is the subdivision into cartridges, which match in number the overlying ommatidia. Several different cells are associated anatomically to form a cartridge. The axons of the six visual cells of type R1–R6 represent the sensory input. The greater part of the afferent output consists of five monopolar neurons (L1–L5). There are cells with lateral branches, which probably mediate interactions between the cartridges. Finally, one finds endings of axons from cells in the second optic ganglion, which may constitute efferent inputs. Detailed descriptions of these anatomical relationships have been published by Trujillo-Cenóz (1965), Boschek (1971), Strausfeld (1971), Braitenberg and Strausfeld (1973) and Strausfeld (1976).

Unfortunately, the methods of electrophysiology have not yet been capable of revealing the function of all the cell types involved in a cartridge. The limitations of present-day experimental techniques have restricted investigation to two cell types, the sensory inputs R1–R6 and the afferent outputs L1–L2. However, despite this paucity of results it is possible to draw qualified inferences about neuronal processes in the lamina.

One ought first to mention certain pecularities of the dipteran eye. The compound eye of a fly, unfortunately, is so constructed that it is impossible - or, more precisely, possible only with very elaborate apparatus - to stimulate the receptors individually. In our experiments we have used a point source of light at a distance of 3 cm from the eye. In this situation a number of receptors must be stimulated simultaneously, since their receptive fields overlap partially or completely (Autrum and Wiedemann, 1962). Closer examination has revealed that there is in fact a nearly complete overlap of the receptive fields of certain subsets of the receptor cells (Kirschfeld, 1967); each such subset consists of eight receptor cells (two of these, R7 and R8, are located one behind the other in a given ommatidium, and the others, R1 to R6, are specific receptors in the six surrounding ommatidia which are oriented optically in the same direction). When a point source of light is used, then, each such set of eight receptors is stimulated similarly; maximal stimulation of one cell corresponds to maximal stimulation of all eight cells. Other sets of receptors are also stimulated, of course, at some fraction of the maximal intensity, depending upon their position.

The axons of six of these eight receptors (R1–R6) project into the lamina, where they converge precisely upon one cartridge (Trujillo-Cenóz and Melamed, 1966; Braitenberg, 1967). These six identically oriented receptors, from different ommatidia, are the sensory input

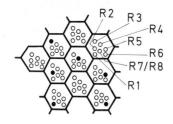

Fig. 1. Schematic diagram of the anatomical organization of the first optic neuropile; the boxes represent the repeated subunits called "cartridges". The array of hexagons at the top of the picture shows the arrangement of receptor cells in the ommatidia. Cells represented by <u>black dots</u>: Type R1-R6 receptors, in six different ommatidia but all "looking" in the same direction, which converge as the inputs to a single cartridge

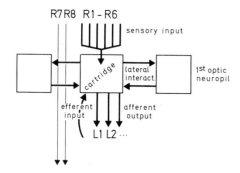

channels of a cartridge (Fig. 1). Their receptive fields are nearly identical, so that the cartridge, rather than the ommatidium, must be regarded as the receptive unit as regards visual space. The axons of the remaining two receptors (R7 and R8) pass through the lamina without synapsing and project into the second optic neuropile, the medulla.

2.4.2 The Responses of Cartridge Inputs R1-R6

Functionally, it is useful to consider a visual cell in terms of two components - the peripheral soma, in which the photopigment is located, and the more proximal axon, which conducts excitation. If the soma of a visual cell is penetrated by a glass micropipette, the response to a light stimulus so recorded is a depolarizing potential with amplitude dependent upon the intensity of the stimulus.

Intracellular recording from the axon of such a cell also reveals a graded depolarization in response to a light stimulus (Fig. 2); we have never seen any sort of spike activity in the visual-cell axons. This means that transmission of the response via the axon takes the form of a wave of depolarization, rather than a volley of spikes.

To demonstrate this we used a technique by which the cell studied could be marked with Procion yellow and the site of the recording determined. Upon completion of the electrophysiological recording the dye was injected into the cell in the usual way (5-10 nA applied for 3 min). 15-30 min thereafter the whole preparation was freeze-fixed by pouring CF_2Br_2 at -160°C over it. The glass pipette, still inserted into the solidly frozen tissue, could then easily be broken off, its tip remaining in place. After the tissue was dehydrated by freeze-drying, the preparation was embedded in wax to cut the block which also included the glass tip of the electrode (Fig. 2).

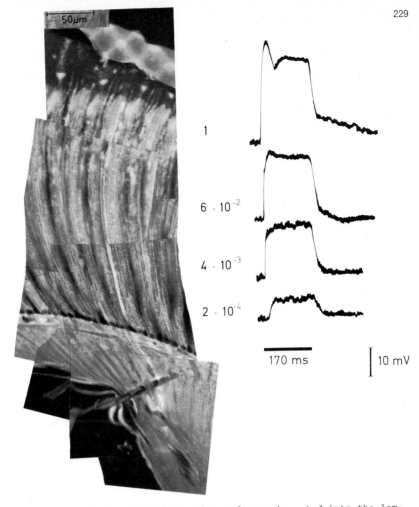

Fig. 2. A marked receptor of Type R1–R6. The electrode was inserted into the lamina and penetrated the axon of the receptor. After dye-injection had been completed, the electrode was advanced somewhat in search of a second cell. On the right are shown the responses of the receptor axon to square-wave light stimuli at different intensities

At first glance, the responses recorded in the soma and in the axon of a visual cell do not differ appreciably. The only obvious difference is that the axon potentials change more smoothly than those of the soma - a phenomenon which suggests that the receptors are capable of lateral interaction. This lateral interaction is more clearly evident when one compares not only the potentials but also the receptive fields measurable from the potential recordings.

When the responses of a visual cell are recorded for various positions of the light source, one always finds a clearly defined position at which the light produces a maximal response. The further the light source is moved from this position (the distance from the eye remaining constant), the smaller the response becomes, since less and less light reaches the cell because of the arrangement of the dioptric apparatus. Although in fact the intensity at the cell is changing, it appears as though the cell is becoming less sensitive as the light source is moved from the optimal position. If the sensitivity of a visual cell is plotted as a function of the angle φ of the light with respect to the optimal position, a bell-shaped curve is obtained (Fig. 3a). This curve is a description of the receptive field, though

in only one dimension. To obtain the complete geometrical shape of
the receptive field, it is necessary to measure several such curves
for a given receptor. The results show - and indeed one would hardly
have expected it to be otherwise - that the receptive field is approximately radially symmetrical (Fig. 3b). This established, measurement of a single curve (3a) now suffices to characterize the receptive
field of a cell. Receptive-field size can be specified in terms of the
angular range $\Delta\varphi_5\%$ within which the sensitivity of a cell is greater
than 5% of the maximal sensitivity, at its center.

We have determined the receptive-field sizes of a number of receptors
of the type R1-R6 (Fig. 3c). Naturally, there is some variation from
cell to cell. The notable point in this regard is that the scatter is
smaller for measurements in the region of the soma than in the axonal
region. The receptive fields measured in the soma region range from
5° to 8°, whereas the range of those measured via axon response is
5° to 15°. Since each such cell was identified as a Type R1-R6 receptor by marking with Procion yellow, there can be no doubt that there
are visual cells with receptive fields smaller in the peripheral part
of the cell than in the proximal part. This would hardly be possible
if each cell functioned as an isolated unit. It must be concluded,
then, that the potentials recorded from a visual-cell axon reflect
more than the response of the corresponding cell; that is, they include components derived from neighboring cells. In other words, at
the level of the receptor axons there must exist a certain lateral interaction not present at the level of the somata. This finding is quite
consistent with the results published by Scholes (1969). Using a very
elegant arrangement for stimulation, he showed that the response of a
receptor axon is a superposition of the potentials of all six receptors
contributing to a cartridge. This discovery provides a ready interpretation for our finding. The receptive field measured in the soma of a
visual cell is the optical field of this single cell, whereas the receptive field measured in the axon of a given cell results from a combination of the fields of all six receptors which form a cartridge.
In the ideal case these six receptors would have identical receptive
fields, so that recordings from soma and axon would yield identical
fields. But in Fig. 3c it is evident that the fields recorded from
axons are often larger than the soma fields, from which we must conclude that the six fields of the receptors forming some cartridges are
not identical. That is, the optical axes of the six receptors in these
cartridges are not precisely parallel, but diverge to some extent.

2.4.3 The Responses of the Cartridge Outputs L1-L2

Fig. 3c also shows receptive fields of the monopolar neurons of
Type L1-L2. These are, on the average, considerably smaller than those
of the receptor axons. This is an astonishing finding, since the receptor axons are the direct presynaptic elements of cells L1 and L2
(Trujillo-Cenóz, 1965; Boschek, 1971). It was, therefore, to be expected that the size distribution of the monopolar-neuron receptive fields
would be the same as that of the R1-R6 axons. Since this is obviously
not the case, there must be other elements presynaptic to the monopolar neurons, by which neighboring cartridges exert an inhibitory influence upon one another. This narrowing of the receptive fields of
the monopolar neurons is thus an indirect demonstration of the existence of lateral inhibition within the first optic neuropile of the
fly (Zettler and Järvilehto, 1972; Laughlin, 1974).

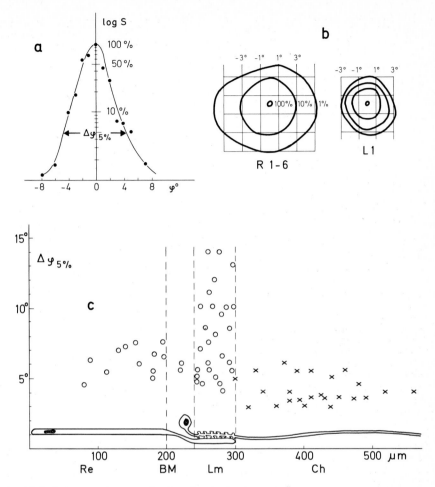

Fig. 3a-c. Spatial intregration at the level of the receptors R1-R6 and the monopolar neurons L1 and L2. a) Relationship between angle of incidence of the light
and sensitivity of the cell. $\varphi = 0$ is the position of a point source of light at
which the maximal response is elicited. The sensitivity to a stimulus at this position is defined as 100%. $\Delta\varphi 5\%$ is the angular range within which sensitivity exceeds
5%. b) Receptive field of a receptor and of a monopolar neuron. The roughly circular
lines connect points of equal sensitivity. The outer circle in the L1 diagram represents 0.1%. c) Size distribution of the receptive fields of 41 R1-R6 receptors and
24 L1-L2 neurons. The abscissa indicates the depth of the recording site, measured
from the distal end of the receptors. All the data points in this graph are taken
from identified cells. The axons of the receptors, on the average, exhibit broader
receptive fields than the somata. The receptive fields measured in the postsynaptic
neurons again become smaller. Re: retinula layer; BM: basement membrane; Lm: lamina;
Ch: chiasma

Our subsequent experiments on lateral inhibition were designed with
two main goals in mind: to obtain more explicit evidence for the existence of lateral inhibition, and to discover the pathways over which
the inhibition is exerted. Annular light stimuli proved especially effective in dealing with the first problem, while an approach to the
second was provided by determining the receptive fields with monochro-

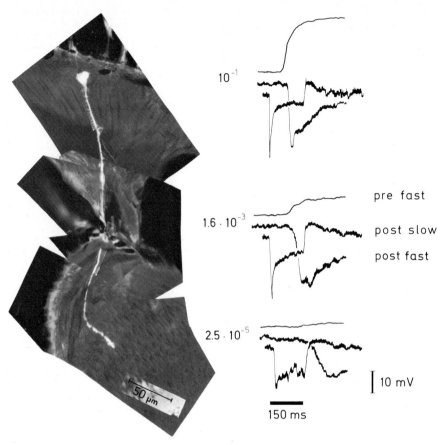

10^{-1}

$1.6 \cdot 10^{-3}$

pre fast

post slow

post fast

$2.5 \cdot 10^{-5}$

10 mV

50 µm

150 ms

Fig. 4. Marked monopolar neuron of Type L1. The electrode penetrated the axon about halfway along its length. A bit of the glass tip can be seen just outside the tissue. On the right are the potentials recorded in response to stimuli at three different intensities, as indicated next to the recordings. The time scale applies to the slow trace; the fast traces are a factor of ten times faster. The presynaptic receptor responses at each intensity are included for comparison

matic visual stimuli. Before describing these experiments, however, some mention must be made of the special nature of the monopolar-cell responses.

The response of the monopolar neurons to stimulation by light is a hyperpolarizing potential, the amplitude of which depends upon the intensity of the stimulus (Fig. 4). These potentials are notable in several points:

1. They are found to have the same shape and size at every position along the axon - that is, they are not limited to the immediately postsynaptic part of the axon in the lamina. This means that the monopolar neurons transmit information not in the form of spike trains, but rather as graded potential fluctuations (Zettler and Järvilehto, 1971). These potential fluctuations, moreover, do not appear to spread passively along the axon according to cable theory. There are some indications that attenuation of the spreading potential is compensated by regenerative processes in the axon membrane (Zettler and Järvilehto, 1973).

2. They are hyperpolarizing - i.e. their polarity is the reverse of that of the presynaptic visual-cell potentials. It might, therefore, at first be thought that they are IPSPs. The error of such an inference in this case is evident, however, since the monopolar neurons produce no spikes at all which could be inhibited. The hyperpolarizing potential itself must be regarded as the basic response of these neurons.

3. They are extraordinarily sensitive to light stimuli. With stimuli so weak that the potentials of the presynaptic visual cells are reduced to the limit of detectability (approx. 0.5 mV), the monopolar neurons still respond with potentials of approximately 10 mV. This result implies a considerable synaptic amplification.

4. The on-phase is very rapid, being completed in a much shorter time than the on-phase of the presynaptic receptor potential. The consequence is that the maximal amplitude of the postsynaptic potential is reached sooner than the presynaptic maximum. It is thus impossible for the presynaptic amplitude to determine that observed postsynaptically. There must be another parameter of the presynaptic potential which determines the amplitude of the postsynaptic potential. This parameter can only be the rate of rise of the initial phase of the receptor potential (Järvilehto and Zettler, 1971).

5. At low stimulus intensities, the potential is tonic in character, and as the intensity is increased the response gradually becomes phasic. That is, at higher intensities a delayed component of the potential appears which counteracts the initial hyperpolarization. This component has positive polarity, a slower time course than the hyperpolarizing component, and a delayed onset. It is very probably generated by a second input, from the neighboring cartridges. The latter are also stimulated, though less strongly than the central cartridge, with the stimulus arrangement used in these experiments. The fact that excitation of the lateral input is weaker offers a natural explanation of the delay of the inhibitory potential component and of its slower time course.

There are two phenomena, then, that indicate the existence of a lateral, inhibitory input to the monopolar neurons: the appearance of the antagonistic potential component just described, and the reduction in size of the receptive fields of the monopolar neurons as compared with those of the receptors. Somewhat more direct evidence of the existence of lateral inhibition can be obtained by using an annular light source as the stimulus. The advantage of an annulus of light as compared with a point is that the former stimulates most strongly those cartridges surrounding the monopolar neuron on which it is centered. Of course, it is impossible even with an annular light source to stimulate only the neighboring, inhibitory elements; the receptors associated with the central cartridge - that of the monopolar neuron being studied - are also illuminated to a certain extent and hence contribute an excitatory (i.e. hyperpolarizing) potential component. The annulus does, however, achieve a stronger relative weighting of the inhibitory potential components.

In the experiments summarized in Fig. 5, the diameter of the annular stimulus was selected such that the stimulus produced a receptor-cell response identical to that produced by a point source at the same intensity. As a result, the response-versus-intensity curves of the receptor are nearly identical, regardless of whether they are measured with the annular or the point light source (Fig. 5a). For these measurements, the point source is located in the maximally effective position, and the center of the annulus is in the same position. By contrast, when response-versus-intensity curves of a monopolar neuron are

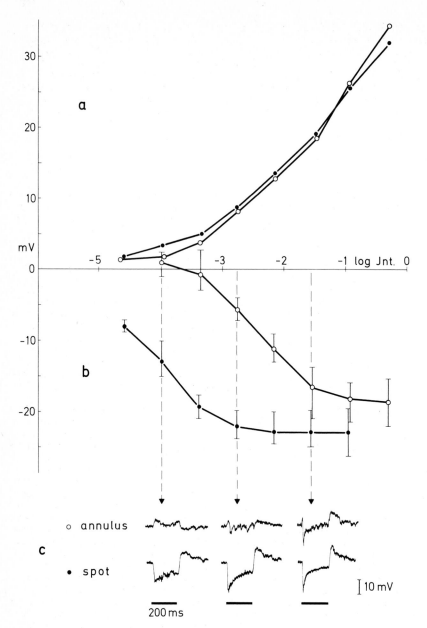

Fig. 5a–c. a) Response-versus-intensity curve recorded from the axon of a Type R1–R6 receptor. ● point stimulus, diameter 1.8°; o annular stimulus; diameter of annulus 12.5°, thickness 1.8°. Each data point represents the average of the responses of three cells. b) Response-versus-intensity curves of monopolar neurons of Type L1–L2. Stimuli as in (a). Each data point represents the average of the responses of four cells, with the maximal deviation indicated by the vertical lines. c) Potentials recorded from a monopolar neuron in response to stimulation at three different light intensities: the point-stimulus and annular-stimulus responses in each column were recorded with the same stimulus intensity; intensities as indicated by the arrows from the abscissae of Fig. 5(a,b)

measured with the same two stimulus arrangements, the curves for the point and annular sources are far from identical (Fig. 5b). In fact, a smaller response is obtained for the annular than for the point stimulus at each intensity. It follows that, since the two stimuli (point and annulus) produce identical responses in the receptors of a cartridge but decidedly different responses in the secondary neuron, the signals from these receptors cannot bear the sole responsibility for the excitation appearing in the monopolar neuron. That is to say, a monopolar cell must receive additional inputs from neighboring cartridges, which are inhibitory in function.

Examination of the time courses of the potentials measured in a monopolar neuron (Fig. 5c) increases the plausibility of this interpretation. At low intensities the annular stimulus actually elicits a positive (depolarizing) potential, so that the antagonistic character of the lateral effects is brought out clearly. At higher intensities the amplitude of the normal, negative component dominates, but the positive component, although of smaller amplitude, can still be seen as an on-effect. The latter phenomenon results from the fact that the annular light stimulates the lateral inputs relatively strongly, so that these respond with a shorter latency than the central inputs.

Once the existence of lateral inhibition has been demonstrated, there remains the rather more difficult question of the neuronal connectivity underlying it. We cannot, of course, provide a complete answer, but we have done a few experiments which we believe at least narrow the range of possibilities. Taking as a starting assumption that the excitatory input channel of a monopolar neuron consists of the six receptors R1-R6 of a cartridge, a first question is whether the inhibitory channel is also driven by neighboring receptors of this same type or, alternatively, by one of the two other receptor types, R7 and R8.

To decide between these alternatives, we examined the receptive fields of the monopolar neurons with monochromatic light of different wavelengths. The basic idea is as follows: experiments combining electrophysiology and histology (Meffert and Smola, 1976) and electrophysiology and genetics (Stark, 1975) have shown that R7 and R8 differ with respect to chromatic sensitivity. Moreover, it is known that both R7 and R8 have spectral sensitivities differing from those of Type R1-R6 receptors, and that the latter form a homogeneous group in this regard (McCann and Arnett, 1972). On the assumption that R7 or R8 drives the inhibitory channel, measurement of the receptive field of a monopolar cell with a wavelength to which Type R1-R6 is much more sensitive than R7 or R8 should indicate that the field is relatively large; that is, under such conditions the inhibitory channel would not be strongly excited. Conversely, at a wavelength to which R7 or R8 respond more strongly than R1-R6, the inhibitory channel would be excited to a much greater degree, and hence the receptive field should be relatively small. Fig. 6 shows the receptive fields of three individual monopolar cells, measured in each case at three different wavelengths. (The survival time of the cells was such that no more than three wavelengths could be tested with a given cell.) Examination of a total of 28 cells revealed none in which there was a detectable change in receptive-field size as the wavelength of the stimulus was changed (Zettler and Autrum, 1975). This implies that our initial assumption - that the inhibitory channel is excited by R7 or R8 - must be false. Further, it implies that the excitatory and inhibitory channels are driven by receptors of a uniform chromatic type (R1-R6); only if this were so would it be explicable that the relative weightings of excitation and inhibition, which determine the size of the receptive field, do not depend upon wavelength.

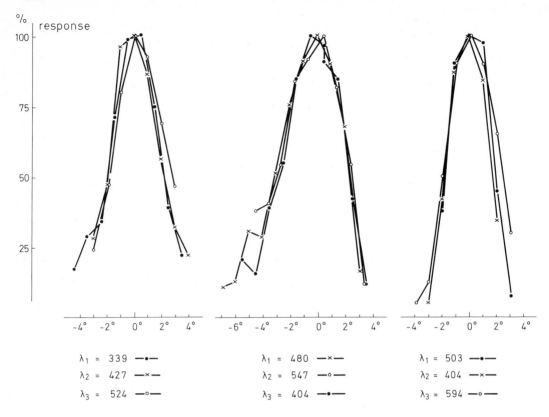

Fig. 6. Receptive fields of three monopolar neurons of Type L1-L2, measured at three different wavelengths in each instance. (From Zettler and Autrum, 1975)

Acknowledgments. This work was supported by the Deutsche Forschungs-gemeinschaft. We would like to thank Dr. Marguerite A. Biederman-Thorson for translating the German text into English and Miss Christa Weyh for technical assistance.

2.4.4 References

Autrum, H., Wiedemann, I.: Versuche über den Strahlengang im Insektenauge. Z. Naturforsch. 17B, 480-482 (1962)
Boschek, C.B.: On the fine structure of the peripheral retina and lamina ganglionaris of the fly Musca domestica. Z. Zellforsch. 118, 369-409 (1971)
Braitenberg, V.: Patterns of projection in the visual system of the fly. I. Retina-lamina projections. Exp. Brain Res. 3, 271-298 (1967)
Braitenberg, V., Strausfeld, N.J.: Principles of the mosaic organization in the visual system's neuropil of Musca domestica L. In: Handbook of Sensory Physiology. Jung, R. (ed.). Berlin-Heidelberg-New York: Springer 1973, Vol. VII/3A, pp. 631-659
Järvilehto, M., Zettler, F.: Localized intracellular potentials from pre- and post-synaptic components in the external plexiform layer of an insect retina. Z. Vergl. Physiol. 75, 422-440 (1971)

Kirschfeld, K.: Die Projektion der optischen Umwelt auf das Raster der Rhabdomere im Komplexauge von Musca. Exp. Brain Res. _3_, 248-270 (1967)

Laughlin, S.B.: Neural integration in the first optic neuropil of dragonflies. III. Transfer of angular information. J. Comp. Physiol. _92_, 377-396 (1974)

McCann, D., Arnett, D.W.: Spectral and polarisation sensitivity of the dipteran visual system. J. Gen. Physiol. _59_, 534-558 (1972)

Meffert, P., Smola, U.: Electrophysiological measurements of the spectral sensitivity of central visual cells in the eye of the blowfly. Nature _260_, 342-344 (1976)

Scholes, J.: The electrical responses of the retinal receptors and the lamina in the visual system of the fly Musca. Kybernetik _6_, 149-162 (1969)

Stark, W.S.: Spectral sensitivity of visual response alterations mediated by inter-conversion of native and intermediate photopigments in drosophila. J. Comp. Physiol. _96_, 343-356 (1975)

Strausfeld, N.J.: The organization of the insect visual system (light microscopy). Z. Zellforsch. _121_, 377-441 (1971)

Strausfeld, N.J.: Atlas of an Insect Brain. Berlin-Heidelberg-New York: Springer 1976

Trujillo-Cenóz, O.: Some aspects of the structural organization of the intermediate retina of dipterans. J. Ultrastruct. Res. _13_, 1-33 (1965)

Trujillo-Cenóz, O., Melamed, J.: Compound eye of dipterans: Anatomical basis for integration - an electron microscopy study. J. Ultrastruct. Res. _16_, 395-398 (1966)

Zettler, F., Autrum, H.: Chromatic properties of lateral inhibition in the eye of a fly. J. Comp. Physiol. _97_, 181-188 (1975)

Zettler, F., Järvilehto, M.: Decrement-free conduction of graded potentials along the axon of a monopolar neuron. Z. Vergl. Physiol. _75_, 402-421 (1971)

Zettler, F., Järvilehto, M.: Lateral inhibition in an insect eye. Z. Vergl. Physiol. _76_, 233-244 (1972)

Zettler, F., Järvilehto, M.: Active and passive axonal propagation of non-spike signals in the retina of calliphora. J. Comp. Physiol. _85_, 89-104 (1973).

2.5 Beyond the Wiring Diagram of the Lamina Ganglionaris of the Fly

V. Braitenberg and W. Burkhardt

2.5.1 Introduction

The recent upgrading of neuroanatomy among neurological sciences has
been due largely to our feeling that fiber patterns may be viewed as
electronic circuit diagrams and in some cases may be directly related
with patterns of behavior. We thought we had found such a case in the
visual system of some insects, where both quantitative behavioral analyses and histology had reached a degree of clarity which seemed to war-
rant fruitful comparisons. We will show how several years of research
in this field have made it increasingly clear that as long as the components of the network are not fully understood, little can be inferred
from the pattern of their connections. As an illustration, we shall
briefly describe four cases, all referring to the first visual ganglion, the Lamina ganglionaris of the house fly.

2.5.2 Gradients in the Thickness of Fibers over the Surface of the Ganglion (Braitenberg and Hauser, 1972)

Very impressively, the thickness of some of the fibers which make up
each of the 3000 periodic subunits of the Lamina varies as a function
of their position in the ganglion. In one case, that of the L3 fiber,
the thickness is covariant with the size of the corresponding lenses
of the compound eye, which also follows a gradient, different in male
and in female flies. Simple mechanical explanations do not suffice.
For example, the supposition that in regions where the lenses are larger, the whole array of the compound eye, including the visual ganglia,
is mechanically stretched and that the fibers are therefore stubbier,
is contradicted by the fact that these fibers remain thick after they
have left the lamina, traversed the chiasma and reached the medulla
in entirely different places. We are forced to assume some functional
relation linking perhaps the greater light-gathering power of the larger lenses with the larger membrane surface of the corresponding fibers, but we must admit defeat when it comes to a real biophysical
explanation. Here obviously the wiring diagram is not the whole story.

Fig. 1 (Braitenberg and Hauser, 1972) shows the thickness of the
L3 fiber in three different regions of the eye, as compared to the
L1 and L2 fibers which do not vary as much in the same regions. They
follow a different gradient (not illustrated here and even more difficult to explain). Possibly the variation of the thickness of these
fibers in different regions of the eye has something to do with variations of optomotor reactions as induced by stimuli in different parts
of the visual field. Again, we are far from even speculative interpretations.

Our lack of understanding is even more fundamental than this. L1, L2
and L3 of one cartridge are all postsynaptic to the same set of axons

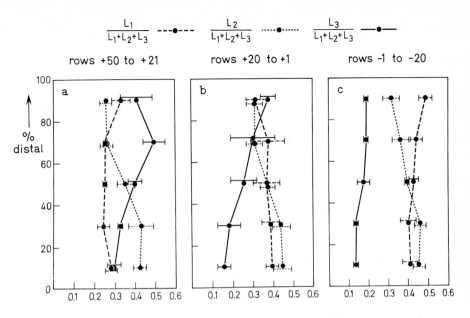

$$\frac{L_1}{L_1+L_2+L_3}$$ --●-- $$\frac{L_2}{L_1+L_2+L_3}$$ ......●...... $$\frac{L_3}{L_1+L_2+L_3}$$ ─●─

rows +50 to +21   rows +20 to +1   rows -1 to -20

Fig. 1a-c. Plots of the thickness of L1, L2 and L3 in three different areas of a
male fly. Vertical coordinate: vertical position in the ganglion. Horizontal coordinate: cross-sectional surface in μ^2. Note areal variation of L3 (<u>solid line</u>)
compared to the other two fibers. <u>Horizontal bars</u>: standard deviations.
(From Braitenberg and Hauser-Holschuh, 1972)

of retinula cells. The electron microscope gives no indication of a
difference in the signals which L1, L2 and L3 receive from the corresponding set of retinula cells. Still, the three fibers very neatly
convey this information to three different levels of the medulla and
we are forced to assume that they serve as different filters. Neither
the shape of the three fibers, nor any known neurphysiological finding
gives any indication of the nature of this differential filtering process.

2.5.3 A Net of Reciprocal Synapses (Braitenberg and Debbage, 1974)

A set of fibers connecting neighboring channels of the compound eye
at a level between the lamina and the chiasma has long ago attracted
our attention because of its striking regularity. Light microscopical
observation of Bodian and Golgi preparations seemed to indicate an
orientation of these fibers (named L4 collaterals) in the two oblique
directions of the hexagonal array (Braitenberg, 1969; Strausfeld and
Braitenberg, 1970; Braitenberg, 1970) but a more careful reconstruction by means of small series of electronmicroscopical sections reveals the array of these fibers as the simplest possible, with the
endfeet of three fibers touching each other in each cartridge (Fig. 2).
The situation is even more symmetrical, when the direction of the signal transmission is taken into account. By all standard interpretations
of pre- and postsynaptic specializations as they appear on electron-
micrographs, it is evident that for each synapse in one direction between one endfoot and the other, there is a synapse in the opposite
direction. The situation seems to indicate signal flow in all directions of the net of Fig. 2.

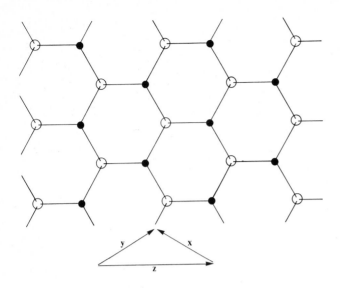

The input to this net presumably stems from the contacts which the L4 fibers have in the upper part of the lamina and which according to Strausfeld and Campos-Ortega (1973) are postsynaptic to α-neurons (see Sect. 4 in this chap.). The output, besides coming from the termination of L4 in the next ganglion (the medulla), is also derived from synapses of the endfeet of the L4 collaterals on the main stem of the L2 fiber. However, the significance of the net of horizontal connections remains obscure. We do not know whether they are excitatory or inhibitory, let alone the spatial and temporal characteristics of the transmission of signals through these synapses. Or course, entirely different macroscopical effects would be obtained depending on the sign and the characteristics of the interaction in the reciprocal synapses.

2.5.4 Synaptic Complexes Postsynaptic to Retinula Cell Axons

This is a preview of a paper by Burkhardt and Braitenberg, now in preparation. It is our contention that all synapses which transmit signals from the axons of the retinula cells 1 to 6 onto the laminar neurons are complex structures involving one presynaptic and four postsynaptic cells. The arrangement is repeated with great constancy throughout the depth and extent of the lamina. There are about 40 such complexes for each of the six times 3000 retinula cell axons entering the lamina of one side.

Fig. 3a,b,c shows the synaptic complex in three electron micrographs oriented at right angles to each other, as indicated in the diagram on the lower left. Fig. 4 serves as an aid in the interpretation of the pictures. There is an elongated presynaptic plate (P) which is supported by an elongated and lobated foot, the presynaptic bar (B), characterized by its high electron-density. The presynaptic membrane is not shown in Fig. 4, while the four segments of postsynaptic membrane facing the presynaptic apparatus, each belonging to a different cell, are labeled I to IV. On the postsynaptic side, the electron micrographs show in addition two pieces of folded membrane, called

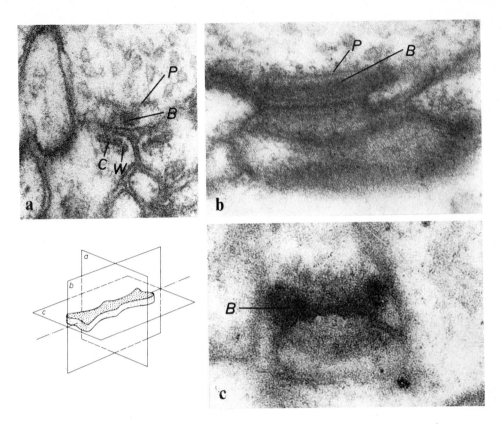

Fig. 3a-c. Electron micrographs from sections through the complex synapse between a retinula cell axon and four postsynaptic elements. The sections are oriented at right angles to each other, as indicated by the inset on the lower left which uses the "postsynaptic bar" as a marker. P: postsynaptic plage; B: postsynaptic bar; C: postsynaptic bag; W: postsynaptic whisker. Scale: 100.000 : 1

postsynaptic bags (C on Fig. 3a), situated symmetrically in the postsynaptic elements number I and II, accompanied by further small stainable densities, called postsynaptic whiskers (Fig. 3a, W). The length of the presynaptic bar varies between 2200 and 3.600 Å, that of the plate is about 5000 Å.

It came as no surprise that the elements in the positions I and II are the L1 and L2 fibers respectively, since it had been observed earlier (Trujillo-Cenoz, 1965; Boschek, 1971) in both light and electron microscopy that these two fibers have the same sort of intimacy with the afferent retinula cell axons. The fiber in position IV was soon recognized as the α-fiber, Cajal and Sanchez' "brush-shaped centrifugal", one of the two elements out of which the basket is woven which evelops each cartridge (the other being a medulla-lamina element called β-fiber, see Sect. 5 of this chap.). The identification of the element in position III raises some problems. As already mentioned, the L3 fiber is also postsynaptic to retinula cell axons. The shape of its dendritic processes, as well as the electron-microscopical appearance of the contact in sporadic observations by Boschek (1971) seemed to indicate the same sort relation to the visual input as that of L1 and L2. Indeed we found evidence of the continuity of the postsynaptic membrane in posi-

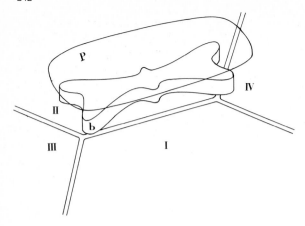

Fig. 4. Diagrammatic view of the complex synapse between the retinula cell axon (presynaptic) and four postsynaptic elements (I to IV). P: presynaptic plate; b: presynaptic bar. The presynaptic membrane is not shown. For the identification of I, II, III and IV see text

tion III with the stem of the L3 fiber, confirming the earlier descriptions and showing at the same time that the relation of L3 to the afferent retinula cell axon is not exactly symmetrical to that of L1 and L2, in the sense in which the latter two fibers enter the postsynaptic complex in indistinguishable ways.

The difficulty arises in the lower half of the lamina, since the L3 fiber, in contrast to the L1 and L2 fibers, has dendritic branches only in the upper part of the lamina and therefore, in the lower half, is not morphologically equipped to make the necessary contacts with all the incoming retinula cell axons. This realization led to the suspicion that in the lower lamina the synaptic complexes are perhaps simpler, with the element in position III missing. However, a topographic study of the shapes of the synaptic complexes did not reveal any differences depending on the level. Everywhere there are four separate pieces of synaptic membrane associated with one presynaptic terminal. Which is the cell which replaces L3 in the lower half of the lamina? To our surprise the cytological characteristics of the element in that position reveals it as being glial rather than neuronal in nature. One of the two types of glia present in the lamina, the so called marginal glia, is in fact only present in the lower half of the lamina: its distribution being complementary to that of the branches of L3, it is plausible that it should replace L3 within its level.

The strange finding that the glia cells have synaptic contacts with neurons will detain us also in the following section.

2.5.5 Gnarls, Capitate Projections, Postsynaptic Glia

We have designated as "gnarl" a synaptic complex involving three cells of the lamina: the brush-shaped centrifugal (α-neuron), the β-neuron and the epithelial glia cell. The number of β-neurons is the same as that of the cartridges of the lamina, each of them forming a basket around one cartridge. Also the number of epithelial glia cells is the same as that of the cartridges. They are housed in the interspaces between the cartridges in such a way that each of them touches three cartridges and each cartridge is touched by three of them (Boschek, 1971). The number of α-neurons is not known but it seems certain that each of them is connected to more than one cartridge. The α-neurons

Fig. 5. Grossly schematic representation of the complex formed by the
brush-like centrifugal cell (α), by
the epithelial glia (γ) and by the
basket cell (β). The glia is invaginated into the β-neuron in places
opposite to where the α-neuron shows
apparent presynaptic specializations
(arrows). The invaginations may be
interpreted as an extreme case of
postsynaptic specialization. La:
lamina; Ch: chiasma; Me: medulla.
It should be obvious that the met-
ric of this diagram is grossly distorted, the microscopical details
being vastly overemphasized as compared to the macroscopical layout

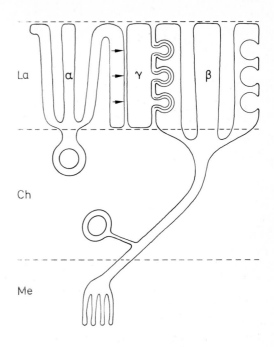

have their nucleus underneath the lamina, their processes stay confined to that ganglion. The β-neurons connect the medulla and the
lamina, their nucleus being situated near the medulla, where they
also have a well ramified termination. The epithelial glia cell stays
entirely within the lamina. The contacts between these three elements
can best be explained with reference to the very diagrammatic drawing
of Fig. 5. There are synapses (i.e. all the structural markers that
we have always interpreted as synapses in fly histology) indicating
signal transmission from the α-neuron onto the epithelial glia cells
(arrows in the diagram). Opposite these synapses the epithelial glia
cell, which is very thin between the α- and the β-neuron, is invaginated into the β-neuron, forming very complicated processes of a shape
not unlike a chain of sausages.

The cleft between α and glia and between glia and β is enlarged and
there is a dark layer interposed between the two membranes. This membrane specialization covers the entire invagination into the body of
the β fiber. If it is to be brought into relation with the apparently
presynaptic specialization within the α fiber, the glia-α-β membrane
complex is the most gigantic postsynaptic structure ever described.
It is interesting to observe that the area of this membrane specialization varies greatly in large (Musca) and small (Drosophila) flies
(Hauser-Holschuh, 1975). The data published in this thesis show that
the area is roughly proportionate to the third power of the linear
dimensions of the flies, indicating a functional relation with some
volume rather than with another membrane surface; but speculation can
hardly go beyond this.

"Capitate projections" have been described by Trujillo-Cenoz (1965)
as small invaginations of the epithelial glia into the retinula cell
axons. They are covered by a darkly stained three-layer membrane complex. Their meaning is even more obscure than that of the other structures described in this paper.

In all cases described, glia cells seem to be synaptically involved with neurons or sense cells as if they were neurons themselves. This is surprising, but could be explained away by giving the glia some sort of supporting role in a synapse which otherwise involves only neurons (e.g. in the synaptic complexes postsynaptic to retinula cells and in the gnarl). But Boschek (1971) has already described instances in which the glia seemed to be postsynaptic to neurons outside of these complicated structures with only the glia on the postsynaptic side. We shall have to reconsider the role of the glia which probably have functions that go beyond the old idea of "support" and "nourishment".

2.5.6 Conclusion

It is obvious that a good deal of biophysics will have to be done before the electron microscopical findings in the fly's visual ganglia can really be explained. In this field, anatomy is perhaps 50 years ahead of physiology as it was around the turn of the century. Cajal described the nervous system with a spatial resolution of about 1 μ at a time when electrophysiologists used electrodes with a diameter of about 1 cm. The electrophysiology reached the resolution of 1 μ only recently with the introduction of microelectrodes and, in fact, rediscovered Cajal. Anatomy meanwhile, however, has reached a resolution of about 100 Å. It will take some time before that resolution can be introduced into electrophysiological techniques.

2.5.7 References

Boschek, B.: On the fine structure of the peripheral Retina and Lamina ganglionaris of the fly Musca domestica. Z. Zellforsch. 118, 369-409 (1971)

Braitenberg, V.: The anatomical substratum of visual perception in flies. A sketch of the visual ganglia. Rendiconti S.I.F. XLIII. Reichardt, W. (ed.). London-New York: Academic 1969, pp. 328-340

Braitenberg, V.: Ordnung und Orientierung der Elemente im Sehsystem der Fliege. Kybernetik 7, 235-242 (1970)

Braitenberg, V., Debbage, P.: A regular net of reciprocal synapses in the visual system of the fly, Musca domestica. J. Comp. Physiol. 90, 25-31 (1974)

Braitenberg, V., Hauser, H.: Patterns of projection in the visual system of the fly. II. Quantitative aspects of second order neurons in relation to models of movement perception. Exp. Brain Res. 16, 184-209 (1972)

Burkhardt, W., Braitenberg, V.: Some peculiar synaptic complexes in the first visual ganglion of the fly, Musca domestica. To appear in Cell and Tissue Research

Cajal, S.R., Sanchez, D.: Contribucion al conocimiento de los centros nerviosos de los insectos. Trab. Lab. Invest. Biol. Univ. Madr. 13, 1-168 (1915)

Hauser-Holschuh, H.: Vergleichend quantitative Untersuchungen an den Sehganglien der Fliegen Musca domestica und Drosophila melanogaster. Ph. D. thesis. Univ. Tübingen (1975)

Strausfeld, N., Braitenberg, V.: The compound eye of the fly (Musca domestica): Connections between the cartridges of the lamina ganglionaris. Z. Vergl. Physiol. 70, 95-104 (1970)

Strausfeld, N., Campos-Ortega, J.A.: The L4 monopolar neurone: A substrate for lateral interaction in the visual system of the fly Musca domestica. Brain Res. 59, 97-117 (1973)

Trujillo-Cenóz, O.: Some aspects of the structural organization of the intermediate retina of dipterans. J. Ultrastr. Res. 13, 1-33 (1965).

2.6 Mosaic Organizations, Layers, and Visual Pathways in the Insect Brain

N. J. STRAUSFELD

2.6.1 Introduction

In this account I want to outline some of the cardinal structural features of the insect visual system which have been resolved from neuroanatomical studies of two species of Dipterous insect, Musca domestica and Calliphora erythrocephala.

The text is divided into 13 sections. After the Introduction, the first illustrates the sequence of channels from the retina to the descending neurons from the brain, into the thoracic ganglia. This section is followed by a brief summary of the numbers of neurons in the visual system and the distribution of visual neuropil within the protocerebrum of the brain. Section 4 describes the blue print of convergence pathways from the columnar neuropil of the optic lobes to dendrites of descending neurons.

The principle features of the optic lobe's first- to third-order neuropils — the mosaics, the precisions of neural elements, the retinotopic projections, the packing of neurons and columnar equivalences — are described in Sections 5-8. Section 9 sketches the types of heterolateral pathways between left and right hand optic lobes and Section 10 classifies the three major groups of tangential elements that structurally interact with large aggregates of the projected retinotopic mosaic.

The third-order neuropil of the lobula plate serves to illustrate the principles of neuronal mappings and structural "summation" of discrete parts of the visual field. Section 11 describes the giant visual neurons of the lobula plate and outlines some associated large field elements that contribute to inputs to channels that leave the brain. The neurons are described against a rather simplified background of behavioural and electrophysiological studies, including Procion yellow investigations by Dvorak et al. (1975) and Hausen (1976a,b).

The large-field lobula plate neurons receive, summate and relay information about the direction of panoramic motion in the visual field. However, it is likely that computations between neighbouring retinotopic channels, leading to the detection of motion - and its directionality and velocity - are performed within the second synaptic region of the medulla. This neuropil is, though, exceedingly complex, being composed of many thousands of neurons comprising at least 60 distinct "species" of nerve cells. Axonal neurons are arranged in columns and their lateral processes, with networks of amacrines, compose at least 24 discrete strata in the medulla. Section 12 attempts to formulate a very simple schema of the medulla architecture and proposes the existence of three "projection modes" from the medulla neuropil on to two functional classes of third-order interneurons. The first of these modes relays the retinotopic mosaic to motion-sensitive units which respond to wide-field events of the visual panorama. The second mode relays the medulla mosaic to small-field convergence pathways

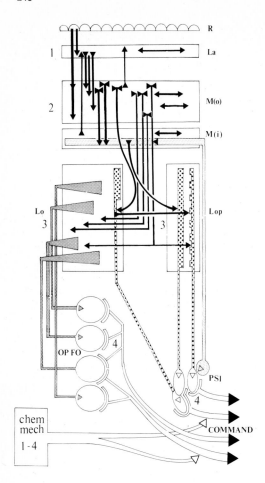

Fig. 1. Summary diagram of the main channels in the optic lobes and brain (see text for explanation and abbreviations)

from the lobula. The third mode is proposed to relay information about small- and/or large-field events to both the lobula and lobula plate.

Finally, by way of a post-script, and in order to hint at the kinds of structural complexities that we should expect at all levels in the brain, I have summarized some synaptic connectivities within one layer of the visual system. For this I have taken our present knowledge of the first synaptic region, the lamina.

2.6.2 Plan of the Visual Neuropils

Fig. 1 schematizes the main regions and channels of the optic lobes into the brain. Each channel (CHAN) represents a major class of elements but does not consider individual types of neurons nor their packing and distribution. These are outlined in the succeeding sections.

Channels in the optic lobes, lateral and through-going, are shown in black. Channels that project from the lobes to the brain are shown stippled. Channels leaving the brain for the thoracic ganglia have open profiles. First- to fourth-order neuropil is numbered 1-4 (Lamina <u>La</u>; Medulla outer layer (<u>M(o)</u>) and inner layer (<u>M(i)</u>); lobula complex (anterior lobula <u>Lo</u>, and posterior lobula plate <u>Lop</u>); protocerebrum (anterior optic foci <u>op fo</u>, posterior slope <u>PSl</u>).

Two channels project from the retina (CHAN, Ra,Rb) which bestow the retinal mosaic on to the lamina and medulla. These are the short and long visual cells, respectively. First-order interneurons relay retinal information, via CHAN 1 to the medulla, parallel with long visual cells. Lateral elements project between parallel pathways within the lamina and centrifugal neurons project back from the medulla to the lamina.

There are two major channels derived from the medulla input, CHAN 2a and 2b. The first serves to link the outer medulla (<u>M(o)</u>) to the inner levels (<u>M(i)</u>). Planar arrangements of neurons subsequently project directly from <u>M(i)</u> to the posterior slope of the protocerebrum. CHAN 2b comprises columnar neurons that pass from <u>M(o)</u> and/or <u>M(i)</u> to the lobula and/or lobula plate. There are three major projection modes of CHAN 2b elements, associated with functional classes of lobula complex neurons (Sect. 12). Subsequent projections from the lobula complex are of two major types. CHAN 3a is composed of columnar third-order interneurons that project to the optic foci of the ventrolateral protocerebrum where they meet dendrites of descending neurons (COMMAND) to the thoracic ganglia. Lobula elements of CHAN 3a are termed "lobular columnar neurons" (LCN). CHAN 3b comprises giant fan shaped visual cells (GVNs), or unique giant visual cells (UGVNs), that project from a special superficial stratum of the lobula and from all levels of the lobula plate. These converge at the posterior slope to meet tangentials derived from M(i) and dendrites of descending neurons. Descending fibres may also receive inputs from channels equivalent to CHAN 3 from chemo- or mechanosensory neuropils.

This summary diagram (Fig. 1) includes only ipsilateral connections. Heterolateral pathways between second-, third- and fourth-order neuropils are summarized in Section 9 of this account. Two features that should be emphasized at the outset are: (1) the division between giant field and small-field neurons which leave the optic lobes (GVNs and LCNs, respectively) and their subsequent partial segregation in pathways from the brain; (2) the absence of lateral intrinsic (amacrine) pathways in the third synaptic regions. Details of the morphology and projections of neurons, from which this schema is derived, are published elsewhere (Strausfeld, 1976).

2.6.3 Numbers of Visual Neurons and Brain Regions Subserving Vision

The total number of neurons in the brain of a female fly (optic lobes, and supra-, suboesophageal ganglia) is approximately 3.4×10^5 (Strausfeld, 1976): Of these 76% can be related to the three regions of the optic lobes. Of the total number of optic lobe neurons 68% invade the medulla, 12% invade the lamina, 18% invade the lobula and 2% are derived from the lobula plate.

Even though the optic lobes appear to be the most dominant structures of the brain it is clear from the projections of lobula derived elements that substantial volumes of protocerebral neuropil also subserve visual integration. These neuropils occupy nearly the whole volume of the ventrolateral protocerebrum and posterior slope. They take up 6.35% of the supraoesophageal neuropil volume and are thus more bulky

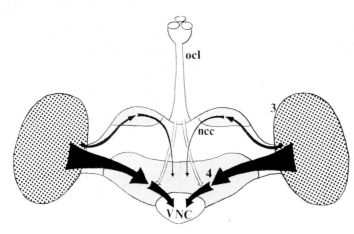

Fig. 2. Divergence of lobula output to central and visceral nervous system (see text) ncc neurosecretory pathways from protorerebrum to retrocerebral complex; VNC: ventral nerve cord

than first-to-fourth-order neuropils of the chemosensory system (Olfactory lobes, Calyces, Alpha and Beta lobes and Inferior Medial Protocerebrum). However, whereas the chemosensory neuropils are derived from 13% of the total number of brain perikarya only 1% of cell bodies give rise to neurons that have dendritic arborizations within the fourth-order visual neuropil of the protocerebrum. These elements are mainly command or commisural neurons which serve, respectively, to leave the brain for the thoracic ganglia or link left and right protocerebra.

Fig. 2 schematizes the protocerebrum, flanked by the lobula complexes. The lower arrows represent lobula complex channels (CHAN 3a,b) to the ventrolateral protocerebrum and subsequent arrows indicate pathways to the ventral cord. Ocellar fibres (ocl) also converge on this output pathway.

There is, however, a second-region of protocerebral neuropil that deserves some comment. This is situated in the dorsal hemispheres and receives terminals from some species of wide-field lobular tangentials, or terminals from small diameter fibres derived from optic foci (Strausfeld, 1976). These terminals arborize amongst interneurons destined either for the neurosecretory pathways to the retrocerebral complex (NCC I and NCC II) or for median bundle pathways destined for the vegetative centre of the brain, the tritocerebrum. Thus, outputs from the lobula complex show a major divergence; to the premotor central nervous system (via the ventrolateral protocerebrum) and to the visceral nervous system.

2.6.4 Convergence of Columnar Pathways on to Noncolumnar Output Neurons

Regular mosaic arrangements of neurons are a characteristic feature of first- to third-order neuropils of the visual system as well as second- and third-order neuropils of the chemosensory system (Strausfeld, 1976). But at the level of fourth-order regions all resemblance to columnar structures is lost. Fig. 3 schematizes this structural "summation" of columnar neurons at the level of the lobula. Inputs to the lobula (LO) from the medulla (M) are shown as a regular palisade that projects across the second optic chiasma. In the medulla the columns represent the retinotopic projection from the retina and

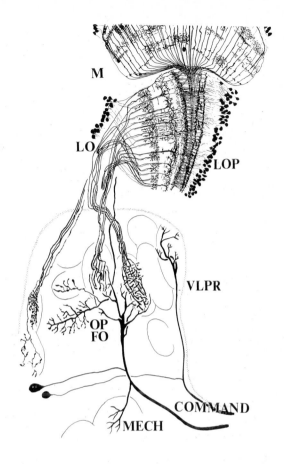

Fig. 3. Convergence of columns from lobula to protocerebrum (see text). LOP: lobula plate; VLPR: ventrolateral protocerebrum; Op.Fo.: optic foci of small-field lobula-derived terminals

lamina. Two "species" of lobula neurons are shown here, also arranged as a regular palisade. Note that the spacing of the axis fibres (axons) of each type of third-order interneuron coarsens the retinotopic periodicity of the lamina and medulla. However, the dendrites of lobular neurons extensively overlap and do in fact cover the whole of the projected mosaic from more peripheral levels. Axons of lobula columnar neurons (LCNs) do not confer the mosaic any deeper in the system. Instead, all axons of one species of LCN converge and pass as a coherent bundle to special and characteristic regions of the ventrolateral protocerebrum. These glomerular regions of neuropil are called optic foci (OP FO). In this diagram three foci have been sketched, two of which are invaded by a large fibre-diameter element that projects to the ventral cord (COMMAND). This sequential arrangement and convergence of lobular neurons is characteristic of all orders of insects. Note too, that the visual pathways to the thoracic ganglia also interact with other sense-modality neuropils (MECH).

2.6.5 Mosaics

Sections cut tangential to the axes of through-going interneurons show that at any level between the incoming fibres from the retina and out-

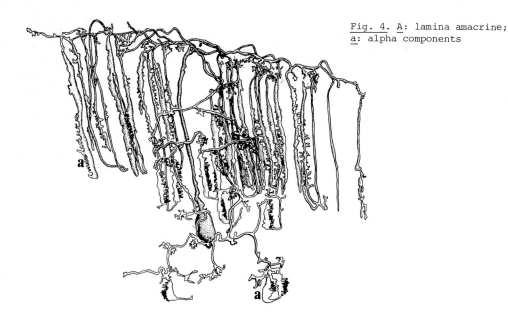

Fig. 4. A: lamina amacrine; a: alpha components

going fibres from the lobula complex the neuropil is organized as an extremely regular array of mosaic components. These are the cross sections of axons that define each column and the regular network of lateral fibres derived from them (Fig. 6). In addition to through-going elements there are both anaxonal neurons (amacrines) and tangential fibres (Sect. 10). And although the former appear individually to have rather asymmetric relationships with columns, the sum of a particular type of amacrine cell does in fact provide a very regular system of fibres that are homogenously arranged throughout a planar stratum. For example, although the amacrine cell of the lamina (Fig. 4) appears to invade a nonsymmetric constellation of columns (its alpha processes are mapped in Fig. 5) en masse Golgi impregnations (Sect. 8) show that all the lamina amacrines together have a homogenous distribution in all columns and at all loci in the first synaptic neuropil. This distribution was originally proposed from electron microscope investigations, where each cartridge was observed to have a dotation of six amacrine components (Campos-Ortega and Strausfeld, 1973). This example may serve for a proposition: namely, that each stratum is composed of homogenous sets of processes across its planar extent. Structural heterogeneity is discernable through the depth of the system.

2.6.6 Precision of Neural Arrangements

The more or less invariant locations of particular forms of insect interneurons in the brain has been recognized since the early researches of Cajal and Sanchez (1915) and Zawarzin (1913). And, recently, "aimed" impregnation of cell tracts by cobalt chloride iontophoresis has demonstrated that insect neurons have exceedingly consistant relationships with one another even though there may occur errors of growth (see C. Goodman, 1975). Many interneurons of ganglionic chains have been shown to have characteristic mappings in the neuropil which can be observed time and again in many individuals of the same species (Cohen and Jacklet, 1967; Tyrer and Altman, 1974). Iterative arrange-

Fig. 5. The lamina amacrine of Fig. 4 mapped on to optic cartridges: rings indicate distribution in alpha components, from one amacrine, in lamina mosaic (see text). Bielschowsky reduced silver preparation; distance separating each cartridge = 12 um

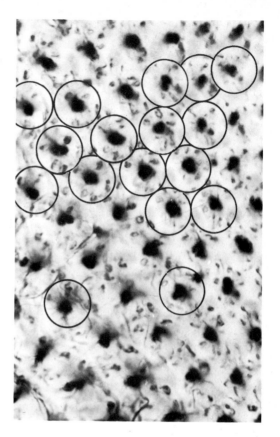

ments of neurons in a single individual were first demonstrated by Cajal (1910) in the optic lobes. And perhaps the best example of neural arrangements, and their precision, is to be found in the structure of the lamina. In the fly there are eleven types of interneurons which invade every optic cartridge and which have precise geometrical relationships with the cyclic order of receptor terminals. In addition, the loci of synapses within one cartridge is representative of the loci of synapses on all cartridges. This remarkably precise neuropil is outlined in the final section of this account.

Other examples of iterative arrangements of nerve cell shape and location, between individuals of the same species, can be referred to the giant visual neurons and associated elements of the lobula plate (Pierantoni, 1973; Dvorak et al., 1975; Hausen, 1976a,b; Strausfeld, 1976) and to the forms and coordinate loci of descending neurons from the brain (Strausfeld, 1976).

2.6.7 Retinotopic Projections and Columnar Mappings

Light and electronmicroscopical observations of the retina and lamina
have shown that each column of the first synaptic region is defined
by the terminals of six "short receptor" endings. These terminals are
the axonic prolongations of a set of outer segments that in the fly
share a common optical axis but are arranged in a rhomboidal pattern
in six different ommatidia (Vigier, 1907a,b; Kirschfeld, 1967; Braitenberg, 1967; Scholes, 1969). Thus, the six short receptors derived from
a single ommatidium project to six different optic cartridges. Electron microscopy observations have also shown that a pair of long visual fibres project from each ommatidium and bypass the set of six terminals in the lamina which have the same optical alignment (Campos-
Ortega and Strausfeld, 1972a). The axons prolongate to the medulla,
across the first optic chiasma, in parallel with (and as part of a
coherent bundle of) first-order interneurons that derive their information from optically equivalent short receptors. Degeneration experiments reveal that there is a strictly retinotopic projection of the
retinal mosaic on to the outer layer of the medulla (M(o)). Diffusion
of Procion yellow into cut retinae also reveal homotopic projections
of long visual fibres and, under certain circumstances, projections
of monopolar neurons from the lamina.

The retinotopic arrangement of long visual fibre terminals and endings
of lamina derived interneurons defines the mosaic of columns in the
medulla. This mosaic is carried through all its strata without permu-
tations by small-field columnar second-order interneurons and short
axoned cells which link the medulla's upper and inner layers (M(o)
and M(i)).

Outgoing axons from medulla columns project across the second optic
chiasma into the lobula and into the lobula plate. There is a chiasma
between the medulla and lobula but none between the medulla and lobula
plate nor between the lobula and lobula plate which face each other.
Sets of three rows of medullary columns contribute to each sheet of
the second chiasma (Braitenberg, 1970), but there is a redistribution
of these aggregates into both neuropils of the lobula complex so that
even here the retinotopic mosaic is well preserved (Preissler, 1975).
The outer surfaces of the lobula and lobula plate lie opposite each
other (Fig. 3) and equivalent columns in both regions are also opposite. These are connected by small-field elements derived from the
outer (UGVN) layer of the lobula which project to two layers of GVN
fibres in the lobula plate (see Sect. 11).

2.6.8. Identicality between Columns

The implicator elements of each lamina and medulla column, the visual
fibres, define at least one order of columnar identicality. Structural
investigations of the lamina have shown that each cartridge contains
the same subset of interneurons, centrifugal cells and amacrine processes. And apart from graded differences of structural nuances from
front to back of the lamina - such as the number of dendrites from the
so-called L4 neuron, the diameters of optic cartridges and relative
diameters of monopolar cell axons (Braitenberg and Hauser-Hohlschuh,
1972; Strausfeld and Campos-Ortega, 1973b) - each column represents
any other in terms of the cell types it contains and the functional
contiguities between them (see Campos-Ortega and Strausfeld, 1972a,
1973, 1976; Strausfeld and Campos-Ortega, 1973a,b, 1976). The lamina,

however, is relatively simple compared to medullae. There are only 11 types of elements (not counting receptor terminals or long visual fibre axons) of which six (in each cartridge) are centripetal interneurons. Two are centrifugal terminals that are derived from each column of the medulla, two are wide-field centrifugal tangentials and the eleventh type is the lamina amacrine.

In the medulla, however, there are 60 forms of neurons, including through-going centripetal cells and amacrines. Electron microscopy observations of medulla columns have resolved at least 34 axon profiles in each column in M(o) and there are scores of satellite profiles associated with each column axis at any level in this region (Campos-Ortega and Strausfeld, 1972b).

It is beyond the capacity of the observer to detect anything more than a general orderliness of medulla columns from electron microscopy. And without sampling very many Golgi impregnated neurons by electron microscopy it is not possible, at this level of resolution, to determine whether or not one column is representative of all columns. However, there are other procedures which will at least give a very good indication if the medulla neuropil is characteristically homogenous across its lateral extent.

All neurons can be strictly classified according to the shape and in-depth relationships of their branching patterns as well as from their lateral spreads through columns; the kinds of specializations that emanate from their processes and axons; and the forms and depths of terminals - in the case of through-going elements - in the lobula complex.

If enough brains are treated by the Golgi method the same range of neuronal "Gestalts" or "Species" will be encountered time and again. This means that there is little or no overlap between the structural types of neurons, and it is exceedingly rare to detect a chimeric gestalt which demonstrates a structural ambivalence between two forms of previously classified cells. This feature was recognized long ago by Cajal (1909) and he was, to my knowledge, the first worker to recognize both the enormous variety of cell forms as well as the finiteness of this variety. The whole gamut of medullary neurons, both through-going as well as tangential and intrinsic cells, has been documented for one species of insect, <u>Musca domestica</u> (Strausfeld, 1976).

More difficult than drawing up lists of cell species is to obtain some reliable information about their populations in the neuropil, with reference to the columns. One way of approaching this is to be able to recognize a neuron species by a staining technique whose selectivity is not "random" but which relies on the presence or absence of some organelle such as neurofibrillae or neurotubuli. One method is the Bodian reduced silver procedure which selects argyrophilic components in neurons and which, under special conditions, can be adjusted to reveal only a proportion of the stainable neurons, such as all cells with large diameter dendrites, or all neurons with wide fields in the lobula. These can be resolved as standing out against the weak background coloration "noise". L4 neurons in the lamina were the first to be recognized as having a one-to-one relationship with columns from Bodian identification (Strausfeld and Braitenberg, 1970). Similarly, lobula columnar elements and some through-going medullary neurons have been separately resolved (see Figs. 6, 8, 9). Thus, by a comparison of Bodian preparations and, say, a hundred Golgi-treated optic lobes it is possible to determine the percentage "hit-chance" by the Golgi procedure on to certain cell types. Similarly, reconstruc-

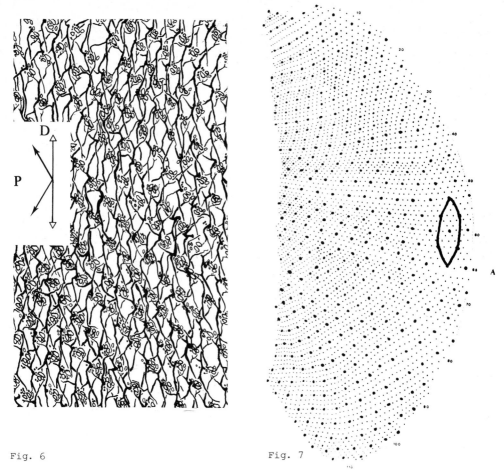

Fig. 6   Fig. 7

Fig. 6. Planar section, parallel to a medulla stratum, showing the regular mosaic arrangement of axons and dendrites of class 2 transmedullary cells. Dendrites oriented vertical and $45°\pm$ to horizontal axis of retinotopic mosaic. D: dorsal; P: oblique posterior axes of the mosaic of columns

Fig. 7. Planar map of retinotopic input groups to lobula complex (each small dot indicates a group of terminals derived from a single medulla column) and the regular array (large dots) of one type of LCN (1:6 columns)

tions of large diameter neurons, such as lobula plate GVNs (Pierantoni, 1973) from semi-thin serial sections, will give precise information about the number and forms of some neurons in the brain. Again, by searching through Golgi-impregnated brains and counting how many of these cell types were impregnated (among the total number available for impregnation) it is possible to estimate "hit-chance" by the Golgi procedure. With single impregnations the worst marksmanship by the Golgi method is into lamina amacrines (hit-chance of 0.004%) and the best marksmanship is into large diameter neurons, such as descending fibres[1] (hit-chance of 0.05%). The average hit-chance for small diameter medulla or lobula interneurons is 0.02%. Assuming that each medulla (two per brain) contains 2.800 columns (equal to the number of facets of the compound eye), and assuming that any one type of through-going cell occurs in each column anywhere in the medulla's

---

[1]There are, in fact, only unique pairs of large descending neurons in each brain (Strausfeld, 1976).

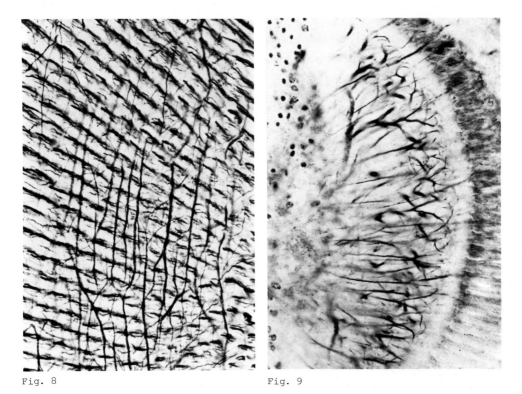

Fig. 8   Fig. 9

Fig. 8. Vertically oriented, supraperiodic, small field tangentials in M(i), destined for the posterior slope

Fig. 9. Selective Bodian preparation of two species of LCNs (see text)

lateral extent, then given a hit chance of 0.02% there should be impregnated 112 cells of the same type in 100 brains.

The majority of through-going neuron species have, on average, been observed one per optic lobe, or slightly more often. Thus, it seems that the medulla should have a homogenous lateral structure. This impression is strengthened by observations of brains that have undergone a triple recycling through the Golgi procedure, thereby increasing the total number of impregnated cells per lobe by a factor of 100! This is called en masse staining from which it is relatively simple to observe, firstly, if a species of neuron can be revealed anywhere across the medulla's lateral extent and, secondly, if the minimal distance between two neurons of the same species is consistently equal to, or a multiple of, the distance separating adjacent columns. It turns out that only eight types of small-field through-going neurons to the lobula are spaced one to every n-columns (1:2 or, alternatively, 1:6). And two of the short axoned cells which link M(o) to M(i) are spaced one to every two columns. These coarse spacings are, however, regular through the medulla's lateral extent. The advantage of en masse impregnations is that they clearly indicate that the great majority of small-field through-going interneurons are, in fact, arranged one per column and also show that T-cells between the medulla and lobula plate, or between the UGVN layer of the lobula and lobula plate, are arranged as pairs per column (see Sect. 12).

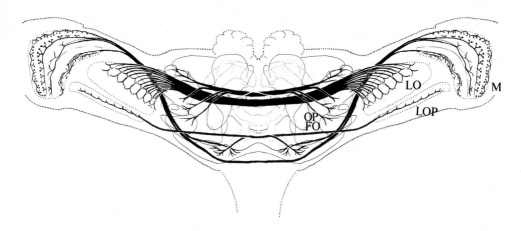

Fig. 10. Heterolateral pathways of the optic lobes and ventro-lateral protocerebrum (see text). Abbreviations as for Fig. 1

In summary, the general rule for the medulla is that small-field in-terneurons have a synperiodic arrangement (1: column) whereas lobula columnar elements (LCNs) and wide-field medulla interneurons have supra-periodic arrangements of one per n columns, where n = 2 or, more usually, 6 (cf. Figs. 6, 7 and 9). A further discrepancy between the medulla and lobula is that the latter region gives rise to far fewer cell types and represents the first stage of a massive conver-gence of parallel pathways to descending fibres from the brain.

2.6.9 Heterolateral Connections of the Visual System

The most peripheral heterolateral pathways between the two optic lobes are via wide-field neurons of the medullae. The most outstanding form of these cells is a pair of tangential elements whose dendritic fields extend through an entire stratum just beneath the level of L2 monopo-lar cell terminals from the lamina. The axons of these cells traverse the brain, via the posterior optic tract, and in the contralateral medulla each gives rise to a giant telodendritic field at the surface of its neuropil. The projection of their tract is shown in Fig. 10 and their dendrites and terminals are shown at level M(o) in Fig. 11.

Section 2 drew attention to medulla tangentials that projected from M(i) to the ipsilateral posterior slope of the protocerebrum. There are also sets of M(i) tangentials which project to the contralateral posterior slopes to meet contralateral terminals of lobula plate-derived GVNs (Strausfeld, 1976). Also shown in Fig. 10 are fourth-order interneuron pathways between optic foci of the ventrolateral protocerebrum and at least one species of lobula columnar neuron which projects a coarsened retinotopic mosaic from the ipsilateral lobula in to the contralateral region.

Heterolateral connections mediated by wide-field lobula tangentials to the protocerebrum have been observed rarely; the majority of these give rise to dendrites in the contralateral protocerebrum and termi-nate as diffuse endings in the ipsilateral lobula. Possibly such ele-ments may represent input into the lobula derived from another sensory

modality and may be the cell types that structure the lobula in eyeless insects, such as worker termites (which lack lamina and medulla),
or in nearly eyeless "sine oculis" Drosophila (where the laminae and
outer layer of the medullae are extremely reduced).

Lastly, both lobula plates are linked by heterolateral neurons. These
cells, which have been exclusively classified from Procion yellow
electrophoresis (Hausen, 1976), serve to connect two characteristic
layers of the lobula plate. The first type invades its outer layer at
the level of "horizontal motion sensitive" neurons (Dvorak et al.,
1975; Hausen, 1976) and projects across the front of the brain to the
same level in the contralateral lobula plate. The second type invades
the deeper layer at the level of "vertical motion sensitive" neurons
and projects posteriorly across the brain into the contralateral plate.
In addition, both these functional levels (see Sect. 12) receive wide-
field terminals or dendrites of cells that are derived from or project
to the contralateral posterior slope at the level of "horizontal" or
"vertical" motion detectors (GVNs). These too have been classified
from Procion yellow diffusion (Hausen, 1976a,b).

2.6.10 Tangential Cells

Wide-field neurons serve to relay information from, or project information to, whole or very large-field aggregates of the projected retinotopic mosaic. Tangentials can be classified into three major groups:
(1) those that serve to link optic lobe regions heterolaterally, such
as between the lobula plates or between the outer levels (M(o)) of the
medullae; (2) cells that link central neuropils of the brain centri-
fugally to the lobula or outer stratum of the medulla; and (3) cells
that serve structural "summatory" pathways from the inner layer of the
medulla (M(i)), or the lobula plate, and project to the posterior
slopes of the protocerebrum. The last neurons are classified as third-
order visual interneurons and, like columnar elements from the lobula
(to optic foci), represent the first stage in convergence of parallel
pathways from optic lobes to descending fibres from the brain. Fig. 11
illustrates some typical forms of tangential arborizations in the
medulla and lobula complex. Note that in the lobula some forms of tangentials appear to have complex asymmetric fields. These are confined
to local regions of the projected mosaic and serve here as the first
example of planar inhomogeneities across the columnar structure of
the lobes. In Fig. 11 the lobula plate is represented by giant tangentials (GVNs). These, and similarly shaped unique giant visual neurons
(UGVNs), provide us with the first functional correlations between
neuronal architectures and behaviour of the insect visual system.

2.6.11. The Giant Visual Neurons: Tangential Cells of the Lobula Plate

2.6.11.1 The retinotopic mosaic

In Section 7 it was stated that the retinotopic arrangement of columns
was carried through the medulla and into both the lobula and the
lobula plate by second-order interneurons. In Fig. 7 each small point
represents one column of the projected medulla mosaic. The whole array
of columns is a precise map of the lobula plate of Musca domestica.
And since it lies opposite the lobula it can be used for mappings of
large field or supraperiodic lobula elements as well. In Fig. 7 one
species of LCN has been mapped on to the mosaic: its distribution is
such that each cell coincides with every third column along a recti-

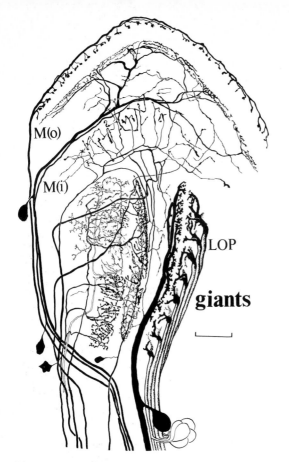

M(o)

M(i)

LOP

giants

Fig. 11. Tangentials of optic lobe neuropils (see text)

linear coordinate. For this species of LCN neuron it can be seen that the whole lateral extent of the lobula (approximately 500 μm x 350 μm) contains a regular distribution of axons. However, this is not the case for the very large outgoing elements from the lobula plate; and it is among this class of cells that the first indications of "column redundancy" become apparent.

2.6.11.2 Giant Visual Neurons: Shapes and Mappings

The lobula plate is a layered neuropil whose strata are defined by two major cell types. The first of these lie superficially in the neuropil and give rise to a system of dendrites that fan out postero-anteriorly across columnar inputs. Reconstructions from serial 2-3 μm sections have demonstrated that there are three such "horizontal" cells and that the axons of these elements project to the posterior slope of the ipsilateral protocerebrum (Pierantoni, 1973). Golgi studies on the same cell types in Musca also reveal the presence of three of these neurons and have resolved the entire spread of their dendrites. Fig. 12 illustrates the upper (north) and lower (south) horizontal neurons and outlines the dendritic spread of the middle (equatorial) element. Their distribution in the lobula plate may, at first sight, not appear remarkable.

However, scrutiny reveals that their dendritic spread interacts only with a proportion of the projected retinotopic mosaic. A posterior strip of columns, running from north to south, is outside their domains. This "redundancy" of columns is also exhibited by a fourth form of giant neuron in Fig. 13 which is a unique element, meaning that there is only one in each lobula plate. The neuron invades a frontal, vertical, strip of projected columns which, if mapped back on to the compound eye, roughly coincides with the anterior quarter of the retina,

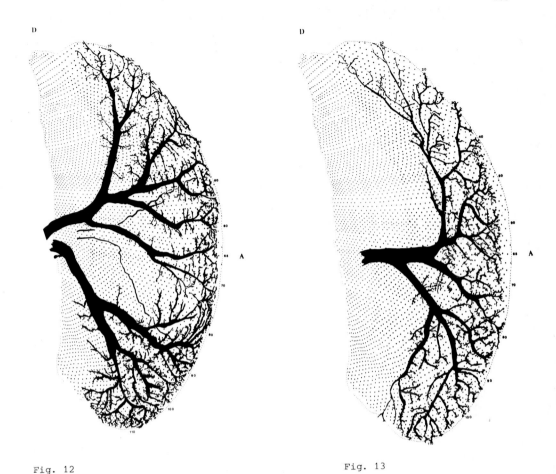

Fig. 12

Fig. 13

Fig. 12. North and south giant horizontal tangentials mapped on to retinotopic pattern of the lobula plate inputs from medulla columns

Fig. 13. A lobula plate "unique" giant visual neuron (see text)

including the region of binocular overlap. One other interesting feature of this neuron is that it shows a density gradient of dendrites which does not match the packing density gradient of the lobula plate columns. Thus it might be supposed that there are more presynaptic sights on to the cell in the lower anterior segment of the projected mosaic than in the dorsal anterior segment. It is worthwhile comparing this gradient with the structures shown in Fig. 12. Here, too, there appears to be a dorso-ventral gradient; however, differences in the packing of dendritic branches can be referred to the closer packing of columns beneath the equator of the lobula plate.

The second layer of the lobula plate is defined by a set of "vertical" neurons, each of which covers a small dorso-ventral strip of the projected mosaic. Reconstructions from Musca (Pierantoni, 1973) showed that each lobula plate contained nine vertical neurons. In Calliphora eight vertical neurons have been resolved which are shown to the right of Fig. 14. Together, the set of vertical elements covers the whole of

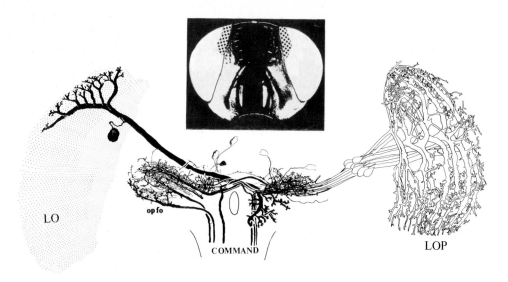

Fig. 14. A lobula UGVN, eight lobula plate vertical giant neurons, and descending neuron dendrites with cobalt sulphide. The lobula UGVNs have been mapped bilaterally on to the retina (see text)

the projected mosaic. However, the eight cells can be subclassified into three subsets: three anterior cells, three medial cells and a pair of posterior cells. Typically, each of the anterior cells is mapped on to a true vertical strip of columns (Fig. 15, left hand diagram, right hand map). The medial cells, however, are mapped on vertical rows of columns only in the lower half of the lobula plate; in the upper half they are mapped on to a hooked-shaped distribution of columns, recurved posteriorly. One of these is schematized in Fig. 15 (left hand diagram, left hand map). The posterior elements also cover vertical strips of columns in the lower half of the plate. However, in the upper half they are mapped on to a hook-like distribution of columns, recurved anteriorly. In addition, each of the eight vertical cells shows a gradient of dendritic density that is reminiscent of the unique neuron in Fig. 13 (mapped in the right hand diagram of Fig. 15).

The axons of the vertical neurons also pass to the posterior slope. There, both the axon terminals of vertical and horizontal cells are met by a giant telodendrion derived from the outer stratum of the contralateral lobula, termed the "unique giant neuron layer". The neuron in question is shown against the lobula columns in Fig. 14, left hand optic lobe. Only one of these elements exists in each lobula and assuming there is no lateral permutation of the relayed retinal mosaic, via projection mode 1 (Sect. 12), then the dendrites can be referred back to the dorsal region of overlap between the two compound eyes. This is mapped in the central inset of Fig. 14.

The giant and unique giant neurons are the largest cells of the visual system. They represent the most abrupt convergence pathway from third-order neuropil on to the dendrites of descending fibres to the thoracic ganglia. Some of these are included in Fig. 14 ("COMMAND"): the set shown here has been derived from retrograde diffusion of cobalt chloride, introduced into the dorsal neuropil of the thoracic ganglia.

Fig. 15. Maps of anterior and medial vertical giants (<u>left hand diagram</u>) and the lobula plate unique giant (<u>right hand diagram</u>)

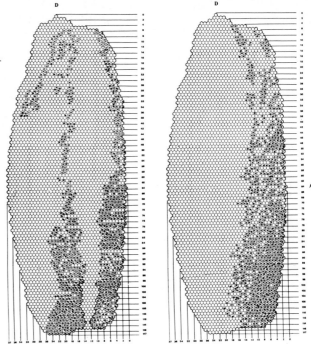

2.6.11.3 Giant Visual Neurons: Some Functional Correlates

Two major types of motor reactions contribute to the course-control behaviour of short bodied insects, such as the fly. The first of these is called the optomotor turning response (see Reichardt, 1970) which can be elicited in flight, or during walking, by rotatory or asymmetric translatory movements of the visual panorama. This reflex enables the fly to maintain forward course; or, in response to vertical motion, its altitude. The second type of behaviour is that of fixation of a simple pattern in the visual field (Reichardt, 1973; Poggio and Reichardt, 1973; Land and Collett, 1974). This mode of behaviour can also be elicited by motion up-down (Wehrhahn and Reichardt, 1975).

Both types of behaviour have been described in terms of phenomenological theories that are able to predict the reaction of the fly to more complex stimulus configurations than moving gratings or simple geometrical forms, and which can be used for defining a hypothetical behavioural repertoire, under certain limiting conditions (Poggio and Reichardt, 1975). Both types of behaviour have also led to the formulation of specific models for minimal interaction channels between parallel retinotopic pathways from the sensory periphery (Reichardt, 1961; Thorsen, 1966; Götz, 1968). These fall into two classes, each composed of wiring diagrams of various complexity: the first class includes symmetrical interactions between pairs of neighbouring pathways which, for example, contain a delay and multiplication station and function by cross-correlation of both channels (Reichardt, 1961). The second class includes asymmetric interactions between parallel channels, which use "and" or "and not" gates, based on the Barlow-Levick model (1965) for motion detection in the rabbit retina (see Collett and Blest, 1966; Kien, 1975). Variations of this model, which can account for the absence of inhibition in the null direction, have been proposed from studies of Locust motion detecting neurons (Kien, 1975). However, both classes of models can account for turning responses and the minimal distance between their parallel channels will depend upon the minimal sizes of the receptive fields of motion detector units. Up to the present time non-directional motion detector elements in the medulla of the fly have been elicited by restricted motion over a neighbouring pair of ommatidia (Mimura, 1975) and behavioural experiments have shown that neighbouring of subadjacent channels are sufficient to release a turning response (Kirschfeld, 1972).

Numerous fibre patterns have been resolved by anatomical investigations of the lamina and medulla (Strausfeld and Braitenberg, 1970; Strausfeld, 1976) which, given the appropriate synaptic connectivities, would meet not only the connectivity demands of the cross-correlation motion detection model (Götz, 1968; Reichardt, 1970) but could be fitted to almost any model that has been hitherto envisaged. And even if the proposed connectivities (or at least part of them) are shown to exist, this alone cannot prove that one particular set of neurons computes the response. A case in point is the lamina (Sect. 13) where appropriate structural connectivities do exist for cross-correlation but, according to the functional characteristics of identified lamina neurons, there seems to be no motion computation at this level.

Both intra- and extracellular recordings in the optic lobes have, however, demonstrated that the structural substrates for many types of motion detection - direction insensitive, direction sensitive, transitory, sustained, and so on - must occur in the second synaptic region as well as in the lobula regions (see Collett, 1970, 1972; Mimura, 1972, 1975). One "complex" motion detection element has been resolved by intracellular recording and procion marking in the fly: namely, a Y-cell that projects to both the lobula and lobula plate (DeVoe and Ockleford, 1976). A similar unit, which projects to the topologically equivalent neuropil regions in the bee, has been identified by Menzel and Kien (pers. comm.) as a complex colour-coded unit subserving a rather wide field of the receptor mosaic.

Although correlation studies of function and form of columnar elements are still in their infancy one may make some rather simple assumptions about the general architectures of neuropils that compute the optomotor response or mediate fixation. Firstly, if the optomotor reflex can be evoked by the appropriate stimulus over the whole eye, and supposedly in any quadrant of its visual field, it is unlikely that lateral pathways which compute this detection would be situated in levels where mosaics show abrupt inhomogeneities. Secondly, since the strength of the optomotor response is proportional to the number of facets which receive the moving pattern (McCann and McGinitie, 1965), it is likely that columnar motion detectors are summed together on to wide-field elements in the optic lobes or directly on to descending neurons to the ventral cord, via columnar projections into the brain. Thirdly, if the initial optomotor computations are performed by neighboring channels then, according to Götz (1968), interconnections would be ideally arranged if they were oriented along the horizontal or vertical axes of the retinotopic mosaic or, obliquely, 45° mirror symmetric to the horizontal axis. The first and third assumption could be fulfilled in the medulla; but the second assumption is difficult to reconcile with levels and forms of tangentials in M(o) - though not in M(i) - that serve to connect second-order neuropil heterolaterally. It is more likely that convergence of elementary directionally sensitive detectors occurs at the level of descending neurons, via LCNs, or onto giant visual neurons in the lobula plate.

Fixation reactions to horizontal (Wehrhahn and Reichardt, 1975) or vertical patterns (Reichardt, 1973) demands computation of positional information. This response does not demand interactions between neighbouring channels but may be computed by single channels (Poggio and Reichardt, 1975). Also, the fixation response can only be elicited by the lower half of the eye (see refs. above), beneath its optical equator (Kirschfeld, 1967). This implies that (1) computations of positional information may possibly be referred to antero-posterior gradients of columnar elements rather than to patterns of cross-linkage between columns, and that (2) some basic up-down asymmetry is likely to exist at some level of the projected mosaic.

In the previous section it was suggested that columns had equivalent structures across the lamina and across the medulla. The only clear evidence for asymmetries at the light microscope level could be referred to gradual changes of size or dendritic number of one or another species of neuron. For example, graded differences of cross sectional diameters of lamina cartridges and some monopolar cell axons (Braitenberg and Hauser-Hohlschuh, 1972); or, graded differences of the number of L4 monopolar neuron dendrites (and thus, graded differences of post-synaptic sites to amacrines) from front to back of the lamina (Strausfeld and Campos-Ortega, 1973b). There are more frontal inputs to L4 than posteriorly, or dorsally, or ventrally: the gradient vector finds its sink anteriorly at, or just beneath, the eye's equator. Another example of a gradient is found in the medulla, where one stratum of line amacrines (Strausfeld, 1970) diminishes in size from the front to the back of the projected mosaic.

In the third synaptic region we see, for the first time, clear evidence of abrupt inhomogeneities that can be referred to some third-order interneurons. Whereas the whole medulla mosaic is projected onto the lobula plate, there is at least one unique giant neuron that gives rise to a dense dendritic subfield which interacts only with the frontally projected columns and is biased beneath the equator of the eye (Fig. 13). This means that its dendritic density is maximal from the fourth horizontal row of columns downwards. Extrapolated back on to a map of the compound eye (Fig. 15, right hand map) the upper borderline of the subfield fits quite well to the upper borderline of the eye region which mediates optimal fixation; that is, 18° below the equator and within the anterior quarter of the visual field (Reichardt, 1973; Wehrhahn and Reichardt, 1975).

This rather crude match between a behavioural mapping and a single neuron is not intended to claim that this type of UGVN necessarily mediates the fixation response. The example merely serves to illustrate that initial columnar computations, such as those for fixation, could be performed beneath the whole retina but only a special fraction of them may impinge upon an outgoing wide-field element as exemplified by the mapping of this UGVN onto the retinotopic mosaic. However, the above example at least demonstrates one point; at the level of the lobula plate there are anatomical features which can be selectively investigated in order to formulate propositions about peripheral columnar computations. Moreover, the mappings and forms of these neurons can help us in the design or interpretations of electrophysiological experiments. Let us now turn to some concrete examples.

Extracellular recordings from the lobula plate vicinity have revealed several "classes" of spiking neurons that respond to moving gratings and which fulfill some of the characteristics of theoretical channels in the optomotor response. These elements characteristically showed pronounced directional selectivity to motion and their response was modulated with respect to the angular velocity of the pattern, its contrast and its spatial wave length (Bishop and Keehn, 1967). These elements were also detected centrally in the brain, near the posterior protocerebrum, and at least one class of motion "detectors" was also recorded from the contralateral lobe (McCann and Dill, 1969; McCann and Foster, 1971).

Recently, intracellular recordings, followed by injection of Procion yellow, have revealed both the shapes of several species of lobula plate neurons <u>and</u> their response characteristics. The following is a rather condensed and necessarily incomplete summary of the functional organization of interneurons that link lobula plate with the ipsi- or contralateral posterior slope or link both left and right lobula plates. All the elements outlined below have large field dendritic or telodendritic spreads in the third-order visual neuropil. The term "wide-field" refers to a specific cell type observed in <u>Phaenicia sericata</u> and <u>Sarcophaga</u>. The summary is derived from papers, and personal communications, of Dvorak et al. (1975) and Hausen (1976a,b).

1. Most lobula plate neurons revealed by Procion yellow have been subsequently resolved by Golgi methods or transynaptic cobalt fillings (Strausfeld and Obermayer, 1976). Procion fillings, silver impregnations, and serial reconstructions have resolved two major sets of giant visual interneurons; these are the set of (three) giant horizontal cells (H-cells) in the superficial layers of the lobula plate and the set of 8-9 giant vertical cells (V-cells) deep in the lobula plate.

2. Other large-field neurons that invade the lobula plate neuropil are structurally associated with two main strata containing H and V cell dendrites (termed H and V layers).

3. The response characteristics of H and V cells cannot be simply summarized because some discrepancies exist between results obtained from <u>Phaenicia</u> (Dvorak et al.) and <u>C. erythrocephala</u> (Hausen). Dvorak et al. describe H-cells as exhibiting epsps, ipsps and also spike activity. Dvorak (pers. comm.) was not able to correlate a specific anatomical type with a particular receptive field and directionality except to classify the neurons, in a very general sense, as horizontal motion sensitive or, in the case of V-cells, as vertical motion sensitive. On the other hand, Hausen finds that H-cells exhibit epsps, ipsps, and conduct by graded potentials although H-cells show some spike-like activity in axon terminal collaterals. These spike-like signals are superimposed on membrane DC-shifts (pers. comm.) and are not regenerative propagating action potentials. Hausen also has demonstrated that the three H-cells (north, equatorial and south) of <u>Calliphora</u> are directionally sensitive: ipsilateral progressive motion (from front to back) elicits depolarization, regressive motion elicits hyperpolarization. H-cells also show a response to contralateral regressive motion and the binocular reaction is thus to a turning panorama. Relays from the contralateral lobula plate are via heterolateral elements.

4. V-cells of <u>C. erythrocephala</u> (Hausen, 1976a,b; Hengstenberg, pers. comm.) were usually found to selectively respond (by graded potentials) to ipsilateral downward motion. Homologous neurons in <u>Phaenicia</u> also exhibited noisy graded potentials but showed ambivalent responses to directional movement of gratings (Eckert; Dvorak, pers. comm). However, although V-cells are likely to be involved in motion detection, a second system of small diameter "mimetic" vertical elements were resolved in <u>Phaenicia</u> (v-cells) that exhibited spiking axon potentials triggered by ipsilateral downward motion (Dvorak et al., 1975; Dvorak, pers. comm).

5. Although the giant H and V neurons are the outstanding landmarks in the neural architecture they may represent two of possibly as many as fifteen different species of large-field motion sensitive elements in the lobula plate. The mimetic v-cells are a case in point and their response characteristics (see Dvorak et al.) are almost identical to a class of centrifugal heterolateral neurons (CV-cells) that project from the terminals of V-cells, across the rear surface of the brain, into the contralateral lobula plate where they give rise to vertically oriented wide-field telodendria (Hausen, 1976a,b).

6. In addition to the ipsilateral H-cells, Dvorak et al. (1975, Fig. 5A,B) describe wide-field horizontal sensitive neurons (WF-cells) that branch throughout the greater part of the projected retina mosaic. Similar homolateral neurons are known from Golgi studies of Musca (Strausfeld, 1976) but have not yet been found in C. erythrocephala.

7. Associated with H-cell terminals are a pair of mossy dendritic arborizations that are linked, by rather large diameter axons, to two telodendritic fields at the outer surface of the lobula plate and which, together, cover the three H-cell dendritic fields (see Hausen, 1976a; Fig. 1A). According to this author CH-cells conduct by graded potentials; contralateral progressive motion hyperpolarizes, motion in the reverse direction depolarizes. Dvorak et al. also illustrate horizontal motion sensitive elements (h-cells) that mimic the giant H-cell dendritic pattern but which are centripetal and which have smaller axon diameters. These authors claim that there are three such elements (h-cells) in each lobula plate; and though their reconstructions are at variance with the structure of the CH-neuron (cf. Dvorak et al., Fig. 10), it cannot yet be excluded that there may be interspecific structural variations between close relatives. For example, in the hover fly Eristalis tenax there appear to be at least 4 H-cell profiles and 3 CH elements (Strausfeld and Hausen, in prep.).

8. Hausen's studies of C. erythrocephala have classified the forms, projections, and response characteristics of several heterolateral neurons which conduct by spiking axon potentials (one of these, the CV-cell, has already been mentioned in paragraph 5). Some examples of these include horizontal motion sensitive neurons that link both lobula plates and which have dendrites and terminals in the neuropil at the same level as H-cell dendrites. A pair of these neurons (H1-cells) have been shown to reside in the brain; their dendrites cover most of the lobula plate's mosaic, omitting a small ventral area, and the telodendria cover the anterior part of the projected retinal mosaic of the contralateral plate. These neurons show spiking axon potentials and respond to regressive ipsilateral motion of gratings, but are silent to contralateral motion in either direction. They correspond to heterolateral elements of Phaenicia (the class II - 1 units) recorded extracellularly be McCann and Dill (Hausen, 1976a). An analogous heterolateral neuron subserves the vertical layer of the left and right lobula plates. In addition to the H1-neurons, two additional types of horizontal motion sensitive neurons should be mentioned: these cells (H2 and H3) serve to link the lobula plate to regions of the contralateral posterior slope at the level of H-cell endings and CH-cell dendrites. One of them (H3) most likely acts as an inhibitory input onto the CH-cells (Hausen, 1976b).

The relative positions of giant H-cells and V-cells (see Pierantoni, 1973) indicate that layers of lobula plate neuropil may possibly be specialized for functional separations of "horizontally" and "vertically" components of motion detected in, or of, the panorama. The discrete segregation of all large-field neurons into two major structural layers (Hausen, 1976b) and the correspondence between layer and direc-

tional sensitivity (i.e. horizontality or verticality) adds credance
to the suggestion that more peripheral elemental small-field motion
detectors, such as those detected by DeVoe and Mimura, project retino-
topically into the lobula plate and there be segregated into two functional levels. Examination of the columnar inputs to the lobula plate,
in particular the T-cells, reveals three levels of termination; at the
H-stratum, at the V-stratum and between H and V (see Fig. 19, sect. 12;
and Strausfeld, 1976). Interestingly, one large-field neuron, the V2-
cell linking both lobula plates, resides in this intermediate layer
and responds to ipsilateral upward and forward motion (Hausen, 1976b).

Although it is not yet known if the computations for verticality and
horizontality are performed between retinotopic channels in the medulla, rather than by elements in the lobula plate, it is clear that
mappings of the retinotopic mosaic onto dendrites of giant visual
neurons and associated large-field elements correspond quite well to
some of the response features of procion filled cells. For example,
the three H-cells are mapped onto quasi-horizontal rows of columns.
And at least the most frontal triplet of the V-cells map rather accurately on to vertical rows of the projected columnar mosaic. The
same holds for the mimetic v-cells (Dvorak et al.). The more posterior
V-cells, whose upper dendrites are mapped almost horizontally, show
ambivalent responses to moving gratings that are reminiscent of H-cells
but are unaffected by contralateral horizontal motion (Hausen, 1976a).

Elements that are localized to a fraction of the projected mosaic, and
which have precisely oriented dendritic fields, might provide suggestions for future experiments. For example, are the horizontal giant
neurons less effected by gratings moved across the back part of the
retina which, according to the columnar projections, should only
sparsely interact with H-cell dendritic mappings? Or do posteriorly
projected gratings evoke as strong responses as the same width of pattern projected on to anterior ommatidia? If they do then it would seem
that retinotopic mosaics are either topographically smudged or the
elementary motion detectors are spaced further apart than in adjacent
columns. Let us take another example. What are the possible relationships between ipsilateral horizontal and vertical giant visual neurons
and the lobula unique neuron (Fig. 14)? These cell types have very interesting relationships: the unique cell terminal meets contralateral
endings of H- and V-cells and all three cell-types end among one set
of descending neurons destined for thoracic ganglia. Cobalt chloride
will (in very specific experimental conditions) migrate from dendrites
of retrograde filled descending fibres into axon terminals of these
three types of visual neurons (Strausfeld, 1976) and in many examples
of second-order diffusion all eight vertical giants, all three horizontal giants and the unique lobula giant were shown in their entirety
in both the optic lobes (Strausfeld and Obermayer, 1976). Although
second-order diffusions still present difficulties of interpretation
it may be proposed that these three sets of elements interact together,
at least onto a common set of descending neurons. Possibly the unique
lobula element serves to modulate some relays from the H- and V-cells
in response to very specific stimuli perceived by a restricted area
of the dorso-frontal retina (see inset, Fig. 14).

2.6.12 Projection Modes on to Convergent Pathways of CHAN 3

The evidence from en masse Golgi impregnations of Musca and Calliphora
suggests that their medullae are, in a general sense, constructed homogenously across their lateral extent (Sect. 8). Although not every

species of medulla neuron has been related to all the columns there
is no evidence to suggest that the architecture of this neuropil radically changes from one segment of the project retinotopic mosaic to
another. This is in contrast to some Orthopterous optic lobes.

From the number of neuronal species it would be expected that if each
cell type was represented, at some level, in each column of the medulla
the structure of a column should be very complex indeed. This is the
case, even at the level of the light microscope (Campos-Ortega and
Strausfeld, 1972b), and it might be asked, with good reason, whether
or not a step by step study of the medulla at higher resolution is a
desirable way of resolving functional contiguities. This kind of study
has already been performed on the first synaptic neuropil (the lamina)
by means of electron microscopy of Golgi-impregnated neurons (see
Campos-Ortega and Strausfeld, 1972a,b, 1973, 1976). However, the number of elements in the first neuropil is relatively small (Sect. 13).

In principle, it would be possible to trace each type of neuron in the
medulla and map all of its postsynaptic sites with other profiles by
means of Golgi electron microscopy. Ultimately the investigator would
be in possession of pieces of an immensely sophisticated jigsaw puzzle
and these would have to be fitted to each other in some logical fashion.
But without some functional clues the procedure of fitting these pieces, from periphery to centre, will almost certainly lead to an intel-
lectual impasse at every stage of the research. Thus, the selection of
neurons which are to be assayed by electron microscopy, or which are
to be fitted into a connectivity schema, should be strictly limited
to sets of few elements. The selection of these sets should also be
based on some functional integer.

One criterion for an initial selection is suggested here. It refers
to the types of columnar components that project on to the dendrites
of specific response-type elements that leave the lobula complex for
the brain. An alternatively selection, which will not be discussed,
it to examine the connectivities of neurons that coincide with large-
field tangentials, situated peripherally in the medulla, which respond
to movement of gratings in the visual field and which project not to
descending neurons but to equivalent levels of the contralateral neuropil (see Collett, 1970).

The following description of the "skeletal" medulla defines three major (and one minor) projection modes (MPM) that can be subjected to
anatomical and physiological correlation. Fig. 16 illustrates some
basic components of the medulla: columnar elements transected by stratified elements, most of which are amacrines. Fig. 17 summarizes the
major forms of amacrines which contribute to at least 24 well defined
strata (Campos-Ortega and Strausfeld, 1972b). These strata can be
coarsely grouped into three major amacrine levels which coincide with
the receptive parts of neurons that belong to some or all of the projection modes. Two levels occur in M(o), the outermost coincident with
L2 and L3 monopolar cell terminals from the lamina, the inner coincident with terminals of L1, L4, L5 and long visual cells.

Fig. 18 illustrates the amacrine levels alongside the major classes
of through-going columnar neurons. Class i are intrinsic cells that
either connect the outer amacrine level of M(o) with M(i), or the
outer and inner amacrine levels of M(o) with M(i). For the sake of
clarity, further discussion of i cells will be omitted. Class 1-3 are
transmedullary (Tm) cells. Class 1 (Tm 1) is the only type of through-
going element to have a lateral dendritic spread equal to, or less
than, the distance separating two columns. It is also the only Tm
cell to terminate exclusively in the superficial UGVN stratum of the

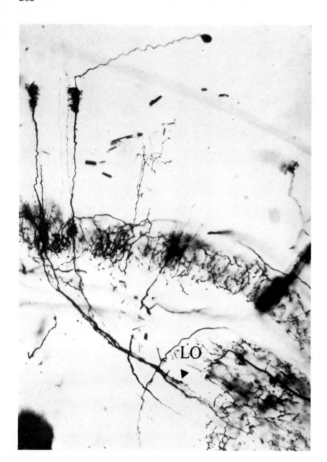

Fig. 16. Pair of Tm1 neu-
rons of MPM 1, with ter-
minals in the UGVN layer
of the lobula (<u>Pieris</u>
<u>brassicae</u>): see text

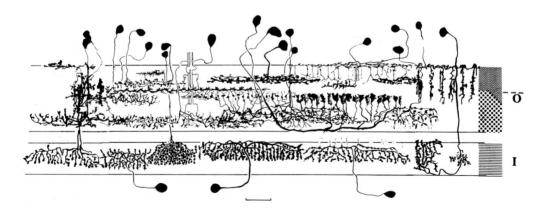

Fig. 17. Forms of amacrines and major amacrine levels of the medulla. Level O (outer
2/3 of medulla) contains two major amacrine layers at the terminals of L2 monopolars
from the lamina and at the terminals of long visual fibres and L1 and L3 monopolars

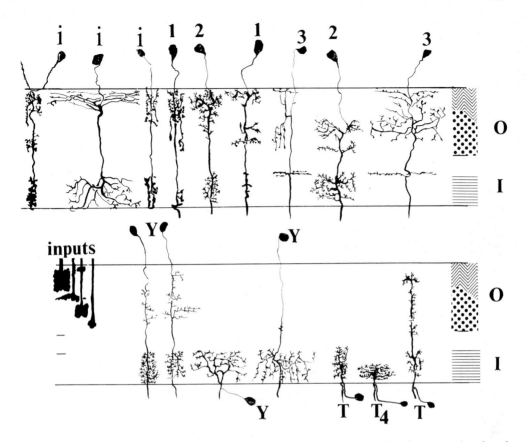

Fig. 18. Dendritic trees of columnar through-going neurons and major amacrine levels. Inputs are, from left to right, terminals of L2, L3, L1 (from lamina) and R7 (from retina). Transmedullary cells to lobula (Tm) are numbered types 1-3, corresponding to their projection modes (Fig. 19). T-cells to lobula are labelled <u>T</u>; to lobula plate T4. Y-cells project to the lobula and lobula plate. <u>i</u>: columnar intrinsic cells

lobula. Each Tm 1 gives rise to a simple varicose axon collateral at the inner face of the medulla and gives rise to a simple swollen terminal in the lobula. Pairs of T5 cells project from each terminal into the V and H layers of the lobula plate. Pairs of T4 cells project from Tm 1 axon collaterals to the same levels. The dendrites of Tm 1 cells receive at least part of their input from L2 monopolar endings and are situated exclusively within the most peripheral amacrine layer of the medulla. The columnar set of neurons, L2, Tm 1, T4 (x2) and T5 (x2) is termed the medulla projection mode 1 (MPM 1) and is schematized in Fig. 19, top left.

Class 2 Tm cells have dendritic spreads that extend through a surround of at least six columns. Their arborizations in the medualla may be mono- or multistratified, but their dendrites are invariably spiny and reside within both amacrine levels of M(o). Stratified class 2 Tms give rise to an inner cluster of arborizations at M(i). All class 2 Tms project to the LCN levels of the lobula. This is schematized in Fig. 19, top right, and is termed the MPM 2.

The second projection mode is twinned by a third, much in the same way as first-order centripetal neurons from the lamina (L1, L2, L3) are twinned by the L4 and L5 monopolars: the former derive their input directly from receptors, the latter derive their input from receptors via amacrines (see Sect. 13). Like L4 and L5 neurons, the class 3 Tm cells of MPM 3 give rise to thin varicose processes peripherally which do not appear to derive their inputs from incoming centripetal elements, but via amacrines. Also, like the L4 monopolar cells of the lamina, class 3 Tms give rise to a regular network of axon collaterals which are distributed peripherally in M(i). And, finally, the analogy between class 3 Tms and L4, L5, can be taken a stage further; the MPM 3 terminals in the lobula either mimic the collateral network in the medulla (cf. L4 terminals, axon collaterals; Strausfeld and Campos-Ortega, 1973b) or they appear to be mimetic to large diameter terminals derived from MPM 2 (cf. the terminals of L5 with L1 and with L2; Strausfeld and Campos-Ortega, 1976).

Several other forms of neurons are shown in Fig. 18 which pass to the lobula complex. Two of these are T-cells that arborize outwards through M(i) or through M(i) or M(o) and which project parallel to MPM 2 and MPM 3. These elements cannot be conveniently ordered to either mode and are, for the present, omitted from the discussion. It suffices to draw the analogy between these T-cells and the centripetal T-cell from the lamina to the medulla (see Sect. 13). The other elements in Fig. 18 are Y-cells. These must be considered separately since they project to both the lobula plate and to the lobula, in that order. This projection mode (MPM Y, Fig. 19, lower right) is derived from all levels of M(o) and M(i), or exclusively from M(i). The axon collaterals of Y-cells in the lobula plate extend into both its functional layers indicating that horizontal and vertical giant neurons may receive some identical Y-inputs.

The projection modes of Fig. 19 represent the simplest schema of the medulla neuropil and its projections. And even though the elements in each mode may, individually, be demonstrated to have very different response characteristics one might expect that the set of neurons which contribute to one mode may share some common functional parameters. For example, do the neurons of MPM 1 relay simple messages about direction and velocity of the moving panorama to the H and V layers? Or, do all MPM 2 elements respond to more complex small-field motion events? One feature is certain: MPM 1 and MPM 2 are carried through into the brain by two very different structural pathways (Fig. 1, Sect. 2) via two very different types of convergent elements, the GVNs (and UGVNs) and LCNs, respectively. In general, these lead separately to two populations of neurons which descend into the ventral cord.

2.6.13 The Blue print of a Neuropil: the Lamina

2.6.13.1 Introduction

Lattice arrangements of neuropil consist of sequentially arranged strata derived from the lateral processes of columnar elements and networks of amacrine processes. Connectivities are basically two dimensional matrices which are serially arranged through the depth of the neuropil. The planar strata are essentially homogenous until the level of the lobula complex. Different connectivity networks, at different levels, define the inhomogeneity in the third dimension.

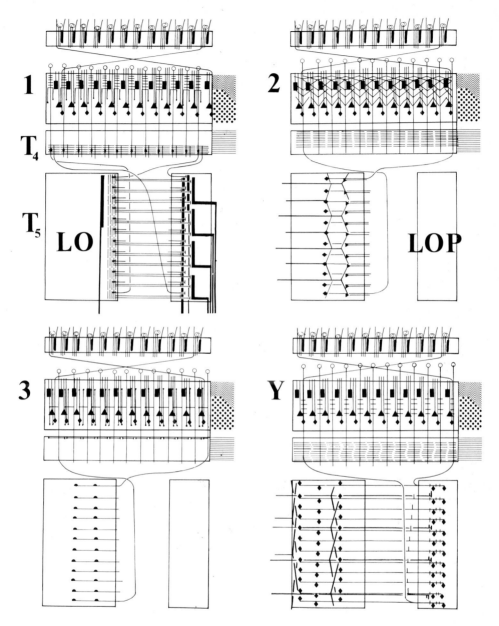

Fig. 19. Medulla projection modes (MPM; 1,2,3,Y). Major amacrine levels shown to the right of each diagram. Inputs to the lamina and, from the lamina, to the medulla, are the same for each mode. Medulla inputs shown here are L2 (<u>outer</u>), L1 and R7 (<u>inner</u>). See text

This section very briefly outlines some strata arrangements of one neuropil, the lamina (which is an embryonic outgrowth of the medulla anlage), and summarizes the Golgi-electronmicroscopy studies of J.A. Campos-Ortega and myself (see references).

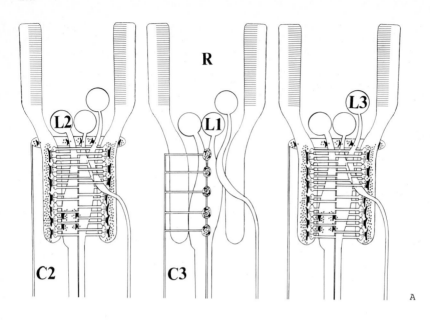

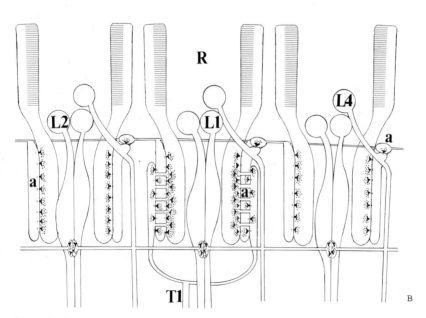

Fig. 20A,B. (A) Intracartridge connectivities; (B) Intercartridge connectivities
of the lamina. Note cardinal position of T1, postsynaptic to receptors as well as
pre- and postsynaptic to amacrine alpha components. Note also the covert stratum
of L2 presynaptic to receptors and L1 in (A)

I shall first outline the basic cellular organization, with reference
to Figs. 20A and 20B. Secondly, I will describe the arrangement of
strata and, thirdly, list the known synaptic connections as a set of
connectivity channels.

Finally, to end this account, I shall briefly summarize the known functions of identified neurons (Procion studies) in order to demonstrate that at least some meet the requirements of the connectivities, and vice versa. Others, including responses from unmarked cells, can be fitted to structural contiguities. The physiological data is taken from Zettler and Järvilehto (1971, 1972), and from studies of the dragonfly lamina by Laughlin (1973-1975, 2.1 of this vol.).

2.6.13.2 Structure

Receptors with identical optical axes end at a single cartridge and form a hollow column of terminals. This is the basic component of an optic cartridge in Dipterous laminae. A pair of first-order interneurons project axially through the column (L1, L2; the former ends deep in the medulla, the latter ends superficially). These two neurons receive their primary inputs from receptors via radial arrangements of dendritic spines throughout the depth of the lamina. Although L1 and L2 appear to be identical there is a remarkable difference between them: L2 is presynaptic to receptors and to the L1 monopolar cells within the inner part of the lamina (Campos-Ortega, 1972; Campos-Ortega and Strausfeld, 1976). This important reciprocal channel will be dealt with in more detail below.

A third first-order monopolar cell, L3, is also postsynaptic to all receptor terminals of a cartridge, but only within the outer third of the lamina. At this level L1, L2 and L3 appose single presynaptic sites in receptors to form triads of profiles. Elsewhere L1 and L2 appose single presynaptic sites in receptors as dyads.

Fig. 20A summarizes these centripetal connections and illustrates the two types of small-field centrifugal terminations from the medulla which invade each cartridge (wide-field centrifugals are not considered here). The first type of centrifugal (C2) is presynaptic to L1, L2 and L3 above the level of their dendrites. The second (C3) is presynaptic to L1 and L2 (and some receptor endings) at the level of their dendrites.

Fig. 20B illustrates interconnections between cartridges. These are mediated by amacrine cells (a) and L4 monopolars. Also shown at the center cartridge is a basket-like dendritic arborization of a centripetal T-cell (T1): each cartridge receives six swollen tendrils derived from three amacrine cells. These, and the receptor terminals, are embraced by the basket T-cell arborization. Basket dendrites of T1 are postsynaptic to receptors terminals of a single cartridge and receive inputs from a surround of receptor terminals in other cartridges (and inputs from their parent cartridges) via wide field amacrines. Amacrines also contribute to a second type of lateral interaction with the L4 monopolar cells: they derive inputs from receptors and are subsequently presynaptic to dendrites of L4. L4 monopolars give rise to axon collaterals which are presynaptic to L1 and L2 of their own and two adjacent cartridges. L4 collaterals are also pre- and postsynaptic to each other. The collaterals form an orderly rectilinear latticework - whose coordinates are oriented 45° mirror symmetric to the horizontal axis of the retinotopic mosaic - beneath the whole of the lamina.

There is a third form of lateral interaction, via the midget L5 cell, which has been omitted here. In principle this neuron receives a sparse amacrine input at the surface of the lamina, and an input from the second form of wide-field centrifugal (also not shown). L5 cells do not interconnect cartridges.

2.6.13.3 Strata

The synaptic arrangements, which are summarized in Fig. 20, occur in each cartridge and at the same precise levels. There are five strata, and two modes of synaptic interaction are common to four of them. These are the postsynaptic relationships of L1 and L2 to receptors and the C3 centrifugal ending, and rows of reciprocal synapses between T1 amacrine tendrils, and receptors.

Four strata can be recognized from light microscopy of Golgi-impregnated cells. The first is a plexus of amacrine and wide-field centrifugal processes which coincide with the minute dendritic spines of the L5 monopolar cells (not shown, see Strausfeld and Campos-Ortega, 1972) and the terminals of C2 small-field centrifugals; these all reside above the level of monopolar dendrites. The second stratum is at the level of L3 and L4 dendrites and the level of presynaptic bags of the first type of wide-field centrifugal (not shown) that is presynaptic to L1, L2, L3 and receptors. The third stratum resides between the inner limit of L3 dendrites and the network of L4 collaterals, which make up stratum five. The fourth stratum is covert and can only be detected by electron microscopy: it is, namely, the layer of reciprocal synapses between receptors and L2, and presynaptic sites of L2 on to L1. This important distinguishing feature between the axial monopolars (Campos-Ortega, 1972) was not recognized by other authors who described L1 and L2 as being identically connected.

At first sight it seems that the strata are so arranged as to provide channels for sequential operations on the receptor input to L1, L2 and L3. This proposal finds support from the types of connectivity pathways and the electrophysiology of the lamina elements.

2.6.13.4 Connectivity Channels

These can be conveniently listed as intra- and intercartridge pathways (1-5 and 6-10, respectively).

1. Convergence of receptors to interneurons: homogenous receptor set of six elements on to a centripetal monopolar.

2. Divergence from a set of identical receptors (CHAN Ra; R1-R6): R1-R6 presynaptic to L1, L2, L3 monopolars and T1 (parallel projection of quadruple lines to the same medulla column).

3. Feedback pathways from the medulla, from single columns, via C2 and C3: C2 presynaptic to L1, L2, L3 above their level of receptor input. C3 presynaptic only to L1 and L2, at the level of receptor input.

4. Reciprocal interactions within a single cartridge: R1-R6 presynaptic to L2, L2 presynaptic to receptors. R1-R6 presynaptic to T1, T1 pre- and postsynaptic to amacrines.

5. Cross-talk between monopolar cells of a single cartridge: L2 presynaptic to L1.

6. Lateral interaction pathways: Receptors to amacrines, amacrines to L4. L4 presynaptic to L1 and L2, via axon collaterals. L4 to L4.

7. Interactions between input to a single cartridge element and input from lateral cartridges: R1-R6 to T1; R1-R6 to T1 from surround, via amacrines.

8a. Serial synapses in lateral pathways: amacrines to amacrines.

8b. Reciprocal synapses in lateral pathways: amacrines to amacrines: L4 to L4 collaterals.

9. Wide-field centrifugals: Lam tan 1 to L1, L2, L3. Lam tan 2 to L5 monopolars (also amacrines presynaptic to L5). Set 9 not shown in Fig. 20.

2.6.13.5 Electrophysiology and Connections of Lamina Neurons

Considered here are only those findings that either confirm structural projections (A), or combine intracellular recordings with procion dye injection, or are unambivalently derived from intracellular recordings of unmarked neurons known to be first-order interneurons (monopolars) on the basis of their synaptic delays (1-2 ms) (see Järvilehto and Zettler, 1973; Laughlin, 1974, 1975).

1. Confirmation that receptors which share the same visual field converge to a single cartridge was derived from intracellular intracartridge recordings of _Musca_ lamina (Scholes, 1966) and from large monopolar cells of Odonatous laminae (Laughlin, 1973). Convergence to L2-type neurons has been shown for both dragonflies and Diptera, and recordings from these cells, in both species, are here generalized.

2. At low light intensities, convergence of signals from several receptors (usually a set of six) on to a single L2-type monopolar cell are averaged postsynaptically; this serves to increase the signal-to-noise ratio. The signal-to-noise amplification has been correlated with the structural arrangement of rows of synapses in receptors which appose rows of dendrites of L2-type neurons (Laughlin, 1973). Apart from reducing receptor noise at low intensities signal amplification enhances sensitivity and increases acuity at low light intensities (Laughlin, 1973). As yet the L3 neuron has not been recorded and identified by procion. However, due to the paucity of dendrites (and hence postsynaptic sites) it might be expected that this cell is more likely to monitor receptor noise to the medulla.

3. L2-type monopolars have relatively short dynamic ranges which can be shifted from a dark to a light adapted state over a range of 4 log units (Laughlin, 1975). The plateau of the triphasic L-cell response decrements to the adapting light but can be held more or less constant over a 2 log unit intensity range. Laughlin also showed that contrast changes which induce modulations of the L-potential were independent of background intensities. This author pointed out that the net result of the neural adaptation enables L2-type monopolars to detect contrast changes at many possible ambient intensities and thus standardize information about contrast intensity to higher order medulla interneurons.

4. Evidence for lateral inhibitory (inter- rather than intracartridge inhibition) have been derived from Odonata and Diptera (Zettler and Järvilehto, 1972; Järvilehto and Zettler, 1973; Laughlin, 1974).

Recordings from Dipterous cartridges demonstrate that in the outer cartridge levels receptor elements have visual fields of between 5° and 15° (see Zettler, this vol.) whereas recordings from L2 or L1 monopolars near the inner margin of the lamina, or in the chiasma, show acceptance fields smaller than single receptors. In addition, the only recording and marking of a T1 neuron showed that this centripetal element had a visual field less than receptors or L1 and L2 neurons. In dragonflies angular sensitivity of L2-type monopolars is also found

to be enhanced by a process of lateral inhibition from neighbouring
cartridges (Laughlin, 1974), whereby off-axis stimulation diminishes,
or abolishes (depending on the angular divergence), the plateau phase
of the L-cell response. The net result of a lateral inhibitory mechanism is to increase the visual acuity of the retinotopic relays from
receptors to medulla interneurons.

Laughlin showed that neural control mechanisms could be related to
responses of L2-type monopolars: one intracartridge, maintaining constant contrast efficiency over intensity ranges of at least 2 log units
the other intercartridge, mediating lateral inhibition. From the anatomy of the lamina, and with respect to only the L1 and L2 neurons,
there are five major substrates that could possibly mediate intracartridge inhibition; reciprocal feedforward of L2 to receptors (and
L1); feedback presynaptic inhibition, via C2 and wide-field centrifugals from the medulla to the "necks" of L1, L2 (+ L3) and receptors;
feedback via exclusively small-field centrifugals (C3) to L1 and L2
at the level of receptor inputs. The amacrine substrate is likely to
mediate lateral inhibition via L4 collaterals to neighbouring cartridges and it would be expected that since L4 dendrites receive inputs
from several amacrines (with large domains) then illumination of receptors some distance (more than subneighbours) from the recorded
L2 neuron should produce strong inhibitory effects. A comparison can be
made between electrophysiology-derived adaptation schema (Chap. 2.1;
Fig. 8) and synaptology. Clearly the lamina provides intriguing possibilities for structural-functional studies, even though the neurons are
rather small! For example, only one small field centrifugal pathway
(C2) is shared by L1, L2 and L3; and should L3 show adaptive mechanisms
identical to L1 and L2 then one possible way to interpret this would be
that C2 (or the wide-field centrifugal that is also shared by L1-L3)
mediates this mechanism. The T1 basket centripetal and amacrine cell
relationships are also interesting since analogous neurons can be found
in several other orders of insects - including Odonata, Hemiptera, Cole-
optera and Lepidoptera. These cells coincide at the same levels in the
lamina and have fibre specializations that are strongly reminiscent of
the same elements in Diptera. In _Musca_ the basket dendrites of T1 appose rows of presynaptic sites in receptors, as do the tendrils (alpha
processes) of amacrines. Also, there are rows of reciprocal sites between T1 and amacrine cells (Campos-Ortega and Strausfeld, 1973). Thus,
one might speculate that the relationships between T1 and amacrines
could possibly mediate an adaptation pathway both within a cartridge -
via L4 to L1 and L2, and from amacrines to T1 - as well as between cartridges, again by the L4 neurons but via their axon collaterals to
neighbouring cartridges.

Acknowledgments. I am very grateful to Claus Hausen for comments and
criticisms of initial drafts of this account, in particular sections
11 and 12. I also thank David Dvorak for additional information about
some properties of lobula plate neurons in _Phaenicia_ and _Sarcophaga_;
and Gisela Pouvatchy for preparation of this manuscript.

2.6.14 References

Barlow, H.B., Levick, W.R.: The mechanism of directionally selective units in
    rabbit's retina. J. Physiol. _178_, 447 (1965)
Bishop, L.G., Keehn, D.G.: Neural correlates of the optomotor response in the fly.
    Kybernetik _3_, 288-295 (1967)
Braitenberg, V.: Patterns of projections in the visual system of the fly.
    I. Retina-lamina projections. Exp. Brain Res. _3_, 271-298 (1967)

Braitenberg, V.: Ordnung and Orientierung der Elemente im Sehsystem der Fliege. Kybernetik 7, 235-242 (1970)

Braitenberg, V., Hauser-Hohlschuh, H.: Patterns of projection in the visual system of the fly. II. Quantitative aspects of second order neurons in relation to models of movement perception. Exp. Brain Res. 16, 184-209 (1972)

Cajal, S.R.: Nota sobre la estructura de la retina de la Mosca. Trab. Lab. Invest. Biol. Univ. Madr. 7, 217-257 (1909)

Cajal, S.R., Sanchez, D.: Contribucion al conocimiento de los centros nerviosos de los insectos. Parte I. Retina y centros opticos. Trab. Lab. Invest. Biol. Univ. Madr. 13, 1-168 (1915)

Campos-Ortega, J.A.: Postsynaptism in R1-R6: A gain control mechanism. N.W. Biocyb. Bull. 4, 1-3 (1973)

Campos-Ortega, J.A., Strausfeld, N.J.: The columnar organization of the second synaptic region of the visual system of Musca domestica L. I. Receptor terminals in the medulla. Z. Zellf. 124, 561-585 (1972a)

Campos-Ortega, J.A., Strausfeld, N.J.: Columns and layers in the second synaptic region of the fly's visual system: The case of two superimposed neuronal architectures. In: Information Processing in the Visual System of Arthropods. Wehner, R. (ed.). Berlin-Heidelberg-New York: Springer 1972b, pp. 31-36

Campos-Ortega, J.A., Strausfeld, N.J.: Synaptic connections of intrinsic cells and basket arborisations in the external plexiform layer of the fly's eye. Brain Res. 59, 119-136 (1973)

Campos-Ortega, J.A., Strausfeld, N.J.: The synaptic organisation of second order and centrifugal neurons in the 1st synaptic region of the fly's visual system (Manuscript) 1976

Cohen, M.J., Jacklet, J.W.: The functional organisation of motor neurons in an insect ganglion. Phil. Trans. Roy. Soc. London 252B, 561-572 (1967)

Collett, T.S.: Centripetal and centrifugal visual cells in medulla of insect optic lobe. J. Neurophysiol. 33, 239-256 (1970)

Collett, T.S.: Connections between wide-field monocular and binocular movement detectors in the brain of a hawk moth. Z. Vergl. Physiol. 75, 1-31 (1971)

Collett, T.S., Blest, A.D.: Binocular, directionally selective neurons, possibly involved in the optomoter response of insects. Nature (London) 212, 1330-1333 (1966)

DeVoe, R.D., Ockleford, E.M.: Intracellular responses from cells of the medulla of the fly, Calliphora erythrocephala. Kybernetik 23, 13-24 (1976)

Dvorak, D.R., Bishop, L.G., Eckert, H.E.: On the identification of movement detectors in the fly optic lobe. J. Comp. Physiol. 100, 5-23 (1975)

Götz, K.G.: Flight control in Drosophila by visual perception of motion. Kybernetik 4, 199-208 (1968)

Goodman, C.: Anatomy of locust ocellar interneurons: constancy and variability. J. Comp. Physiol. 95, 186-201 (1974)

Hausen, K.: Struktur, Funktion und Konnektivität bewegungsempfindlicher Neuronen in der Lobula Plate der Schmeissfliege, Calliphora erythrocephala. Doctoral Dissertation. Eberhard-Karls-Universität Tübingen. Augsberg: Blasaditch 1976a

Hausen, K.: Functional characterization and anatomical identification of motion sensitive neurons in the lobula plate of the blowfly Calliphora erythrocephala. Z. Naturforsch. 1976b (in press)

Järvilehto, M., Zettler, F.: Electrophysiological-Histological studies on some functional properties of visual cells and second order neurons of an insect retina. Z. Zellforsch. 136, 291-306 (1973)

Kien, J.: Motion detection in locusts and grasshoppers. In: The Compound Eye and Vision of Insects. Horridge, G.A. (ed.). Oxford: Clarendon 1975, pp. 410-422

Kirschfeld, K.: Die Projektion der optischen Umwelt auf das Raster der Rhabdomere im Komplexauge von Musca. Exp. Brain Res. 3, 248-270 (1967)

Kirschfeld, K.: The visual system of Musca: studies on optics, structure and function. In: Information Processing in the Visual System of Arthropods. Wehner, R.: ed.). Berlin-Heidelberg-New York: Springer 1972, pp. 61-74

Land, M.F., Collett, T.S.: Chasing behaviour of houseflies (Fannia canicularis): A description and analysis. J. Comp. Physiol. 89, 331-357 (1974)

Laughlin, S.B.: Neural integration in the first optic neuropile of dragonflies. I. Signal amplification in dark-adapted second-order neurons. J. Comp. Physiol. 84, 335-355 (1973)

Laughlin, S.B.: Neural integration in the first optic neuropile of dragonflies.
III. The transfer of angular information. J. Comp. Physiol. 92, 377-396 (1974)

Laughlin, S.B.: Receptor and Interneuron light-adaptation in the dragonfly visual
system. Z. Naturforsch. 30C, 306-308 (1975)

McCann, G.D., Dill, J.C.: Fundamental properties of intensity, form and motion perception in the visual systems of Calliphora phaenicia and Musca domestica.
J. Gen. Physiol. 53, 385-413 (1969)

McCann, G.D., Foster, S.F.: Binocular interactions of motion detection fibres in
the optic lobes of flies. Kybernetik 8, 193-203 (1971)

McCann, G.D., McGinitie, G.F.: Optomotor response studies of insect vision.
Proc. R. Soc. London 163B, 369-401 (1965)

Mimura, K.: Neural mechanisms subserving directional selectivity of movement in
the optic lobe of the fly. J. Comp. Physiol. 80, 409-437 (1972)

Mimura, K.: Units of the optic lobe, especially movement perception units of
Diptera. In: The Compound Eye and Vision of Insects. Horridge, G.A. (ed.).
Oxford: Clarendon 1975, pp. 426-436

Pierantoni, R.: Su un tratto nervoso nel cervello della Mosca. In: Atti Della
Prima Riuniore Scientifica Plenaria (Camogli, dicembre 1973). Soc. Ital. Biofis.
Pura e Applicata, 231-249 (1973)

Poggio, T., Reichardt, W.: A theory of the pattern induced flight orientation of
the fly Musca domestica. Kybernetik 12, 185-203 (1973)

Poggio, T., Reichardt, W.: Nonlinear interaction underlying visual orientation
behaviour of the fly. Cold Spring Harb. Symp. Quant. Biol. 40, (in press, 1975)

Preissler, M.: Struktur des inneren Chiasma im Sehsystem der Fliege Musca domestica.
Manuscript: Diploma Dissertation, Eberhardt-Karls-Universität zu Tübingen (1975)

Reichardt, W.: Autocorrelation; a principle for the evaluation of sensory information by the central nervous system. In: Sensory Communications.
Rosenblith, W.A. (ed.). M.I.T. John Wiley, pp. 303-318

Reichardt, W.: The insect eye as a model for analysis of uptake, transduction,
and processing of optical data in the nervous system. In: The Neurosciences 2.
Schmitt, F.O. (ed.). New York: Rockefeller Univ. 1970, pp. 494-511

Reichardt, W.: Musterinduzierte Flugorientierung. Verhaltensversuche an der Fliege
Musca domestica. Naturwissenschaften 60, 122-138 (1973)

Scholes, J.: The electrical response of the retinal receptors and the lamina in the
visual system of the fly Musca. Kybernetik 6, 149-162 (1969)

Strausfeld, N.J.: Golgi studies on insects. Part II. The optic lobes of Diptera.
Phil. Trans. R. Soc. London 258B, 175-223 (1970)

Strausfeld, N.J.: Atlas of an Insect Brain. Berlin-Heidelberg-New York: Springer 1976

Strausfeld, N.J., Braitenberg, V.: The compound eye of the fly (Musca domestica):
connections between the cartridges of the lamina ganglionaris.
Z. Vergl. Physiol. 70, 95-104 (1970)

Strausfeld, N.J., Campos-Ortega, J.A.: Some interrelationships between the first
and second synaptic region of the fly's (Musca domestica L.) visual system.
In: Information Processing in the Visual System of Arthropods. Wehner, R. (ed.).
Berlin-Heidelberg-New York: Springer 1972

Strausfeld, N.J., Campos-Ortega, J.A.: L3, the 3rd 2nd order neuron of the 1st visual ganglion in the "neural superposition" eye of Musca domestica.
Z. Zellforsch. 139, 397-403 (1973a)

Strausfeld, N.J., Campos-Ortega, J.A.: The L4 monopolar neurone: a substrate for
lateral interaction in the visual system of the fly Musca domestica.
Brain Res. 59, 97-117 (1973b)

Strausfeld, N.J., Campos-Ortega, J.A.: Synaptic relationships underlying adaptation
and lateral inhibition beneath the insect retina. In prep., 1976

Strausfeld, N.J., Obermayer, M.: Resolution of intraneuronal and transynaptic migration of cobalt in the insect visual and central nervous systems.
J. Comp. Physiol. In press, 1976

Thorsen, J.: Small-signal analysis of a visual reflex in the locust.
Kybernetik 3, 41-66 (1966)

Tyrer, N.M., Altman, J.S.: Motor and sensory flight neurons in a locust demonstrated
using cobalt chloride. J. Comp. Neurol. 157, 117-138 (1974)

Vigier, P.: Mécanisme de la synthèse des impressions lumineuses receuillies par les
yeux composés des Diptères. C. R. Acad. Sci. Paris 63, 122-134 (1907a)

Vigier, P.: Sur la réception de l'exitant lumineux dans les yeux composés des
    Insectes, en particulier ches les Muscides. C. R. Acad. Sci. Paris 63,
    633-636 (1907b)
Wehrhahn, C., Reichardt, W.: Visually induced height orientation of the fly
    Musca domestica. Biol. Cyber. 20, 37-50 (1975)
Zawarzin, A.: Histologische Studien über Insekten. IV. Die optischen Ganglien
    der Aeschna Larven. Z. Wiss. Zool. 108, 175-257 (1913)
Zettler, F., Järvilehto, M.: Decrement-free conduction of graded potentials along
    the axon of a monopolar neuron. Z. Vergl. Physiol. 75, 402-421 (1971)
Zettler, F., Järvilehto, M.: Lateral inhibition in an insect eye.
    Z. Vergl. Physiol. 76, 233-244 (1972).

2.7 Structure and Function of the Peripheral Visual Pathway in Hymenopterans

R. Wehner

2.7.1 Introduction

Sensory systems provide an animal with an appropriate internal representation of the outside world. To understand what this internal representation looks like the most straight forward strategy would be to unravel the neurophysiological mechanisms by which sensory information is processed. The visual systems of insects are favorable objects for such a neurophysiological approach. The positions viewed by each individual receptor cell can be readily defined according to the highly-ordered geometry of receptors within the compound eyes. Furthermore, in the peripheral visual systems of insects, the photoreceptor cells are well separated anatomically from the other classes of higher order neurons and are not included into a single retina as in vertebrates. Finally, insects contain four to five orders of magnitude fewer receptor cells than does man, although the highly developed visual system of a fly or a bee is capable of abstracting information on position, movement, contrast, spectral and spatial frequencies and electric vector of linearly polarized light from the complex visual world of the insect's environment.

Until now, the transfer of information from the receptor cells to the second order neurons has been most thoroughly studied in flies. By the extensive and beautiful work on the anatomy of the retina-lamina projection (muscoid and calliphoroid flies: Musca, Calliphora, Lucilia, Sarcophaga; Trujillo-Cenoz, 1965, 1972; Braitenberg, 1967; Strausfeld, 1970a, 1971, 1975; Boschek, 1971) the first visual neuropile of the fly has become the most completely described neuropile of all organisms. It may well be that it is only in the cerebellar cortex of vertebrates that neural wiring is as well known as in the fly's lamina. In Hymenoptera, e.g. in bees and ants, however, our knowledge about cell types and connectivity patterns in the peripheral visual system is far less advanced.

Nevertheless, considerable information can already be drawn from the pioneering work of Cajal and Sanchez (1915) and the more recent papers of Strausfeld (1970) and Ribi (1974, 1975a,b). In these investigations using reduced silver staining and Golgi impregnation techniques, an extensive descriptive catalogue of laminar neurons has been prepared, but it was not until very recently that combined Golgi and EM techniques (Sommer and Wehner, 1975; Ribi, in prep.; Meyer and Rothenbach, unpubl.) allowed one to speculate about the topographical relationship between receptor terminals and second order neurons. However, because of our insufficient knowledge on the synaptology of the lamina (for the only information available see Varela, 1970), until now all anatomical investigations have not led by themselves to a general hypothesis of laminar functions.

The situation becomes even more unsatisfactory, when the physiology of the peripheral visual pathway is considered. Here also much more information is available on the fly's lamina than on that of the bee.

Although in bees, and more recently in ants, receptor potentials have been recorded intracellularly (Naka and Eguchi, 1962; Autrum and von Zwehl, 1962, 1963, 1964; Shaw, 1967, 1969; Baumann, 1968, 1975; Menzel and Snyder, 1974; Menzel, 1975b), until now there is only one paper dealing with intracellular recordings from second order neurons (worker bee: Menzel, 1974). Outside of this important study, neither intracellular nor extracellular electrophysiological investigations have been performed in the lamina of any Hymenopteran species. This situation contrasts severely with the extensive work on the physiology of the fly's lamina drawing on single unit recordings (Scholes, 1969; Mote, 1970a,b; Autrum et al., 1970; Zettler and Järvilehto, 1971, 1972; Arnett, 1971, 1972; Smola and Gemperlein, 1972; Gemperlein and Smola, 1972; Rehbronn, 1972, see Chapt. 2.4 of this vol.).

Why then should one study the visual systems of Hymenoptera? However, the visual systems of Hymenoptera are especially interesting in that natural selection has forced the Hymenoptera more than any other group of insects to develop highly efficient foraging strategies which are mainly based on visual cues. The most primitive Hymenoptera must have looked like sawflies - clumsy flyers which feed on plants. The most typical representatives of this order, however, are wasps, fast flying, predatory hunters with highly developed visual systems. There are many groups of wasps, e.g. digger wasps, spider wasps, mason wasps, cuckoo wasps, paper wasps and hornets, which all provision their nests with insect or spider prey. From two different lines the most sophisticated of all Hymenoptera have evolved: the bees and the ants. Bees are simply flower-visiting wasps which have abandoned predation, but many ants have maintained their ancestral feeding habits. Some bees and all ants are social insects, and as the consequence of their social organization they have evolved what can be readily called the most complex behavior among invertebrates. Especially in the flying bees, but also in many of the fast-running ants, there is strong need for highly developed visual orientation, because foragers often travel to feeding places over large distances and have subsequently to retun to small and hidden nest entrances. As early as 1898 the Peckhams gave an elaborate description of the rich receptoire of visual orientation performances in wasps. More recently, it has been reported that South American bees (Euglossini), which are long-distance pollinators of tropical plants, return to their nests from as far as 23 km. Each day, an individual bee visits the same set of plants in the same order along a feeding route (Janzen, 1971).

Although in behavioral complexity, bees and ants can well compete with flies, in our knowledge on the visual pathway they cannot. However, the ingenious and elaborate electrophysiological work of Laughlin (1973, 1974a,b,c, 1975a,b,c, and 2.1 of this vol.) on neural integration in the lamina of dragonflies (Odonata) might help to understand the homologous laminar functions in bees, because the retina-lamina projections are similar in Hymenoptera and Odonata (insects with apposition compound eyes) and are different from the projection patterns in Diptera, the compound eyes of which belong to the neural superposition type (Kirschfeld, 1967).

Among the truly limiting factors in our knowledge on the information which a bee abstracts from its visual world, is our complete lack of knowledge concerning the passage of information beyond the level of the lamina. It is not yet possible to reconstruct the exact steps in the neural processing of any visual cue important for the insect, e.g. the hue of color or the direction of polarization. At least against the background of our present knowledge one may argue that the neurophysiological approach will lead more readily to an understanding of general strategies in neural integration rather than to the specific model of the world built up in the brain of a given insect species.

However, the internal model that an insect uses has been shaped by natural selection according to the ecological needs of the species during its evolutionary history. The question what information is transferred and what is discarded by a nervous system can, therefore, be rephrased as follows: What information has been incorporated into a specific nervous system by the selective pressures of the environment? As the animal has to cope with the challenge of its environment primarily through behavioral responses, the analysis of these behavioral responses should indicate the cues processed by the nervous system. According to this strategy, which by necessity has to be a black-box approach, color coding, analysis of E-vector and other discriminatory capacities have been thoroughly studied in Hymenoptera.

In the following pages I shall (1) propose a hypothesis, which has been encouraged mainly by behavioral work (Sect. 2), (2) describe the principal feature of the peripheral visual systems of bees and ants (Sect. 3), and (3) present some arguments which heavily draw on the anatomical and physiological data presented in the preceding chapter and with which one might be able to judge the proposed hypothesis (Sect. 4).

2.7.2 Outline of a Hypothesis: Separate Visual Subsystems

Any nervous system has to compute simultaneously information from different sets of stimuli. The strategy used by the nervous system will depend on how early in the integrative processes information from the different sets of stimuli are transferred to separate neural pathways. In an extreme case, information from each set of stimuli could be processed by its own neural subsystem fed by its own set of receptors. Whether a nervous system can afford the luxury of using a variety of exclusive subsystems, will be determined by the number of different classes of stimuli to be processed. In addition, if each receptor views a fraction of the environment, an increase in the number of receptor types exclusively used for one subsystem, decreases resolution, i.e. increases grain of the picture.

The compound eye of a bee is composed of 5.500 subunits, called ommatidia. Each ommatidium contains nine visual cells. Consider the hypothetical case that the nine visual cells must analyse two parameters: the hue of color and the direction of polarization at a given point of the bee's environment. As to the economy and efficiency of neural integration the question arises, whether within the nine receptors, specific cells are used for color coding and others for E-vector analysis or whether all nine receptors simultaneously operate as input channels for the color and E-vector detecting subsystems. In the latter case, the information from both sets of visual stimuli has to be separated at a later stage of neural integration.

In that context I would like to propose the following hypothesis which draws heavily on behavioral analyses in bees and ants: originally, the visual systems of insects have evolved two rather independent subsystems - one fed by the ultraviolet receptors and the other by green receptors. Both systems are already found in rather primitive insects such as cockroaches, where ultraviolet and green receptors occur in the same ommatidium (Mote and Goldsmith, 1970, 1971; Butler, 1971). In general, the same is true for Hymenoptera.

The two subsystems are characterized by different types of receptors, different types of second order neurons, and - most important of all - by the different modes of behavior they trigger. In bees and ants, the UV system has proven to be the only system involved in the perception of polarized light in the sky (E-vector detection) and, hence, in celestial navigation (Duelli and Wehner, 1973; von Helversen and Edrich, 1974). On the other hand, the subsystem fed by the green receptors, is the main system mediating optomotor responses (Kaiser and Liske, 1972; Kaiser, 1975). The optomotor system, however, which monitors movement of the environment caused by the insects own locomotion provides some means for stability of direction, i.e. for maintaining a straight course by use of terrestrial cues.

In short: (1) The UV receptors alone mediate celestial navigation. Hence, they control principally the direction of course in long-distance orientation. (2) The system triggered by the green receptors stabilizes the course by processing the apparent movement of terrestrial objects through the insect's visual field.

Let us push this model a little further. Of course, each system by itself is incapable of detecting color, and the very first insects might not have been able to discriminate colors. Color coding most likely evolved concomitantly with the need for identifying objects in the terrestrial environment more efficiently. The color of a target is a cue that improves short distance orientation, and hence is used in deciding the direction of a course rather than in maintaining and stabilizing its straightness. For color vision a cross-talk between the two subsystems previously mentioned had to be established. In addition, a third type of receptor, the blue receptor, evolved for trichromatic color vision, for which we have compelling evidence in bees (Daumer, 1956; Menzel, 1967; von Helversen, 1972). I suppose that the establishment of color vision has been a rather recent event in the evolution of the bee's visual system and has most likely been part of a coevolutionary process with the radiation of the flowering plants in middle Cretaceous times.

The concept of the primary visual subsystems may seem too facile, and one may argue that outlining this hypothesis is more like hazarding a guess than proposing a theory, but it is out of such deliberate oversimplification that the beginnings of a theory are made.

2.7.3 The Retina-Lamina Projection

The compound eyes of insects are composed of structural subunits, called ommatidia. In Hymenoptera each ommatidium contains nine retinular (photoreceptor) cells. Each bundle of nine retinular cells, called a retinula (Grenacher, 1879), views the environment through its own dioptric apparatus consisting of a distal corneal lens and a proximal crystalline cone. It is the hexagonal array of these corneal lenses that at first glance characterizes the faceted compound eyes (Fig. 3). The photoreceptor membranes of the retinular cells are folded into densely packed hollow tubes, the microvilli, which are the sites of phototransduction. All microvilli of a given cell form a rhabdomere, which is functionally equivalent to the outer segment of the vertebrate photoreceptor. In Hymenoptera as in most other insects, e.g. Odonata and Orthoptera, within each retinula the rhabdomeres of all photoreceptor cells fuse to form a single central rhabdom. Light is focused on the distal tip of the rhabdom by the dioptric apparatus. As the refractive index of the rhabdom (n_r) is higher than that of

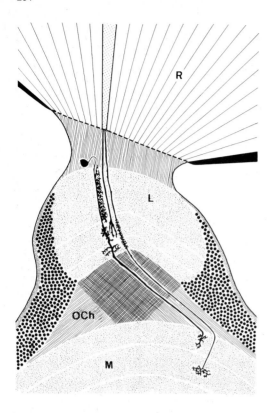

Fig. 1. The peripheral visual system of the worker bee, Apis mellifera. Horizontal section. A short visual fiber (svf), a long visual fiber (lvf) and a monopolar cell (type L1) are shown (not drawn to scale). R retina (one retinula is indicated by shaded area), L lamina (first visual neuropile), OCh outer (first) optic chiasma, M medulla (second visual neuropile)

the surrounding cytoplasmatic medium (n_c), the rhabdom acts as a light guide (worker bee: n_r = 1.365, n_c = 1.339; Varela and Wiitanen, 1970; Stavenga, 1975): by total internal reflection at the boundary of the rhabdom with the cytoplasm of the visual cell, light is contained within the rhabdom.

In vertebrates, the photoreceptor cells and the first synaptic neuropiles are all included into a single retina. In insects, the retinular cells and the different higher order neurons are confined to their own distinct regions in the visual pathway: retina, lamina (first visual ganglion), medulla (second visual ganglion), and lobula (third visual ganglion). In some insect groups, e.g. Diptera and Lepidoptera, the latter is divided into two separate neuropiles. In Hymenoptera, however, the lobula is a uniform structure.

The following presentation only deals with the retinular cells and the projection patterns of the retinular axons. Similar to the retina, the lamina is composed of subunits, the cartridges or "neurommatidia". There are as many cartridges in the lamina as retinulae in the retina. In Hymenoptera as in all insects with the fused-rhabdom type of eye, all retinular cells of one ommatidium project to the exactly underlying cartridge (Horridge and Meinertzhagen, 1970; Strausfeld, 1970b; Meinertzhagen, 1971; Ribi, 1974, 1975a). Whereas six photoreceptor cells of each retinula give rise to six short visual fibers terminating on second order neurons in the lamina, the three remaining photoreceptor axons (long visual fibers) project through the lamina to the medulla. The medulla also is composed of distinct subunits, called

Fig. 2. The peripheral visual pathway of the worker bee. Frontal section (perpendicular to the plane of the cross-over of fibers in Ch1 and Ch2). Interference contrast micrograph. R retina, Fl fenstrated layer, CBL cell body layer, EPL external plexiform layer (first visual neuropile, laminar neuropile), Ch1,2 first and second optic chiasma, M medulla (Ribi, 1974)

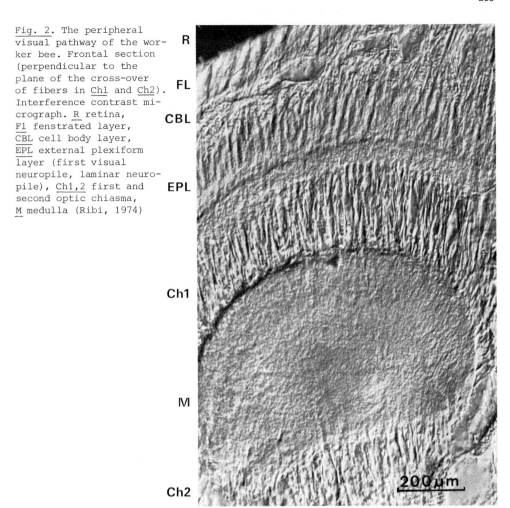

columns, the number of which coincides with the number of laminar cartridges and hence with the number of ommatidia. The projection patterns from the laminar cartridges to the medullar columns (Horridge and Meinertzhagen, 1970; Meinertzhagen, 1971) guarantee a one-to-one relationship between the bee's ommatidial fields of view and the sampling stations in the medulla.

In principle, this one-to-one relationship was already suggested by some light microscopists of the last century, although its existence remained controversial for some time (Hickson, 1885; Kenyon, 1897). Viallanes (1892), who first studied the optic centers of Hymenoptera by serial section techniques, introduced the name "neurommatidia" for the distal laminar subunits he had discovered in 1887. He found that each neurommatidium received its input from the spatially corresponding "retinula", named so by Grenacher (1879). Furthermore, Patten (1887) in investigating the development of the visual system of wasps (Vespa cabro) concluded that the number of subunits in the medulla coincided with the number of neurommatidia, which he called "nerve spindles".

2.7.3.1 Retina: the Photoreceptor Cells

2.7.3.1.1 Structure of Retinula and Rhabdom

Predatory and fast flying insects are characterized by a large number
of retinular cells. Correspondingly, many aculeate Hymenoptera including wasps, ants and bees have large eyes composed of hundreds and thousands of ommatidia: 10.621 ± 134 S.E. in drone bees and 5.519 ± 60 S.E. in
worker bees (Apis mellifera)[1], 600-1.200 (depending on body size;
Menzel and Wehner, 1970) in the carnivorous desert ant Cataglyphis
bicolor, 400-700 in the wood ant Formica rufa (Bernstein and Finn,
1971), about 750 in Formica polyctena (Menzel and Lange, 1971) and
about 250 in Myrmica ruginodis (Vowles, 1954).

The geometry of the hexagonal array of ommatidia in Hymenoptera is
given in Fig. 3.

The first attempt of a three-dimensional reconstruction of the bee's
ommatidium has been made by Jan Swammerdam in the middle of the 17th
century (his famous and elegant illustration of the bee's compound
eye, which he had partly dissected, was posthum published by the Dutch
physician Boerhaave in 1737). However, in contrast to this picture and
to most schematic drawings of ommatidia, as they are usually presented
in textbooks, ommatidia of bees are very slender structures. In the
worker bee the ratio of length to width amounts to about 20 to 1
(Fig. 4). In absolute terms, the retinular cells are rather large
(400 µm long, 10 µm wide) in the drone bee (Perrelet, 1970), so that
they can be easily impaled by microelectrodes, but are much smaller
in worker bees and especially in ants, where they are only about
100 µm long: 80-90 µm in F. polyctena (Menzel, 1972), 100 µm in C. bicolor (Wehner et al., 1972), and 125 µm in Myrmecia gulosa (Menzel
and Blakers, 1975). Hence, the first successful intracellular recordings from the visual cells of any Hymenopteran species have been performed in drone bees (Autrum and von Zwehl, 1962; Baumann, 1968).

Large and Small Retinular Cells. To understand integrative functions
of compound eyes, e.g. color coding and E-vector detection, one must
first number the nine retinula cells within one ommatidium individually. In many groups of insects the receptor cells of one retinula differ markedly from one another even in gross morphology. In Hymenoptera,
however, clear-cut structural differences are difficult to define. In
that respect, the most striking feature is the occurrence of a small
basal receptor cell. Whereas eight cells contribute to the rhabdom
along its total length (long visual cells), the ninth cell is short
and restricted to the proximal part of the retinula (worker bee:
Phillips, 1905; drone bee: Perrelet and Baumann, 1969; Gribakin, 1972;
Grundler, 1972; wood ant, F. polyctena: Menzel, 1972; desert ant:
C. bicolor: Wehner et al., 1972; bulldog ant, M. gulosa: Menzel and
Blakers, 1975). A short basal retinular cell already occurs in the
most primitive Hymenoptera, the sawflies (Tenthredinidae), although
their retinulae have been reported to be composed of only eight photoreceptor cells (Corneli, 1924). Wolken (1975) briefly mentions that
in a few ommatidia of the wasp, Vespa maculata, a small distal ninth
cell occurs. In dorsal regions of the worker bee's eye (10th to 30th
z-row), from which we have clear evidence that they are involved in
the perception of polarized light, the basal ninth cell is only 40 µm
long, compared to a total length of the retinula of 220 µm at that

---

[1]Even in more recent papers on the anatomy of the honey bee's retina, the number of
retinulae has been underestimated: 4.000 (Varela and Porter, 1969) and 4.300
(Skrzipek and Skrzipek, 1971) in worker bees, 8.000 (Perrelet, 1970) in drones.

Fig. 3. Replica of the corneal surface of a worker bee's left eye (eye bristles removed before preparation). The numbers on the left refer to the numbers of the horizontal z-rows within the hexagonal array of ommatidia (x-y-z system). The distribution of interommatidial angles along the horizontal and vertical axes of the eye is given in Baumgärtner (1928) and Del Portillo (1936). During free flight the dorsoventral axis of the eye is inclined by 70-75° to the horizontal (Wehner and Flatt, in prep.)

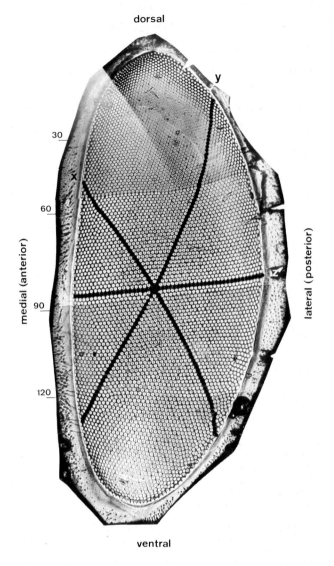

level (Wehner et al., 1975). Correspondingly, in frontal and ventral parts of the eye, where the retinulae are up to 300 μm long, values of 50-100 μm are reported for the length of the ninth cell (Menzel and Snyder, 1974; Grundler, 1974; Gribakin, 1975). Among the eight long visual cells of drone bees and ants two small cells lying opposite to each other, and six large cells can be distinguished at any level of the retinula, but such a difference cannot be consistently found in worker bees (compare the corresponding structure of the rhabdoms in Fig. 5). However, as first noted by Gribakin (1967a,b), the eight long visual cells of the worker bee can be grouped into three classes (type I, II and III cells) according to the vertical position of their nuclei. Using thin and ultra-thin cross sections, the mean

288

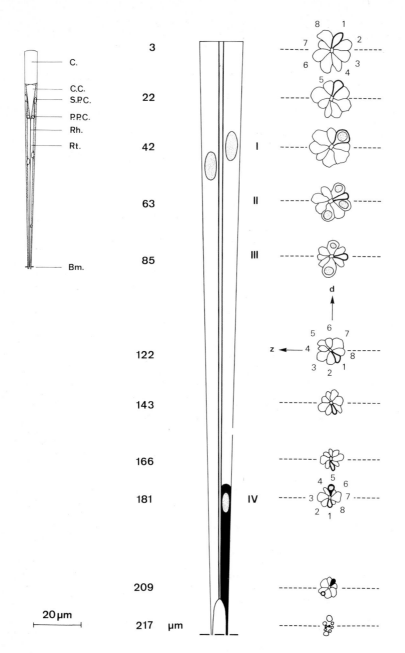

Fig. 4. Geometry of the worker bee's retinula (dorsal part of the eye, z-row 15-25). The longitudinal section (middle part) and the series of cross sections (right part) are drawn according to the same scale. In the cross sections, the distance underneath the distal tip of the retinula is indicated at the left-handed side of the longitudinal section. Cell no. 1 is marked by heavy outline, cell no. 9 is shown in black. I, II, III and IV nuclear layers, d dorsal, z direction of z-axis. Inset figure: longitudinal section of one ommatidium drawn on scale (according to Phillips, 1905). C. corneal lens, C.C. crystalline cone, S.P.C. secondary pigment cell, P.P.C. primary pigment cell, Rh. rhabdom, Rt. retinula, Bm. basement membrane

positions of the nuclei underneath the tips of the crystalline cones were calculated and compared with similar data in ants (Table 1). Near the basement membrane, the nuclei of the ninth cells constitute a forth nuclear layer, which had already been described - together with the most distal nuclear layer - by Phillips (1905). In the retinulae of two ant species, only six (<u>Camponotus herculeanus</u>, Wolken, 1975) resp. seven retinular cells (<u>M. ruginodis</u>, Vowles, 1954) have been reported. In both cases, however, it cannot be excluded yet that short, small retinular cells have been overlooked. In Hymenoptera, the cross-sectional area of a particular rhabdomere does not vary dramatically along the total length of the rhabdomere, as it is described for other compound eyes of the fused rhabdom type (e.g. <u>Periplaneta americana</u>, Trujillo-Cenoz and Melamed, 1971).

I would like to add a short note on the discovery and rediscovery of the ninth cell in the worker bee. Its existence had been first proved by Phillips (1905) in his famous paper on the development of the honey bee's eye. However, he did not recognize the ninth cell as a short, particular cell, but merely described it as the cell that had its nucleus nearest to the basement membrane. Furthermore, he could not decide whether the occurrence of the ninth cell was a common or a rare phenomenon. He reports that occasionally nine retinular cells can be found in one cross section - as indeed had already been demonstrated by Grenacher (1879) - but concludes that the presence of this proximal nucleus in all ommatidia cannot be accountable for the rare occurrence of nine retinular cells as seen in cross sections. The logic conclusion would have been that there are nine visual cells per ommatidium, but that in most cases the ninth cell does not extend over the full length of the retinula. However, there is no evidence that this conclusion had already been drawn. Since the first electron micrograph of a cross section through the worker bee's retinula undoubtedly showed eight retinular cells (Goldsmith, 1962), in the years to come it was generally accepted that the bee's ommatidium was composed of eight photoreceptors. So strong was this belief that Varela and Porter (1969) having found nine receptor cell axons in each retinular nerve bundle explained the occurrence of the spare axon by the dichotomy of a particular axon. Hence, one receptor cell should give rise to two axons. Although one cell, unfortunately termed "eccentric" cell (see Ratliff et al., 1963) by Varela and Porter (1969), did show a nucleus in the very proximal part of the retinula, the authors did not realize that just at this level one retinular cell (cell no. 1 or no. 5 according to the classification in Fig. 8 and Table 1) had been replaced by the short ninth cell. This fact became established by the EM studies of Gribakin (1972) and Grundler (1972). There is only a very small distance (in worker bees, at the maximum 3 μm; Menzel and Snyder, 1974; Sommer and Wehner, 1975; in the ant <u>C. bicolor</u>, 15 μm; Herrling, 1975), along which the ninth cell and the long visual cell that is replaced by it both contribute to the rhabdom. In most parts of the eye it is only there that nine receptor cells can be seen simultaneously.

<u>Rhabdom Geometry</u>. <u>Hymenoptera</u> are characterized by fused rhabdoms, i.e. all retinular cells of one ommatidium contribute with their photoreceptor membranes (microvilli) to a central light guide and light absorbing structure, the fused rhabdom.

The fact that the microvilli of different receptors possessing different spectral sensitivities and different polarization sensitivities are packed together in a unified structure, is of uppermost importance for the visual cells to act as efficient color and E-vector receptors (Snyder et al., 1973; Snyder, 1975). Given specific optical densities and cross-sectional areas, the rhabdomere can increase its absolute sensitivity by increasing its length. Therefore, highly developed

Table 1. Types of visual cells in the retina of the worker bee (Apis mellifera) and the bulldog ant (Myrmecia gulosa)

Type of cell[a]	Cell no.	Apis mellifera				Myrmecia gulosa		
		Cell no. used in previous publications[b]	Position of nucleus[c]	Spectral sensitivity λ_{max}[d]	Axon terminal[f]	Cell no. used in previous publications[g]	Position of nucleus[c]	Spectral sensitivity λ_{max}[h]
I	1, 5	1, 2	0.19	350 nm	lvf	1, 5	0.12	UV
II	2, 6	8, 7	0.29	530 nm	svf (2b)	8, 4	0.26	G
	4, 8	4, 3		440 nm	svf (2a)	6, 2		G
III	3, 7	6, 5	0.39	530 nm	svf (1)	7, 3	0.40	G
IV	9	9	0.82	UV[e]	lvf	9	0.80	?

[a] Classification introduced by Gribakin (1967a,b)

[b] Wehner et al. (1975); for further reference to classification schemes previously used by a number of authors see Table 1 of Sommer and Wehner (1975)

[c] Mean distance between nucleus and distal tip of retinula divided by the total length of the retinula (220 µm in worker bees, dorsal part of the eye, z-row 15-25; µm in bulldog ants). According to unpubl. data (bee) and Menzel and Blakers (1975; ant)

[d] In bees, spectral sensitivity types of the long receptor cells have been correlated with types of axon terminal by intracellular recordings and dye injections (Menzel and Blakers, in prep.)

[e] Menzel and Snyder (1974)

[f] Correlation between axon terminals and retinular cells according to Ribi (unpubl.); svf short visual fiber (retinular axon terminating in the lamina; for types of svf see Fig. 19), lvf long visual fiber (retinular axon terminating in the medulla). In ants, Cataglyphis bicolor, cells nos. 1 and 5 also give rise to lvf

[g] Menzel and Blakers (1975)

[h] UV and green (G) sensitivities of the 8 long visual cells have been determined by use of selectively induced screening pigment responses

photoreceptors often show considerably elongated light obsorbing
structures. In <u>Hymenoptera</u>, the maximal length of a rhabdomere has
been found in <u>drone bees</u> (400 µm). In such a long rhabdomere characterized by a high amount of absorption, spectral and polarization
sensitivities of the corresponding cells should be considerably smaller than the spectral extinction and the dichroic ratio of the photopigment, respectively. This holds, because light passing down the
rhabdomere is selectively absorbed with respect to wavelength and
E-vector direction. Due to this self-screening effect a rhabdomere
cannot exceed a certain length without severely affecting its spectral and polarizational characteristics. The largest rhabdoms in insects, found in some species of dragonfly, are twice as long as in
drone bees. In dragonflies, however, the rhabdoms are tiered, i.e.
the receptor cells only contribute to the rhabdom at certain levels.
It may well be that it is because of the considerable amount of self-
screening in such a long light absorbing structure that intracellular
recordings from the most proximal receptor cells of dragonflies reveal
flat spectral sensitivity curves (Horridge, 1969). In a fused rhabdom
where the individual receptor cells differ in their spectral and E-vector absorbing properties, the effects of self-screening can be compensated by the lateral filter effect of adjacent rhabdomeres (Snyder
et al., 1973). Provided that the different spectral and E-vector types
of receptor are properly arranged within the rhabdom, a single receptor cell can combine high absolute sensitivity with both high spectral
sensitivity, which is close to the spectral absorption of its visual
pigment, and high polarization sensitivity, which is close to the
dichroic ratio of the pigment.

In Hymenoptera, the fused rhabdoms are not tiered, but composed of
eight rhabdomeres along their entire length. Although the rhabdoms
of bees and ants considerably vary in size and shape, they can be
grouped into two classes according to the geometry of their microvillar arrangements (Fig. 5). At any cross section the rhabdom consists
either of two sets of microvilli oriented in mutually perpendicular
directions (bees, wasps V. maculata, see Fig. 8.11 in Wolken, 1975;
specialized eye region in the ant <u>C. bicolor</u>), or of three sets of
microvilli including an angle of 60° between adjacent rhabdomeres
·(ants). Very small, but long receptor cells, which contribute only a
few microvilli to the rhabdom (drone bee, some ants) further enhance
the diversity in types of rhabdom (Figs. 11 and 12).

Rhabdom geometry has been most completely described in the worker bee
(Wehner et al., 1975) and the Australian bulldog ant (Menzel and
Blakers, 1975). The former shows two, the latter three microvillar
directions. In both, bees and ants, however, two types of rhabdom can
be discriminated because of symmetry reasons. They are mirror-images
of each other (Figs. 6 and 8) and occur with equal frequencies throughout the whole retina. No specific pattern of distribution between the
two types could be found (see Fig. 9 in Menzel and Blakers, 1975;
Fig. 3 in Wehner et al., 1975). The most striking feature, however,
is that the rhabdoms are not straight, but twisted. On their way down
to the basement membrane one type of rhabdom twists clockwise and the
other counterclockwise. In the dorsal part of the bee's retina the
twist rate is rather smooth and uniformly distributed along the total
length of the rhabdom. It amounts to about 1°/µm (1.13°/µm and 0.99°/µm
for both twist types, respectively; Fig. 9). In the bulldog ant, where
the rhabdoms are shorter and the twist rates smaller (0.54°/µm and
0.61°/µm, respectively), rather sudden changes in microvillar directions are observed. As can be seen in Figs. 10-15 in Menzel and
Blakers (1975), rapid changes of 10-20°/µm are common along distances
of 3-5 µm.

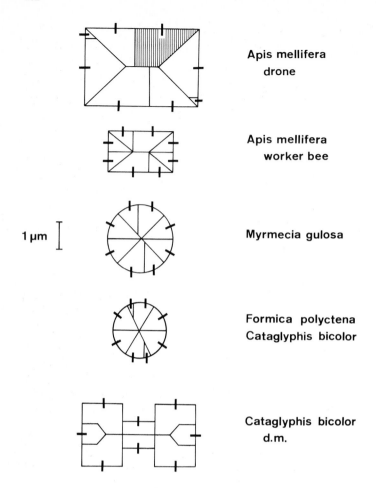

Apis mellifera
drone

Apis mellifera
worker bee

1 µm

Myrmecia gulosa

Formica polyctena
Cataglyphis bicolor

Cataglyphis bicolor
d.m.

Fig. 5. Cross-sectional areas of rhabdoms in bees (Apis) and ants (Myrmecia, Formica, Cataglyphis), drawn to same scale. Based on EM data of Gribakin (1967a,b, 1969a,b); Varela and Porter (1969); Perrelet (1970); Skrzipek and Skrzipek (1971); Menzel (1972); Brunnert and Wehner (1973); Grundler (1974); Wehner et al. (1975); Menzel and Blakers (1975); Herrling (1975) and own unpubl. material. Microvillar directions of rhabdomeres are indicated by heavy black bars. Microvilli are only shown for one rhabdomere of the drone bee's rhaddom. In the species mentioned, the average diameter of the microvilli within one rhabdom ranges between 40 and 70 µm. d.m. dorso-medial part of the eye

In any cross section through the retina, the retinular cells can be individually numbered according to the geometry of the rhabdom and the angular orientation of the rhabdomeres within the coordinate system of the eye (x-, y-, and z-rows of ommatidia). Each particular cell is unambiguously specified when the direction and rate of twist, the level of the cross section and the x-y-z coordinates are known. Until now, in ants, drone bees and worker bees, different authors have used different numbering systems (see Table 1 in Sommer and Wehner, 1975).

Fig. 6. Cross section of the bee's dorsal retina, 60 μm underneath the focal plane of the lens system (nuclear layers of type II and type III cells). In this picture most rhabdoms belong to the type Y retinulae, the two encircled ones to the type X retinulae (see Fig. 8 A1). Large numbers horizontal z-rows (nos. 11-14) of the eye; see Fig. 3 for definition of x-y-z system. Small numbers long retinular cells (nos.1-8), numbered clockwise or counterclockwise according to the type of rhabdom (X or Y). d dorsal, l lateral (posterior), m medial (anterior), v ventral. EM micrograph provided by E. Meyer

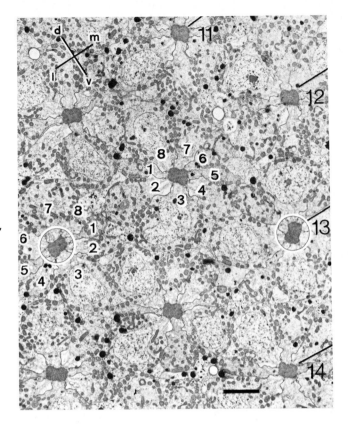

However, as rhabdom geometry is similar in all these species of Hymenoptera, one should try to provide homologous retinular cells with identical numbers. The following system, which can be readily applied to all eyes where rhabdom geometry and the parameters of twist are known, has been worked out together with Menzel: the type I cell that points dorsally at the distal tip of the retinula is called cell no. 1. The other cells are continuously numbered in the direction of twist. Hence, looking into the left eye, the numbers run clockwise in the type Y rhabdoms and counterclockwise in the type X rhabdoms (Fig. 8). For definition of type X and type Y rhabdoms see Fig. 6. In Table 1 this numbering system is compared with the ones previously used.

At least in the dorsal half of the worker bee's eye the distal tips of the rhabdoms all show the same orientation of their eight rhabdomeres. This can be shown by referring to the line where the microvillar tips of two opposite cells meet. This line is called the transverse axis of the rhabdom (TRA, see Figs. 8 and 10) and provides a reference line for the numbering of retinular cells. In the dorsal half of the eye the TRAs of the distal tips of the rhabdoms are perpendicularly oriented to the horizontal z-axis of the eye. In the bulldog ants, Menzel and Blakers (1975) have defined an "ommatidial axis" in a slightly different way as the line running through cells nos. 1 and 5. The angle between this ommatidial axis and the z-axis of the eye significantly differs from $90°$ and between both types of rhabdoms ($71° \pm 12.3°$ and $42° \pm 14.1°$, respectively).

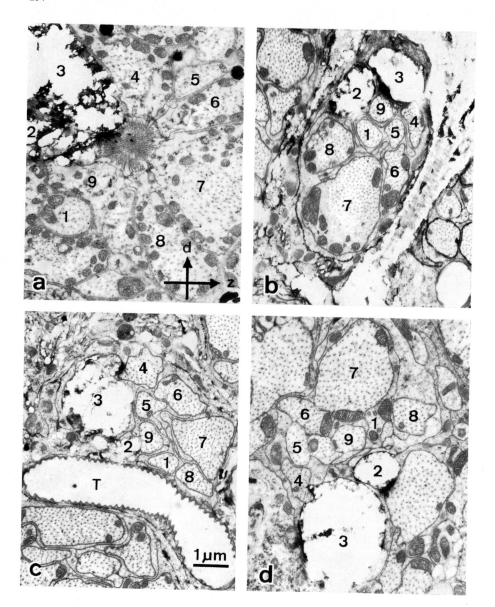

Fig. 7a-d. Correlation between retinular cells (a) and axon fibers in the pseudo-cartridge (b-d). (a) retinula 30-40 μm above basement membrane (bm), cell no. 9 has replaced cell no. 1; (b) pseudocartridge at the level of bm, (c) 40 μm below bm, (d) at the level of the cell body layer. T trachea, d dorsal, z direction of z-axis (Sommer and Wehner, 1975)

In the worker bee, the rhabdoms are differently oriented in the dorsal and ventral parts of the eye (Grundler, 1974). In the ventral parts the TRAs run parallel to the z-axis, i.e. they are perpendicularly oriented to the TRAs in the dorsal ommatidia, if both are measured at the distal tips of the rhabdoms. The retinular cells, however, remain in their absolute positions relative to the x-y-z system of the eye.

Fig. 8. Rhabdom geometry in
the worker bee (Apis mellifera; A1 dorsal retina,
A2 ventral retina) and the
bulldog ant (Myrmecia gulosa;
B, according to Menzel and
Blakers, 1975). Heavy arrows:
direction of twist in both
types of rhabdom (worker bee:
left type X rhabdom, right
type Y rhabdom). In the ventral half of the worker bee's
eye the twist of rhabdom has
not been studied yet systematically. TRA transverse
axis of rhabdom (intra-rhab-
domic line where the microvillar tips of the two large
rhabdomeres meet), OA ommatidial axis (line combining
the cell bodies of the two
small retinular cells, type I
cells, cells nos. 1 and 5),
d dorsal, z direction of
z-axis of the eye (see Fig.3).
The angular orientation of
the rhabdoms within the
x-y-z system of the retina
is given for the distal tips
of the rhabdoms (focal plane)

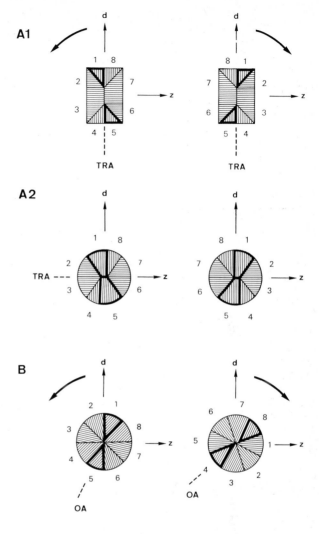

According to this interpretation the large rhabdomeres, which form
the TRA by the tips of their microvilli, belong to cells nos. 3 and 7
(type III cells) in the dorsal part of the eye, but to cells nos. 1
and 5 (type I cells) in the ventral half. This statement is supported
by the observations of Grundler (1974) and Kolb and Autrum (1974) that
cells nos. 1 and 5 are consistently characterized by a smaller number
of pigment granules and sometimes by a denser structure of the cytoplasm. At the tips of the rhabdoms, the microvillar directions of these
cells are always perpendicularly oriented to the z-axis, irrespective
of the orientation of the TRA in both, the dorsal and ventral halves
of the eye. Therefore, the absolute position of a cell within the
x-y-z system is a more reliable criterion than the position of the
cell relative to the TRA. Especially in those regions and at those
levels of the retina, where the cross-sectional area of the rhabdom
is more or less circular in shape, TRAs are not related invariably to
particular receptors. The numbering system given in Table 1 draws on
these arguments.

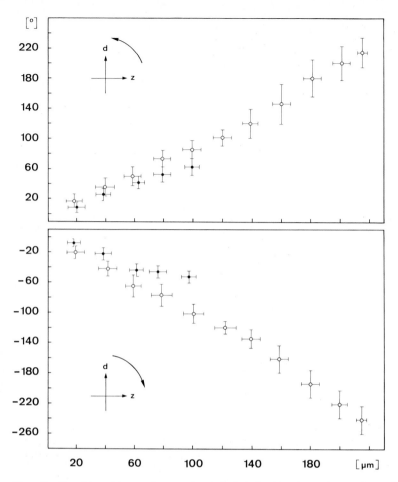

Fig. 9. The direction and amount of twist in the two mirror-imaged types of rhab-doms in the eyes of worker bees (o) and bulldog ants (●). The direction of TRA (ordinate), i.e. the direction of the microvilli of rhabdomeres nos. 1 and 5, is plotted against depth underneath focal plane (abscissa). Means ± S.E. At the level of the distal tips of the rhabdoms (focal plane, O μm) the directions of TRAs are normalized to O°. Bee, Apis mellifera: dorsal part of the eye, 16th-24th z-row, above type X rhabdoms, below type Y rhabdoms (see Fig. 8), according to Wehner et al. (1975). Ant, Myrmecia gulosa: dorso-medial frontal part of the eye; the plotted ●-values are calculated from the data given in Figs. 7a,b of Menzel and Blakers (1975). d dorsal, z direction of z-axis

Until now a continuous twist of the rhabdom has not been described in other insect species, possibly because it can be only investigated by tedious semi- and ultrathin serial sections along distances of hundreds of microns. In the tiered rhabdoms of damselflies (Ishunura senegalensis and Cersion calmorum), the microvilli of four cells smoothly twist along the short distance of 20 to 30 μm (Ninomiya et al., 1969). This small amount of twist, however, is mainly due to the fact that these four middle cells of the ommatidium replace the most distal cell and hence have to rearrange their microvilli on the way from the distal to the middle layer of the rhabdom. The remaining parts of the rhabdom, which in total is 250 μm long, do not twist.

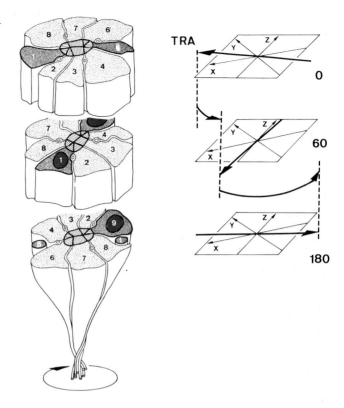

Fig. 10. Three-dimensional model of the twisted retinula of the worker bee. Cross sections are shown for 0, 60, and 180 μm underneath the distal tip of the retinula (focal plane of the dioptric apparatus). The angular orientation of the transverse axis of the rhabdom (TRA) within the x-y-z reference system of the eye is given for the dorsal half of the eye. For definition of TRA see Fig. 8 and text. In the axon bundle, the direction of twist (arrow) is opposite to the direction of twist of the retinula

In the worker bee, Grundler (1974) had already suggested that the rhabdoms twist around their longitudinal axes, because he had observed that they were less orderly arranged in the proximal than in the distal parts of the retina. Horridge and Mimura (1975) and Horridge et al. (1975) conclude from their measurements of polarization sensitivity in the fly's retinular cells nos. 1-6 that the distal and the proximal parts of each of these rhabdomeres are twisted relative to each other.

Because of the twist of the bee's retinula the four cone cell processes running down to the basement membrane between adjacent retinular cells, also have to twist around the rhabdom. These processes, first shown by Goldsmith (1962) in the worker bee (see also Waddington and Perry, 1963; Horridge, 1966; Burton and Stockhammer, 1969), have often been thought to function as a cytosceletal device for supporting the straight alignment of the rhabdom. Armed with the observation that rhabdoms indeed twist, this interpretation can be ruled out. One function of the cone cell processes most likely is to bring the distal tip of the rhabdom exactly in focus with the tip of the crystalline cone, i.e. to fit two light guides together. There is considerable evidence that the proximal part of the crystalline cone acts as a light guide (Bernard, 1975).

It is a challenging problem for developmental biologists to analyze the behavior of individual retinular cells while they perform the actual twist of the retinula during retinal differentiation. However, the data available on the exact sequence of events during the differentiation of any insect retina are rather scanty. Meinertzhagen (1973)

298

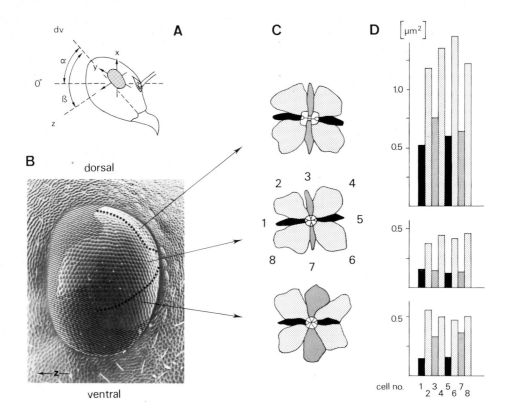

Fig. 11A-D. Distribution of different types of retinulae within the compound eye
of the desert ant, <u>Cataglyphis bicolor</u>. (A) Head and eye position in freely running
ants; α = 46.1° ± 1.9° (according to motion pictures done by R. Weiler),
β = 79.6° ± 1.1°; <u>dv</u> dorsoventral head axis. (B) Scanning micrograph of the eye
revealing the distribution of retinular types. In the <u>hatched area</u> the structure
of the retinulae has not been investigated. (C) Types of retinulae and rhabdoms.
(D) Cross-sectional areas of rhabdomeres as measured 10 μm below distal tips of
rhabdoms. (A) from Wehner (1975), data for (B-D) provided by Herrling (1975)

convincingly remarks that as far as the development of the compound
eyes and optic lobes of insects is concerned, "many of the studies
are either old and obscure or unpublished and preliminary". The first
statement holds for Hymenoptera (wasps: Patten, 1887; bees: Phillips,
1905; sawflies: Corneli, 1924). In the worker bee, just after the
larva is sealed up (semipupa stage), rapid growth occurs in a spindle-
like group of cells which subsequently gives rise to an ommatidium.
The retinular cells become longer, and the formation of the rhabdom
starts at the distal end of the retinula. At that stage each retinular
cell has already developed an axon, which penetrates the basement membrane, but all nuclei lie together at one level of the retinula. As
the formation of the rhabdom progresses toward the basement membrane,
the proximal part of the retinula enlarges and the nuclei move inward
until they reach their adult positions at three different levels
(see Table 1). The twist of the retinula most likely occurs during
this differentiation of the spindle-like group of cells into the long
column of retinular cells - a process, which is accompanied by the
formation of the rhabdom. Furthermore, it would be interesting to

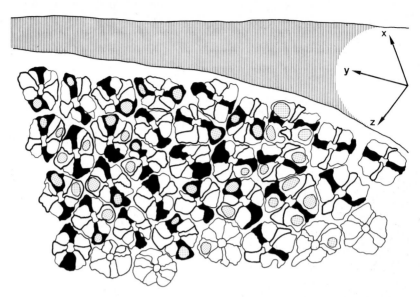

Fig. 12. Oblique tangential section through the dorso-medial retina of Cataglyphis
bicolor (see. Fig. 11). In the uppermost ommatidia near the dorsal eye margin
(hatched) the rhabdoms have dumb-belled cross sections (Fig. 5). In addition, the
transition zone to rhabdoms of circular cross sections is shown. The plane of the
tangential section runs from a more distal level (right) to a more proximal level
(left). The nuclei of the large cells (nos. 2,4,6 and 8) are in a more distal position than the nuclei of the small cells (nos. 1 and 5, black; nos. 3 and 7, heavy
outline. According to Herrling (1975)

know how the twist of the axon bundle, which is opposite to the twist
of the retinula (see Figs. 7 and 10), is related to the formation of
the retina-lamina projection.

The uppermost dorsal part of the worker bee's eye strikes by two significant structural specializations. Within about one hundred ommatidia, all nine visual cells of one ommatidium are of equal length
(about 150 µm) and contribute to the rhabdom along the total distance
from the focal plane of the dioptric apparatus down to the basement
membrane (Schinz, 1975). In addition, the retinulae characterized by
the nine long visual cells (Fig. 13), are not twisted, but straight
(Wehner et al., 1975, see Fig. 14). As the rhabdom is composed of nine
rhabdomeres, the regular pattern of two mutually perpendicular microvillar directions is distorted. Most commonly, three or four microvillar directions occur at any level of the rhabdom. However, even in
adjacent ommatidia, rhabdom geometry can differ markedly. Until now,
we have not been able to number the nine cells of the straight retinulae consistently.

2.7.3.1.2 Spectral Sensitivity

By elaborate behavioral experiments color vision has been proved in
worker bees (Daumer, 1956; Menzel, 1967; von Helversen, 1972), wasps
(Paravespula germanica: Baier and Menzel, 1972), and ants (F. polyctena: Kiepenheuer, 1968; C. bicolor: Wehner and Toggweiler, 1972).
According to these investigations the color coding system is trichromatic in bees and wasps and at least dichromatic in ants. Corre-

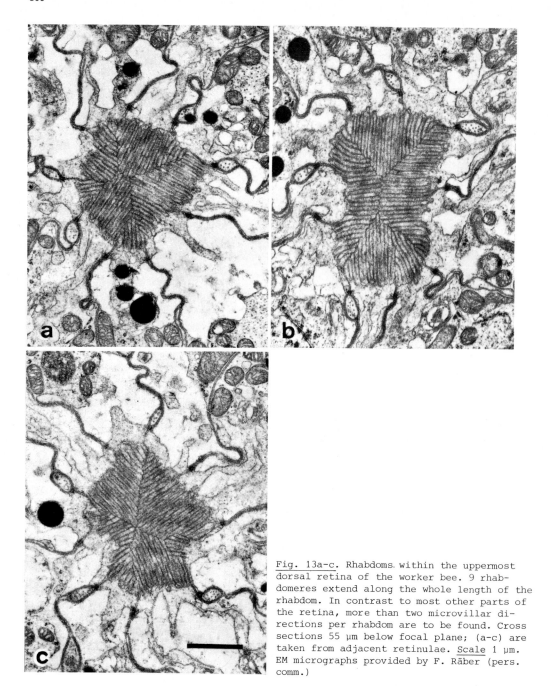

Fig. 13a-c. Rhabdoms within the uppermost dorsal retina of the worker bee. 9 rhabdomeres extend along the whole length of the rhabdom. In contrast to most other parts of the retina, more than two microvillar directions per rhabdom are to be found. Cross sections 55 μm below focal plane; (a-c) are taken from adjacent retinulae. Scale 1 μm. EM micrographs provided by F. Räber (pers. comm.)

pondingly, electrophysiological recordings reveal UV (λ_{max} between 340 and 360 nm), blue (λ_{max} between 440 and 460 nm) and green receptors (λ_{max} between 520 and 540 nm) in worker bees (Autrum and von Zwehl, 1964; Menzel and Snyder, 1974) and wasps (Menzel, 1971), and

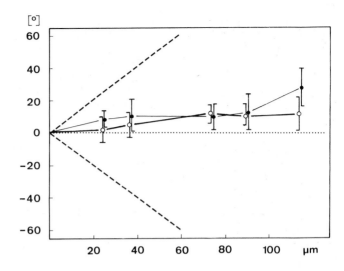

Fig. 14. Microvillar directions in rhabdomeres of the uppermost dorsal retina of the worker bee. In each rhabdom the microvillar directions (<u>ordinate</u>) of all 9 rhabdomeres are determined and plotted against depth underneath focal plane (<u>abscissa</u>). Means ± S.E. In the focal plane (0 μm) all microvillar directions are normalized to $0°$. Hence, all angular values refer to relative microvillar directions. ● means of relative microvillar directions of the 9 rhabdomeres within one rhabdom, o means of 10 ●-curves. <u>Dashed lines</u>: microvillar directions of the type X and type Y rhabdoms (see Fig. 9). According to data of F. Räber (pers. comm.)

UV and green receptors in ants (<u>F. polyctena</u>: Roth and Menzel, 1972; <u>M. gulosa</u>: Menzel and Roth, inpublished). More detailed information on the shape of the spectral sensitivity functions and an excellent discussion of the factors by which these functions are influenced are given in Menzel (1975a).

The distribution of the color receptors within the ommatidium can be most convincingly studied by two methods: either by intracellular dye injections or by applying selective chromatic adaptation, which causes radial movement of screening pigment granules in particular color receptors. In insects with the fused rhabdom type of eye both methods were first used in cockroaches (<u>P. americana</u>), the former by Mote and Goldsmith (1971), the latter by Butler (1971). These two investigations were the first to prove that in compound eyes with fused rhabdoms different color receptors occur within one ommatidium and that they are not confined to separate ommatidia. In the cockroach three UV receptors are always associated with five green receptors. The same number of UV receptors per ommatidium seems to be true for all Hymenoptera.

<u>Selective Chromatic Adaptation</u>. In the two small, but long cells of the ant's retinula (cells no. 1 and 5) the intracellular vacuoles around the rhabdom disappear and the screening pigment granules move close to the rhabdom, when the eye is illuminated with an appropriate intensity of UV light. The six large cells (nos. 2-4, 6-8) remain in the dark adapted state, where the rhabdom is surrounded by the "palisade" of vacuoles. In these cells the radial movement of pigment granules toward the rhabdom is caused by light with $\lambda > 430$ nm (<u>C. bicolor</u>: Wehner et al., 1972; Herrling, 1975; F. <u>polyctena</u>: Menzel and Knaut, 1973; <u>M. gulosa</u>: Menzel and Blakers, 1975; Fig. 15). By measuring the

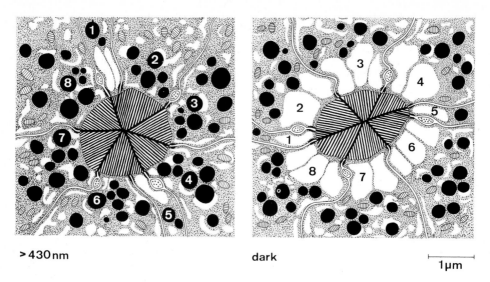

>430nm   dark $\overline{\qquad}$
                                                        1µm

Fig. 15. Cross sections through two retinulae of the desert ant, Cataglyphis bicolor, after dark adaptation (right) and long-wavelength adaptation (left). In the dark adapted state of a cell, the vesicles of the endoplasmatic reticulum form large vacuoles, the so-called "palisade", around the rhabdom (cells no. 1 and 5 of the left retinula, all cells of the right retinula). During light adaptation (cells no. 2-4, 6-8 in the left retinula), the large vacuoles disappear and the screening pigment granules (black dots) move toward the rhabdom. Schematic drawings from EM micrographs. The directions of numbering the retinular cells are chosen arbitrarily.

response/intensity functions of this pupillary response for three different wavelengths, Menzel and co-workers were able to conclude that the only color receptors in the ommatidia of wood ants and bulldog ants are UV receptors (cells no. 1 and 5) and green receptors (cell no. 2-4, 6-8). No blue receptors occur. One has to mention, however, that the latter type of color receptor is more difficult to detect because of the overlapping sensitivity functions of all three receptors in the blue spectral region. On the other hand, the conclusion that all six large cells are green receptors is supported by intracellular recordings as well as ERG measurements in eyes which have experienced selective chromatic adaptation. In desert ants, the presence of a blue receptor cannot be excluded yet. Behavioral experiments on color discrimination (Wehner and Toggweiler, 1972) and unpublished data on the spectral sensitivity functions of phototactic behavior present some evidence that in the desert ant the green receptor is not the only long-wavelength receptor.

Although strong efforts have been made to demonstrate the effects of chromatic adaptation in bees as well, only Kolb and Autrum (1974) succeeded in determining two UV receptors among the eight long retinular cells. According to the considerations on rhabdom geometry, TRA orientation, etc., made in the previous section, one has to conclude that also in bees cells no. 1 and 5 (type I cells) are UV receptors as far as the results of chromatic adaptation experiments are concerned. As the short ninth cell lacks screening pigment, no data on its spectral sensitivity can be obtained by this kind of experiment.

Intracellular Staining. In Hymenoptera there is only one unpublished report on intracellular dye injections into retinular cells (worker bee: Menzel and Blakers, in prep.). These data confirm that among the long visual cells receptors nos. 1 and 5 are UV receptors. The short ninth cell has not been marked yet, but it follows from rare recordings in which the tip of the electrode has been positioned near the basement membrane of the eye (three out of in total 260 intracellular recordings) that the ninth cell, too, is a UV receptor (Menzel and Snyder, 1974). How the green and blue receptors are distributed among the remaining six large visual cells, is tentatively described in Table 1.

Other Techniques. Effects of strong illumination on the ultrastructure of microvilli have been observed in the rhabdoms of crustaceans and insects (Röhlich and Török, 1965; White, 1967; Tuurala and Lehtinen, 1971). According to Gribakin (1969a,b) the microvilli of the worker bee's rhabdom swell, become shorter and disintegrate during light adaptation. Using strong chromatic adaptation and osmic acid fixation, the microvillar distortions were found to be restricted to particular color receptors. From these findings Gribakin concluded that in the ventral half of the eye the two type I cells are UV receptors and the six type II and III cells are green receptors. Furthermore, electron dense osmic staining of specific cells after illumination with long wavelength light ($\lambda > 480$ nm) seemed to confirm these results in so far as in the ventral half of the eye the type II and III cells (in the dorsal half only the type II cells) were stained (Gribakin, 1972). These results, however, could not be confirmed in recent reexaminations (worker bee: Grundler, 1973, 1975; Kolb and Autrum, 1974; bulldog ant: Menzel and Blakers, 1975). Both effects - swelling and disruption as well as osmic staining of microvilli - do occur, but are not consistently related to light adaptation. Irrespective of the state of adaptation, the rhabdomeres of the UV receptors, which can be undoubtedly determined by the screening pigment response (see above), often show heavy osmic staining. This effect is not observed in the microvilli of the green receptors. Furthermore, in the UV rhabdomeres swelling or even destruction of microvilli occurs at low pH of the fixative (pH < 7.4), whereas alkaline fixative (pH > 7.8) affects the microvilli of the green receptors, correspondingly. The latter effect was also observed in our studies in bees, where by this means type I cells and the ninth cell turned out to be the UV-receptors (unpublished).

Therefore, the results obtained by Gribakin (1969a,b, 1972) can indeed be used to discriminate between different classes of receptors, but provide no means for specifying these receptors according to their spectral sensitivities. The argument that osmic acid could react with free aldehyde being deliberated from the bleached photopigment cannot be accepted any longer, since we now know that in rhabdomeric photoreceptors rhodopsin does not bleach (Gogala et al., 1970; Goldsmith, in press). On the other hand, we have not yet understood, how the consistantly observed effects of pH and osmic staining on the structure of the microvilli is to be explained in terms of cell physiology. Until now, they are just used as indicators for the presence of specific visual pigments.

One observation, however, may be of more general significance. In worker bees and bulldog ants, the microvilli of the UV receptors are smaller in diameter (37 and 33 nm, respectively) than the microvilli of the green receptors (50 and 52 nm, respectively; Gribakin, 1967a, Menzel, 1975b). As Laughlin et al. (1975) have pointed out, alignment of pigment molecules in the microvillar membrane has at least partly to result from the tubular geometry of the microvillus and the asymmetry of the photopigment molecule. This effect of microvillar structure on the alignment of the rhodopsin molecules is the more pronounced the smaller the diameter of the microvillus becomes. Hence, all other parameters unchanged, higher dichroism should be expected in UV receptors, because their photoreceptor membranes are bent into narrower tubes than they are in green receptors.

Conclusion. Three of the nine retinular cells in the ommatidia of bees and ants are UV receptors - two long retinular cells (nos. 1 and 5) and the short ninth cell. In the worker bee and the most primitive group of ants (Myrmeciinae), the long UV receptors belong to the type I cells characterized by their nuclei in the distal nuclear layer of the retina. However, in the highly advanced formicine ants (subfamily Formicinae, e.g. F. polyctena, C. bicolor), the nuclei of the long small cells (UV receptors) are lying deeper in the retina than the nuclei of the large cells (Menzel, 1972; Herrling, 1975; Fig. 12).

In at least two of the three ant species studied the six large retinular cells are exclusively green receptors. The same seems to be true for the most ventral part of the worker bee's eye (the lower 40-50 of the in total 145 horizontal rows of ommatidia; Gribakin, 1967b). In the middle and dorsal part, however, blue receptors have been demonstrated by intracellular recordings, but further experiments have to reveal their exact number and location within the retinula. Most likely, two blue receptors occur besides four green receptors (for specification see Table 1). In the drone bee color receptors have not been marked intracellularly, but we know from the early work of Autrum and von Zwehl (1962, 1963) that most parts of the retina are composed of UV and blue receptors and that the green receptors are restricted to the uppermost section of the ventral part of the eye. In wasps, P. germanica, apparently no differences exist between the frequencies of UV, blue and green receptors in different eye regions (Menzel, 1971). Until now there has been no evidence that in any Hymenopteran species the UV receptors are exclusively confined to, or that they are the only receptor types of the dorsal part of the eye, as it is described for other fast flying, visually guided insects (Neuroptera: Gogala, 1967; Odonata: Autrum and Kolb, 1968; Horridge, 1969; Eguchi, 1971; Diptera: Burkhardt and de la Motte, 1972). One should mention, however, that in the worker bee the most dorsal part of the eye is characterized by structural pecularities (see Sect. 3.1.1.2 of this chapt.), but that the color receptors in this dorsal eye region have not yet been investigated.

In all cases where the particular types of color receptor have been localized within the retinula, opposite cells coinciding in microvillar direction, always belong to the same spectral type of receptor. They also give rise to the same type of axon terminal in the lamina (see Sect. 3.2.1 and Table 1 in this chapt.).

2.7.3.1.3 Polarization Sensitivity

At first glance it seemed surprising that in Hymenoptera - and in other insects with fused rhabdom eyes as well - the retinular cells failed to show considerable polarization sensitivities (Menzel and Snyder, 1974; Menzel, 1975b). Especially in the green and blue receptors, from which most data are available, polarization sensitivities (PS) are either low or absent (PS values range between 1 and 2). Candidates for mechanisms reducing PS are electrical coupling between receptor cells that have different directions for maximum sensitivity, and the twist of the rhabdom described above.

Electrical coupling between cells of one ommatidium has been most convincingly demonstrated by Shaw (1967, 1969) in the eye of the drone bee. While simultaneously recording from two retinular cells of one ommatidium with a pair of electrodes, he found very low PS. As, however, PS does not change with light adaptation, i.e. is no dynamic process (Menzel and Snyder, 1974), the possibility cannot be ruled out yet that electrical coupling is an artefact produced by the technique of intracellular recording (Laughlin, 1975b).

Be this as it may, the twist of the rhabdom seems to be much more important in reducing PS. By an optical analysis performed for the worker bee's eye we were able to show that the twist of the rhabdom alone destroys PS of the long visual cells completely (Bernard and Wehner, 1975; Wehner et al., 1975; see also Snyder and McIntyre, 1975). As it is well known from behavioral experiments that only the UV receptors participate in E-vector detection, our calculations specifically refer to the UV sensitive retinular cells (nos. 1, 5 and 9). Taking into account the appropriate anatomical and optical parameters (direction of microvilli, cross-sectional areas and lengths of rhabdomeres, twist rate, spectral sensitivity at λ = 350 nm, optical density, dichroic ratio), one arrives at a clear-cut difference between the long and the short UV receptors. Whereas PS of the long twisted UV cells (nos. 1 and 5) is unity, PS of the short basal UV cell (no. 9) is still high, provided that effective birefringence is restricted to very low values ($\Delta n < 10^{-3}$; compare the measurements of effective birefringence in single rhabdomeres of flies, Kirschfeld and Snyder, 1975).

As two types of rhabdoms occur in the bee's retina - one twisting clockwise and the other counterclockwise - the directions for maximum sensitivity differ between the ninth cells of both types of retinula by 30-40°. For a minimum model of E-vector detection two analysers and one polarizationally insensitive receptor are sufficient. In the bee's retina the two types of ninth cells provide two independent analysers, and the long twisted UV cells deliver polarizationally insensitive signals. Hence, these three types of UV cell within two adjacent ommatidia are most probably the receptors that mediate E-vector detection, which is independent of both, mean light intensity and degree of polarization. The distribution of the two different twist-types of rhabdom within the retina optimizes the E-vector-detecting system for high spatial resolution.

2.7.3.2 Lamina: the First Visual Neuropile

According to the compartmentalization of the retina into retinulae, the photoreceptor axons penetrating the basement membrane of the eye from separate axon bundles, often called pseudocartridges. Each bundle containing nine axons twists clockwise or counterclockwise and opposite to the direction of twist of the corresponding retinula (Figs. 10 and 16), before it enters the first visual neuropile, the lamina ("granular layer" of Berger, 1878; "retinal ganglion" of Patten, 1887; "outer fibrillar mass" of Kenyon, 1897; "retina intermediaria" of Cajal and Sanchez, 1915).

Within the lamina, the short visual fibers of one retinula (svf: axons of type II and type III cells, nos. 2-4, 6-8) terminate within one cartridge, where they project to second-order neurons, the monopolar cells. The long throughrunning fibers (lvf: axons of type I cells, nos. 1, 5 and 9) pierce the lamina, cross the first optic chiasma and terminate in the outer layers of the second visual neuropile, the medulla. In gross morphology, the lamina can be stratified into two

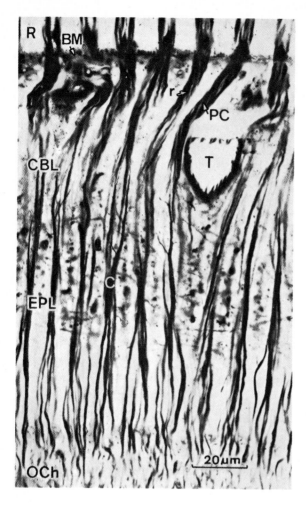

Fig. 16. The retina-lamina projection of the worker bee. Frontal section. Reduced silver staining. BM basement membrane, C cartridge, CBL cell body layer, EPL external plexiform layer (first visual neuropile, laminar neuropile), OCh outer optic chiasma (first chiasma), PC pseudocartridge, R retina, r indicates the twist of the axon bundle (pseudocartridge), T trachea (Ribi, 1975a)

main areas, the outer cell body layer (CBL) and the adjacent plexiform layer. The latter is often referred to as external plexiform layer (EPL), because it constitutes the first (outer) visual neuropile. Both, CBL and EPL, are separated from the basement membrane of the eye by a fenestration layer, rich in tracheae (Fig. 17).

In contrast to the single layer neuropile of Diptera, the laminar neuropile of Hymenoptera is tiered and consists of three strata, numbered A, B and C according to Ribi (1974, 1975a,b). Cajal and Sanchez (1915) did not discriminate between layers A and B and delineated both as "zona plexiforme externa" from the most proximal "plexo tangencial (limitante) inferior" (stratum C). However, stratum B can be clearly recognized as a light band characterized by a preponderence of glial cells and by considerably less lateral spread of neurons than it is due to the inner and outer layer of the lamina. Especially in ants, C. bicolor, neither the retinular cell endings nor the monopolar cells have significant lateral processes in stratum B (Meyer, unpubl.; Ribi, 1975b describes only one type of monopolar cell with short bilateral branches at this level). Furthermore, according to some preliminary histochemical investigations in bees (Weber, pers. comm.), stratum B

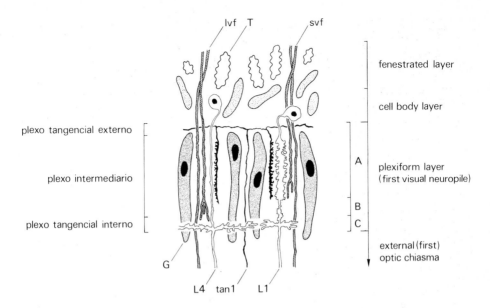

Fig. 17. Schematic diagram representing types and arrangement of neurons in the worker bee's lamina. Left stratification of the laminar neuropile according to Cajal and Sanchez (1915), right classification scheme applied in this paper. G glial cell, L1, L4 monopolar cells (second-order neurons), lvf long visual fibers (axons of retinular cells nos. 1,5 and 9), svf short visual fibers (axons of retinular cells nos. 2-4, 6-8), T trachea, tan 1 centrifugal type of fiber

seems to show high activities of unspecific acid phosphatases and other hydrolytic enzymes, which are supposed to be enriched in glial cells.

What follows is a short survey of types of laminar fibers in bees and ants as revealed by Golgi and Golgi-EM techniques.

2.7.3.2.1 Retinular Cell Axons

In worker bees, three types of retinular cell endings (short visual fibers, svf) can be discriminated by the diameter of the axon, the structure of the terminal branches and the level, at which these branches end in the lamina (Figs. 18 and 19). All types of fibers occur in each cartridge, where they form three pairs of svf. By Golgi-EM serial sections they can be individually correlated with six retinular cells in the directly overlying ommatidium (Ribi, unpubl.).

svf(1), axons of retinular cells nos. 3 and 7, R(d) fibers of Ribi (1974, 1975a): deep fibers terminating in stratum B by tassel-like structures. These thick fibers (diameter 2-3 μm) lack lateral processes in stratum A, but appear pitted along their way through the external plexiform layer. They have been excellently described and portrayed by the early Spanish authors (Cajal and Sanchez, 1915, Fig. 15).

The remaining two types of svf are shorter and hence named shallow fibers (R(s) fibers according to Ribi, 1974, 1975a). They are restricted to stratum A of the laminar neuropile. In contrast to svf(1), they have some lateral spreads, which, however, are confined to one car-

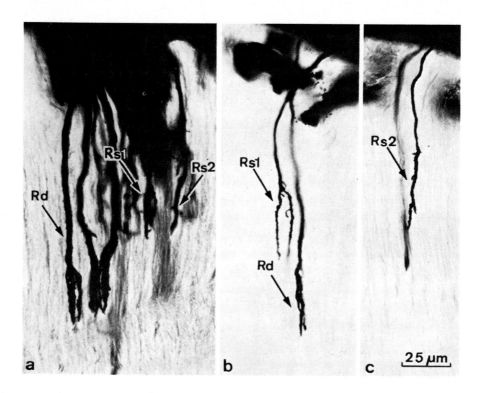

Fig. 18a-c. Types of short visual fibers in the worker bee. Golgi impregnation.
Rd deep visual fiber terminating in stratum B with tassel-like ending: svf(1);
Rs1 and Rs2 shallow visual fibers terminating in stratum A with either forked,
svf(2a), or unforked endings, svf(2b). (Ribi, 1975a). For classification of types
of svf see Fig. 19

tridge. Unpublished data of Ribi confirm that retinular cells nos. 4
and 8 give rise to svf(2a), and retinular cells nos. 2 and 6 to svf(2b).

svf(2a) are characterized by forked endings in the lower part of stratum A. According to Ribi (1974, Fig. 5; 1975a, Fig. 11) their fiber
diameters range between those of the svf(1) and svf(2b) fibers. Although Cajal and Sanchez (1915) do not describe that type of fiber explicitly, they already had found it and shown it in one of their pictures (Fig. 19).

svf(2b) fibers end with slightly tapering, but not branching terminals.
From the representations of these fibers given by Ribi (1974, 1975a,
unpublished data) one has to conclude that these are the most slender
of all short visual fibers. However, the Golgi pictures of Cajal and
Sanchez (1915) reveal these fibers as heavily impregnated structures,
much larger in diameter than the svf(2a) terminals. According to this
early paper they also seem to be impregnated more readily than the
svf(2a) type of fiber, because they are as frequently pictured by the
Spanish authors as the deep retinular fibers svf(1). In Fig. 16A of
Strausfeld (1970b), again svf(1) and svf(2b) fibers, but no single
svf(2a) fiber can be identified. However, in the more quantitative investigations of Ribi (1975a, Fig. 20), no differences in the frequency
of impregnation between svf(1), svf(2a) and svf(2b) could be found. By

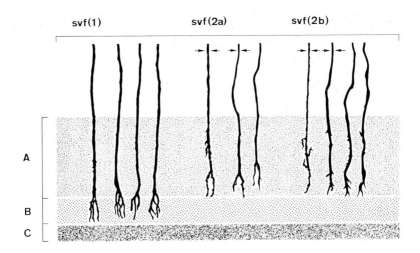

svf(1)   svf(2a)   svf(2b)

A

B

C

Fig. 19. Short visual fibers (svf) in the worker bee's lamina. Golgi impregnations.
The first specimen in each of the three sets of fibers is from Ribi (1974, Fig. 5;
1975a, Fig. 11), the following ones (two or three) are from Cajal and Sanchez (1915,
Figs. 15 and 19); svf(1) deep visual fiber, svf(2) shallow visual fiber. Note the
differences in the diameters of the svf(2) axis fibers (arrows) between the fiber
profiles presented by Ribi and the Spanish authors. The correlation between types
of svf and retinular cell bodies (cells nos. 2-4, 6-8) is given in Table 1

all means, it cannot be ruled out yet that the shallow type of fiber,
svf(2a,b), shows a considerable amount of variation between different
ommatidia or different eye regions.

In ants, C. bicolor, all short visual fibers terminate with stout plug
endings that lack lateral processes. They are all restricted to stratum A and cannot be discriminated by differences in diameter. However,
by variations in their club-like endings and by the levels where they
terminate, Ribi (1975b) has described three different types, whereas
recent Golgi-EM studies of my graduate student E. Meyer only revealed
one type. As long as these fibers and their corresponding retinular
cell bodies are not characterized physiologically, any classification
scheme drawing on pure anatomical data cannot be conclusive. It would
be superfluous to dispute about types and subtypes of fibers, because
we do not know yet the functional significance of variations in branching patterns.

The long visual fibers (lvf, diameter 0.5-0.8 μm) originating from the
three UV receptors of each retinula (cells nos. 1, 5 and 9) terminate
either in the first or in the second outer layer of the medulla. Three
types of lvf have been described in the original papers of Ribi (1974,
1975a), but recent reexaminations (Ribi, unpubl.) reveal only two main
types, which can be readily discriminated in the worker bee: the long
UV cells (nos. 1 and 5) terminate in knot-like endings or in short
lateral spreads in the two outer layers of the medulla, whereas the
short UV cell (no. 9) forms a brush-like terminal in the second medullar layer. One branch of this brush leaves the ramified terminal, passes laterally within the second layer and then ascends back to the
first layer of the medulla. At least in the bee, however, the considerable variation of lvf terminals in the two outer strata of the
medulla has resisted clear-cut classification.

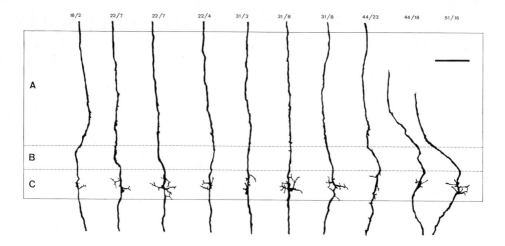

Fig. 20. Long visual fibers within the lamina of <u>Cataglyphis bicolor</u>. Golgi impregnations, 10 specimens arbitrarily chosen. <u>A</u>, <u>B</u> and <u>C</u> strata of laminar neuropile, <u>scale</u> 20 μm. Courtesy of E. Meyer

Within the lamina spine-like structures of the long visual fibers have been mainly observed in stratum B (Ribi, 1975a; Fig. 11). In at least one type of long visual fiber, spines seem also to occur in stratum A, as can be read off drawings in Cajal and Sanchez (1915) and Strausfeld (1970b, Fig. 16A). In <u>Lepidoptera</u>, "spiny long visual fibers" have been observed (Fig. 11 in Cajal and Sanchez, 1915; Figs. 25 and 26 in Strausfeld and Blest, 1970). In these fibers the spines are arranged radially down the total length of the fiber in the external plexiform layer. As described in <u>Lepidoptera</u> (Strausfeld and Blest, 1970), two long visual fibers leaving one retinula are lying closely to one another. This close association between two long visual fibers terminating at two different levels within the medulla has also been observed in bees (Strausfeld, pers. comm.). Most likely these two fibers characterized by their paired arrangement are the axons of the two polarizationally insensitive long ultraviolet receptors. In <u>Lepidoptera</u> pairs of long visual fibers maintain their paired arrangement in the medulla, where their endings are accompanied with a pair of monopolar cell terminals (Strausfeld and Blest, 1970). These arrangements of four endings termed "quads", have not yet been described in bees.

Within the bee's lamina lateral processes of long visual fibers have been demonstrated only in some instances. If they occur at all, they are restricted to the innermost neuropile (stratum C, e.g. see Fig. 16A in Strausfeld, 1970). However, in ants, <u>C. bicolor</u>, all long visual fibers have lateral spreads in stratum C (Ribi, 1975b; Meyer, unpubl.; Fig. 20).

A correlation between retinular cells, their axon terminals, microvillar directions and spectral sensitivities is given in Table 1 and Fig. 28.

2.7.3.2.2 Monopolar Cells

Like the bipolar cells of the vertebrate retina, the monopolar cells of the insect's lamina provide the first relay between the receptor

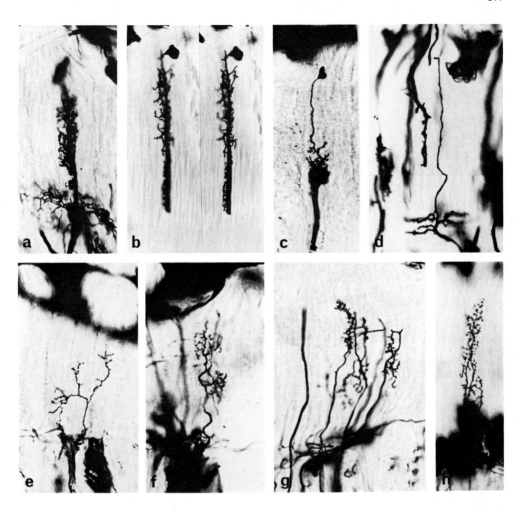

Fig. 21a-h. Monopolar cells (a-d) and centrifugal cells (e-h) in the lamina of the worker bee, _Apis mellifera_. L1a(a), L2(b), L3(c), L4(d). Centrifugal fibers with lateral branches in three cartridges (e), two cartridges (f) and one cartridge (g,h). Courtesy of W.A. Ribi

cells and higher order neurons within the medulla (transmedullary cells). Cajal and Sanchez (1915) described two main classes of monopolar cells in the bee's lamina: the "giant monopolars", defined by lateral processes along the whole length of the axon in the external plexiform layer (EPL, Fig. 22), and the "small monopolars", in which lateral spreads are confined to particular layers. Both classes correspond to the type I and type II cells of Melamed and Trujillo-Cenoz (1968).

Large Monopolar Cells (type I cells, "monopolares gigantes"; axon diameter 4-5 μm). Although A and B have not been distinctly defined by the Spanish authors, it can be read off their figures and descriptions that in the large (giant) monopolars lateral spreads are arranged down the whole length of the axis fiber in stratum A. In addition, wide field, brush-like arborizations ("penacho basal") are regularly found in stratum C (Fig. 22). Hence, in bees the large monopolars are of a stratified appearance. In other groups of insects, e.g. Odonata, they have lateral processes at all levels in the external plexiform layer.

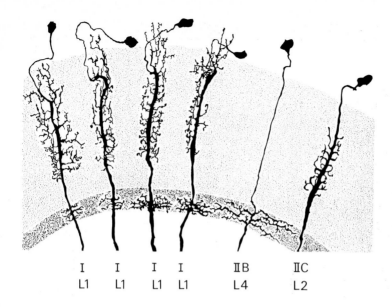

I I I I ⅡB ⅡC
L1 L1 L1 L1 L4 L2

Fig. 22. Monopolar cells in the worker bee's lamina. Golgi impregnations. Fiber profiles selected from Cajal and Sanchez (1915, Figs. 7, 10 and 15). I, IIB, IIC classification of monopolar cells according to the Spanish authors; L1, L2, L4 classifications used in this paper (see Fig. 23)

Small Monopolar Cells (type II cells, "monopolares pequenas"; axon diameter 1-2 μm). In the small monopolars lateral spreads are always less abundant than in the large monopolars and are restricted to certain levels of the lamina. The cell bodies, which often occupy a more distal position than the cell bodies of the large monopolar cells, are connected to the main trunk of the axis fiber by a slender neck (intercalar segment). Three subtypes have been discriminated by Cajal and Sanchez (1915):

II-A. Corpusculo monopolar con penacho inferior: these fibers are mainly characterized by wide-field collaterals at the inner margin of the synaptic layer (stratum C) and by a sudden increase in diameter of the

$\triangleright$

Fig. 23. Monopolar cells of the worker bee, Apis mellifera. Golgi impregnations. Original drawings provided by W.A. Ribi. Types of fiber are characterized by the presence or absence of lateral spreads in strata A, B, and C. L1: lateral spreads at least in strata A and C; L1a corresponds to L1 in Ribi (1974, 1975a), correspondingly L1b = L2, L1c = La (Ribi, 1974) and L3 (Ribi, 1975a); L2 (= Lb): lateral spreads confined to stratum A; L3 (= L3 in Ribi, 1974, and La in Ribi, 1975a): lateral spreads confined to stratum B; L4: lateral spreads confined to stratum C

$\triangleright$

Fig. 24. Monopolar cells of the desert ant, Cataglyphis bicolor. Golgi impregnations. Original drawings provided by E. Meyer. L-fibers are classified according to the scheme outlined for the bee's lamina (Fig. 23). In L1a Ribi (1974b, L1) has found lateral spreads not only in strata A and C, but also in stratum B; the following fibers have been also described by Ribi (1975b): L1b ≈ L2, L2 = L3, L4 = L4. Furthermore, L5-type monopolars have been portrayed by Ribi (1975b, Figs. 4, 5h, and 6a). A, B, and C stratification of lamina, L lamina, M medulla, OCh first optic chiasma

L1a L1b L1c   L2   L3   L4

A

B
C

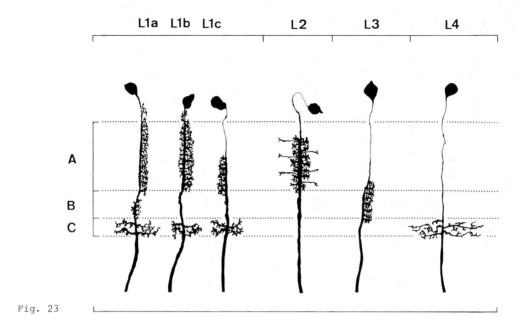

Fig. 23

L1a   L1b   L1c   L2   L4

L   A

    B
    C

OCh

M

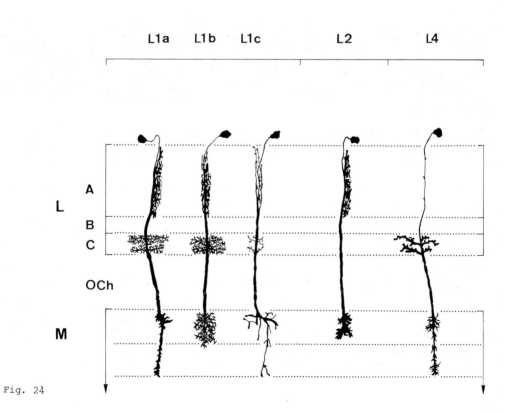

Fig. 24

axis fiber immediately underneath the basal brush-like arborization. The branches leave the axon in a slightly retrograd direction. No other lateral processes or spines occur.

II-B. Corpusculo monopolar diminuto: only minor differences between this and the latter type of fiber are reported by the Spanish authors. The collateral spreads in stratum C are more straight, longer and often spiny. The axis fiber lacks the thickened sigment of the type II-A fiber.

II-C. Corpusculo monopolar con largo penacho descendente: primarily, this type of fiber has been defined by means of its terminal in the outer layers of the medulla, resembling the branches of a weeping willow (see Fig. 15h in Cajal and Sanchez, 1915). However, it is also characterized by its pattern of laminar processes which are restricted to stratum A. Hence, in the papers of the Spanish authors these fibers are the only ones lacking wide-field aborizations in stratum C.

Another classification scheme applicable to all groups of insects has been proposed by Strausfeld and Blest (1970). These authors define as "giant" those monopolar cells that have lateral processes through more than one cartridge. Correspondingly, the lateral spreads of their "small" monopolars are restricted to a single cartridge. Hence, at least in flies the type I cells of Melamed and Trujillo-Cenoz (1968) and the giant monopolars of Cajal and Sanchez (1915) are small monopolars according to Strausfeld's and Blest's (1970) classification.

Most appropriately, monopolar cells should be defined by means of their connection patterns which they form with their input and output fibers (receptor cell terminals, centrifugal fibers, amacrine cells, and transmedullary fibers). In flies, the type I cells receive inputs from all short visual fibers and hence can be clearly discriminated from the other types of monopolar cell. However, as there are only some preliminary data on the synaptology of the laminar cartridge in Hymenoptera (Varela, 1970, see Fig. 26), patterns of arborization have still to be used in classifying monopolar cells. Drawing on the most recent data of Ribi (unpubl., Fig. 21a-d) in bees and Ribi (1975b) and Meyer (unpubl.) in ants, C. bicolor, our present status of knowledge can be most appropriately summarized by applying the following classification scheme (Figs. 23 and 24).

L1 fibers. Because of their lateral spreads in strata A and C, these fibers correspond to the giant monopolars of Cajal and Sanchez (1915). In ants, long collateral processes originating from the axis fiber in stratum A run parallel to the main trunk of the axon and extend towards the soma of the cell. The wide-field arborizations in stratum C mainly occur within two levels. Different subtypes can be discriminated according to the presence or absence of lateral spreads in stratum B and their extent and symmetry of arrangement in stratum A (bees: L1, L2 and La in Ribi, 1974, 1975a; ants: L1 and L2 in Ribi, 1975b).

L2 fibers. Lateral processes are confined to stratum A. These fibers described as Lb in Ribi (1974, 1975a) and L3 in Ribi (1975b) correspond to the small monopolars type II-C of Cajal and Sanchez (1915). In bees, besides short processes, up to 20 μm long lateral spreads occur. They run to the six adjacent cartridges as well as to one more distant cartridge along the yz-axis. The characteristic terminations of this type of fiber in the outer layer of the medulla (see above) have been described by Ribi (1974, 1975a) as well, but are absent in ants.

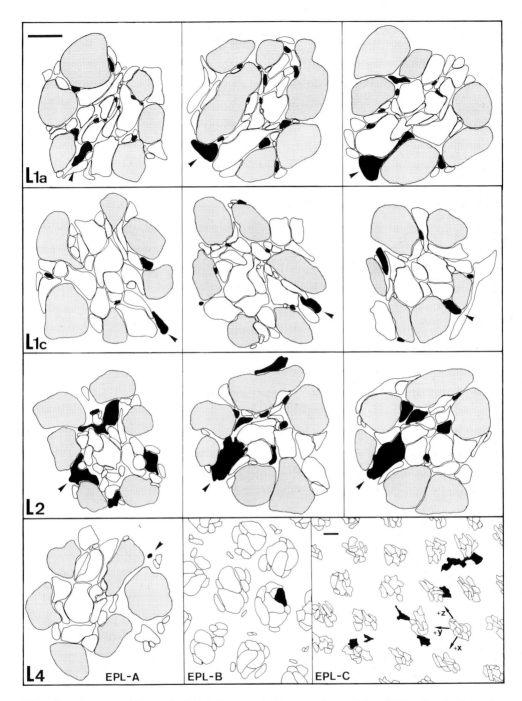

Fig. 25. Cross sections of Golgi impregnated monopolar cells within the lamina of Cataglyphis bicolor. Elements of single L1a, L1c, and L2 fibers are shown at three different levels of stratum A (left distal, right proximal). L4 profiles are given for strata A, B and C of the laminar neuropile (external plexiform layer, EPL). Black profiles of Golgi impregnated monopolar cells, shaded cross sections of short visual fibers, the arrow points at the axis fiber, scale 2 μm. Courtesy of E. Meyer

316

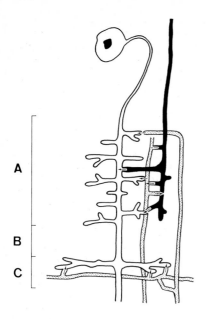

Fig. 26. Synaptic connections in the cartridge of the worker bee, based on data, especially on Fig. 11, of Varela (1970). White monopolar cell (type L1), black short visual fiber (retinular cell axon), shaded centrifugal fibers, A, B, and C stratification of laminar neuropile. Among the different types of synaptic connections (arrows) no presynaptic elements of monopolar cells within the lamina have been reported

A

B

C

L3 fibers. In this type of fiber which has been found exclusively in bees, short lateral processes spread radially in stratum B.

L4 fibers. Long collaterals confined to stratum C are arranged bilaterally along the dorsoventral axis of the eye. In bees, these lateral branches reach six cartridges lying in the (+x), (+x+y), (+y), (-x), (-x-y), (-y) axis. In ants, too, the branches extend in the dorsoventral direction and contact neighboring cartridges (Fig. 25). L4 fibers have already been carefully described and portrayed by Cajal and Sanchez (1915; small monopolar cells, type II-A). In flies, small-field intercartridge connections are provided by a similar type of fiber, termed tripartite monopolar cell (Strausfeld and Braitenberg, 1970; Strausfeld, 1971; Braitenberg and Debbage, 1974). Three collaterals of this cell run to three different cartridges: one collateral to the parent cartridge and the others to the nearest neighbors along the (+y) and (-x) axes.

The axons of the monopolar cells and of the long visual fibers leave the lamina, cross the outer (first) optic chiasma and reach the second synaptic layer, the medulla. As the crossover of fibers is restricted to the horizontal plane of the head, fibers of the anterior portions of the lamina (corresponding to the anterior visual field) terminate in the posterior portion of the medulla and vice versa. Hence, the topology of fibers is reversed in the anterio-posterior direction, but not in the dorsoventral direction.

2.7.3.2.3 Amacrine Cells and Centrifugal Fibers

In addition to receptor axons and monopolar cells, extensive descriptions of fibers branching within the bee's lamina have been provided by the Spanish authors and more recent investigators (Fig. 21e-h). Although different names have been used in describing these fibers, only two types will be distinguished here: (1) <u>amacrine cells</u>, the cell bodies of which are located in the inner parts of the plexiform layer or just underneath that neuropile, and (2) <u>centrifugal fibers</u>, which invade the lamina from more centrally positioned cell bodies. The term "centrifugal" only implies that morphologically the fiber terminals within the lamina grow peripherally from their cell bodies. In general, these cell bodies are placed around the first optic chiasma, i.e. between the lamina and the medulla (Fig. 1), but they may be also positioned within the medulla or even more centrally.

In bees, Cajal and Sanchez (1915) distinguished three types of centrifugal fibers: (a) Fibers which ascend through the whole lamina and branch with their horizontally arranged terminals in the most distal layer of the laminar neuropile, i.e. in the "plexo tangencial externo". The Spanish authors seemed to have supposed that some small monopolar cells had a few lateral processes within that distal layer of the laminar neuropile, but those processes have not been mentioned by later authors. Similarly, at the same level the "midget cells" of flies (Strausfeld, 1971; type <u>L5</u> monopolars) are characterized by a few processes from their axis fibers, as are the tripartite cells (type <u>L4</u> monopolars). (b) Fibers which terminate in the inner plexiform layer (stratum C, "plexo tangencial (limitante) inferior"), and (c) fibers which ascend up to the distal margin of the lamina, where they split into horizontal branches. From these branches "garlands" descend back through the laminar neuropile parallel to the cartridges. The latter fibers have been described in more detail by Ribi (1974, 1975a) and have been christened tangential fibers (type <u>tan (1)</u>, Fig. 27 B1). All types of tangential fiber are characterized by a main axon trunk (linking fiber), which originates from the cell body and then splits into two branches, one ending in the lamina and the other in the distal parts of the medulla (Fig. 27B).

The diversity of amacrine and centrifugal types of fiber by far exceeds the number of different fiber types involved in the direct (vertical) pathway from retinular cells to monopolar cells. However, we are far away from any successful attempt to classify these fibers and to correlate them functionally to receptor cell terminals and second-order neurons. Even morphologically, the branching patterns of many centrifugal fibers have not been traced in full detail, so that they have to be referred to as "incerta sedis". Furthermore, in comparison to the equivalent types of fiber in the vertebrate retina (horizontal and amacrine cells), nothing at all can be said about the functional significance of amacrine and centrifugal fibers in insects, because of the complete lack of physiological data on these fibers. However, it is probably safe to estimate that the integrative functions of the bee's first visual neuropile cannot be understood even in its major features as long as the phsysiology of these local circuit neurons is unknown.

One has to bear in mind that not only amacrine cells, but also some of the so-called centrifugal fibers might primarily provide local integration within the lamina rather than efferent control of laminar neurons. Especially in the tangential fibers described above, the two separate areas of dendritic spread within the lamina and the medulla, respectively, could well act as local circuits. The axon combining the two dendritic fields may actually serve to isolate the two local integrating systems as is supposed to be the case in one type of vertebrate horizontal cell (Fisher and Boycott, 1974). In tentative explanation, the functional role of the tangential fibers in bees could in some way resemble the functional role of the axonal type of horizontal cell in vertebrates (Fig. 27A). However, it is much too early yet to speculate that the laminar dendritic field of the tangential fiber

A

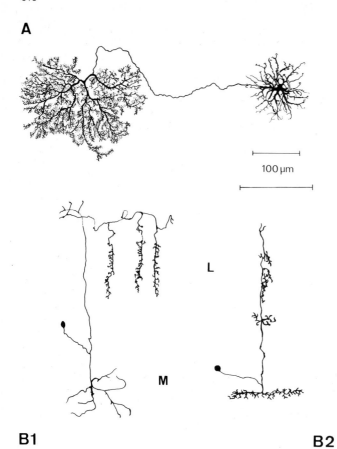

100 μm

L

M

B1   **B2**

Fig. 27A and B. Local circuit neurons in vertebrates (A) and Hymenoptera (B). Golgi impregnations. (A) axonal type of horizontal cell from the retina of the cat (Fisher and Boycott, 1974). (B) centrifugal fibers branching in the lamina (L) and the medulla (M) of the worker bee; (B1) tan(1), (B2) twf (Ribi, 1975a)

receives input from the short visual fibers and the medullar field from the long visual fibers as the two dendritic fields of the vertebrate horizontal cells are exclusively connected to rods and cones, respectively (Kolb, 1974). Whereas in the vertebrate case synaptic interaction between receptor cell terminals, bipolar cells and horizontal cells is proved, nothing is known about reciprocal junctions between the corresponding neurons in insects. Hence, the evidence for analogous elements in the local neuronal circuits of the insect's and the vertebrate's peripheral visual systems is still scanty. These elements may be, nonetheless, not as different between both groups of animals as they at first sight appear.

2.7.3.2.4 Patterns of Connectivity

It is far beyond the scope of this article to draw wiring diagrams for either the projection neurons (receptor cells, monopolar cells, transmedullary cells) or the local circuit neurons (amacrine cells, centrifugal fibers). However, armed with the catalogue of cell types, which

have been consistantly and beautifully portrayed by Golgi and reduced
silver techniques, the raw material for such a synthesis shall be summarized briefly.

Vertical Projections. Synaptic contacts between the short visual fibers
(types svf(1), svf(2a), svf(2b)) and monopolar cells (types L1, L2, L3)
are established in the outer and middle plexiform layer of the laminar
neuropile (strata A and B in bees, stratum A in ants). In the unilay-
ered lamina of Diptera each of the two large monopolar cells within
one cartridge receives input from each of the six short visual fibers
(Trujillo-Cenoz, 1975, 1972). Because of functional as well as morphological reasons, this cannot be proposed for Hymenoptera, where three
different types of short visual fibers are supposed to interact with
a variety of monopolar cell types (L1-subtypes, L2, L3). From spine-
like protrusions of the long visual fibers within strata A and B it
can be furthermore assumed that at least two of the long visual fibers
(axons of cells nos. 1 and 5) form synaptic junctions with cartridge
neurons.

Although the pattern of convergence of receptor cells on monopolar
cells is still unknown in any fused rhabdom type of compound eye, evidence continues to accumulate that the large monopolar cells receive
their major input from retinular cells with the same field of view
(Odonata, Orthoptera; see Laughlin, 1975b). This conclusion heavily
draws on the fact that the angular sensitivity functions of the large
monopolar cells are not broader, but mostly narrower than the angular
sensitivities of the receptor cells. On the other hand, one can conclude from the spectral sensitivities of monopolar cells that all three
spectral types of receptor converge upon an individual large monopolar
cell and supply excitatory (green and blue receptors) as well as inhibitory inputs (UV receptors) to that cell (Menzel, 1974). These two
pieces of evidence together make it quite likely that there is at
least one type of large monopolar cell that receives input from all
types of color receptor within one ommatidium.

Lateral Interactions. In some insects, lateral inhibition has been
found to act on monopolar cells (Diptera: Zettler and Järvilehto,
1972; Arnett, 1972; Odonata: Laughlin, 1974b, 1975c), but the laminar
interneurons providing such an inhibitory mechanism, are still uniden-
tified. Morphologically, lateral processes between receptor cell axons
originating from different ommatidia, have not been found. Among the
different types of monopolar cell, in bees the type L2 neurons establish direct lateral connections to at least the six nearest neighbor
cartridges (stratum A), and in bees as well as ants the L4 neurons
have bilateral spreads to six more cartridges along the dorsoventral
axis of the eye (stratum C). Furthermore, in ants, C. bicolor, the
lateral processes of L1-fibers, stratum C, extend to at least the
nearest neighbor cartridges. At all levels of the lamina, however,
most of the lateral interactions are provided by amacrine cells or
centrifugal fibers. As far as these local circuit neurons have been
described until now, they are horizontal wide-field fibers. Especially in stratum C this horizontal system forms a dense network of lateral processes which may interact with monopolar cells leaving the
cartridges. Type L1 and type L4 cells and - at least in ants - the
long visual fibers show collateral spreads at that level.

If the concept is right that the centrifugal fibers within the lamina
serve as local integration and efferent control systems, concomitantly, besides lvf the monopolar cells are the only output neurons of
the lamina. The retinotopic projection of receptor cell axons to the
monopolar cells and the sharpening of spatial and temporal contrast

by the monopolar cells (Zettler and Järvilehto, 1972; Laughlin, 1974b, 1975c) enable them to transmit a high acuity picture of the outside world to the second visual neuropile. Most likely, it is within this second neuropile that particular properties of single points within the picture, e.g. the hue of color and the direction of E-vector, are determined.

2.7.4 Some Functional Aspects

"An observation that explains nothing is no more a fact than an ex-planation founded upon nothing."
William Patten in "Eyes of Molluscs and Arthropods", 1886.

It should have become apparent from the preceding sections that our knowledge on what happens beyond the level of the bee's or ant's ret-ina is rudimentary, and that it is completely missing if we proceed further beyond the level of the lamina. This substantial lack of phys-iological data on the peripheral visual system of Hymenoptera is the more discouraging, as clear-cut results on visually guided behavioral responses have accumulated during the last years. One way to cope with this unsatisfactory situation may be to shift our attention from the sequence of integrative functions performed by the peripheral pathways to the needs the whole visual system must meet. In following that line, a careful analysis of behavioral performances in Hymenoptera may well provide us with the main traits in the functional layout even of the peripheral neuropiles of compound eye systems.

While foraging, bees and ants rely on visual information mainly in three respects:

1. The insect has to determine the direction of course by means of the pattern of polarized light in the sky (E-vector navigation).

2. During locomotion the insect's course has to be kept constant by optomotor control (course stabilization).

3. Finally, as the foraging course is destined to lead to specific food sources, objects in the environment have to be located and recognized (target detection).

In the following I want to substantiate the tentative idea that (1) navigation and (2) course stabilization are mediated by their own subsystems, each fed by one set of color receptors: by the UV-re-ceptors and the green-receptors, respectively. In evolutionary terms, both these subsystems may well be old systems, which - once established - had not to be further improved substantially. On the other hand, one may suppose that at least in the visual systems of the flower visiting bees, in more recent times natural selection has particularly acted upon (3) target detection, in order to increase the efficiency of foraging. As a consequence of this specific selection pressure, the bee's visual system has developed the high discrimatory capacities for which it is so well known.

Consider the general case in foraging: during flight a bee is steering its course by the E-vector pattern in the sky and is stabilizing this course by optomotor control, mediated by wide-field motion sensitive neurons. At some instant, the bee becomes attracted by a specific vis-ual stimulus. As the bee flies, the system detecting the particular object, has to be motion sensitive, too, but the needs for the detection of an object and for the analysis of the apparent movement of the surround are completely different. The latter system (motion detection for the stabilization of course) should be insensitive to particular features of the floating terrestrial environment, whereas the former (motion detection for the recognition of objects) should instead emphasize those features. Hence, there is no need for color coding in the latter system, but strong need for it in the former system. As experiments on the bee's behavior support these suggestions, one might presume that it is only in the target detecting system that all spectral types of receptor cooperate.

2.7.4.1 E-Vector Navigation

As first suggested by von Frisch (1954) and subsequently proved by Duelli and Wehner (1973) and von Helversen and Edrich (1974), only the UV-receptors are involved in E-vector navigation. At least in the bee, we furthermore know that the only polarizationally sensitive receptors are UV receptors, i.e. the short ninth cells (Menzel and Snyder, 1974; Wehner et al., 1975). Hence it does not seem surprising that besides E-vector navigation other behavioral responses elicited by polarized light are also restricted to the UV range, e.g. it is only by UV light that optomotor turning reactions can be induced in bees by means of moving patterns of polarized light (Kirschfeld, 1973b).

It seems reasonable to assume that originally insects evolved polarizationally sensitive UV receptors for taking advantage of skylight polarization. There is nothing important on earth that is consistantly related to polarized light. What a bee experiences in Kirschfeld's (1973b) apparatus will appear to her as a rather unusual and unexpected situation. Under natural conditions, polarized light is created by reflection of light on terrestrial objects, especially on wet surfaces, but this effect only blurs the image the insect wants to resolve. Polarization sensitivity, so necessary for skylight navigation, is a completely undesirable property for receptors specialized for terrestrial orientation. Hence, the bee should get rid of it in visual cells others than UV receptors. Only in the latter selection pressure has favored polarization sensitivity.

As in fused rhabdoms high alignment of rhodopsin dipoles with the microvillar axis provide the rhabdom with maximum absorption of unpolarized light, it seems that dichroism is merely a secondary consequence of microvillar structure that primarily has been designed for maximum absolute sensitivity (Snyder and Laughlin, 1975). Subsequently, polarization sensitivity caused by the dichroism of the microvillus has proved advantageous only in the UV receptors. By twisting the rhabdoms the bee has completely abolished polarization sensitivity in the green and the blue receptors, but also in the two long UV receptors. As a consequence of that twist the only polarizationally sensitive UV receptors are the short UV cells, of which only one occurs in each ommatidium.

Given a certain rate of twist and a certain concentration of visual pigment, it is only the shortness of the ninth cell that provides polarization sensitivity of this UV receptor: the shorter the rhabdomere the higher the polarization sensitivity (PS) of the corresponding cell, compared with the dichroic ratio of the microvilli. For reasonable values of specific optical density and dichroic ratio, Gribakin (1975) has calculated the optimal length of a straight (untwisted) rhabdomere by maximizing the difference in absorption with E-vector parallel and perpendicular to the microvilli. He arrived at an optimal length of 80 μm. In fact, however, PS increases continuously with decreasing length of rhabdom, and it is only by optimizing PS together with absolute sensitivity that intermediate values of rhabdomere length are favored. In dorsal regions of the worker bee's eye, which are used efficiently in E-vector navigation, the length of the twisted ninth cell amounts only to about 40 μm. According to our calculations, even in such a short cell incorporated into a long rhabdom the maximum sensitivity to polarized light is still 26% of that of an overlying long UV cell, and PS is 4.2 compared with dichroic ration $D_{\parallel}/D_{\perp} = 5.0$ of the microvilli (twist rate $1^{\circ}/\mu m$; Wehner et al., 1975). Hence, the short UV cells of bees serve as efficient analyzers.

The high polarization sensitivity of a short retinular cell is not only due to the low rhodopsin concentration (low absorption), but also to the small amount of twist. In dragonflies, too, the short distal UV cell is the only retinular cell showing high polarization sensitivity (Laughlin, 1975b). If the short cell occupies a proximal instead of a distal position within the retinula, overlying cells may provide the basal cell with a crossed-polarizer effect (ninth cell of ants, Menzel and Blakers (1975); eighth cell of flies). However, as in the bee's basal ninth cell the overlying long UV cells have completely lost their polarization sensitivities by twisting, the latter act as density filters, but not as polarizers (Wehner et al., 1975).

There are several possibilities by which short UV cells can design an E-vector detecting system. Provided that E-vectors shall be detected independently of the mean intensity and the degree of polarization and that E-vector detecting is done instantaneously (without temporal scanning), three receptors are necessary: either three polarizationally sensitive receptors (three analyzers with three different directions for maximum sensitivity; Kirschfeld, 1972, 1973a), or two analyzers and one polarizationally insensitive receptor (Wehner et al., 1975). The second possibility, which is most likely realized in the bee's visual system, is advantageous because of a number of reasons: only two analyzers are needed, which are provided by the ninth cells of two adjacent ommatidia twisting in opposite directions. As the polarizationally insensitive UV receptors are incorporated into the same ommatidium, only two ommatidia have to cooperate for the detection of E-vector. The visual fields of two ommatidia markedly overlap (Laughlin and Horridge, 1971; Eheim and Wehner, 1972), so that E-vectors can be detected in rather small areas of the sky. It is not yet know how many E-vector detectors, i.e. pairs of ommatidia twisting in opposite directions, have to be stimulated for a precise determination of E-vector, but from preliminary behavioral experiments it can be concluded that seven ommatidia - one ommatidium and its six nearest neighbors - are sufficient (Helversen, pers. comm.; own data). Another advantage of the proposed E-vector detecting system consists in the fact that the long twisted UV cells can be used for both the analyses and E-vector.

2.7.4.2 Course Control

By an ingenious experimental device Kaiser and Liske (1972, 1974) and Kaiser (1974, 1975) were able to measure the spectral sensitivity of the optomotor turning reaction in flying and walking bees. Stripes illuminated by one monochromatic beam moved in front of a background of different color. In the experiments no color-specific responses were found. The turning reactions always disappeared at a sharply defined ratio of intensities of the two colors. From these matching intensities an accurate spectral sensitivity function can be obtained.

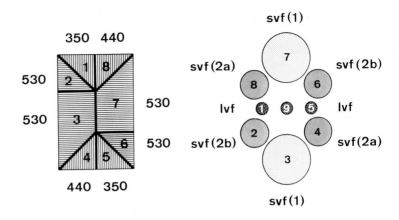

Fig. 28. Correlation between rhabdom structure and spectral sensitivity of retinular cells (left) and arrangement of retinular cell axons within the pseudocartridge and axon terminals in the lamina (right). 1-8 numbers of retinular cells, 340, 440, 530 λ_{max} of spectral sensitivity function, svf, lvf types of short and long visual fibers (see Table 1, Fig. 19 and text). According to Sommer and Wehner (1975), Ribi (in prep.), Menzel and Blakers (in prep.)

As this spectral sensitivity function exactly coincides with the spectral sensitivity of the green receptors (Autrum and von Zwehl, 1964), by approximation it can be concluded that the wide-field motion detecting system receives input only or at least predominantly from sampling stations of the green receptors.

This conclusion might be supported by recordings from motion detecting units, first investigated by Kaiser and Bishop (1970) in the anterior optic tract between lobula and protocerebrum. There are four sets of directionally sensitive motion detectors with their preferred directions pointing at anterior, posterior, dorsal and ventral directions. Both, Kaiser (1972, 1975) and Menzel (1973) have recorded from these units and have arrived at the same results: the neurons cease to react to wide-field movements, when the brightness contract between the two colors is eliminated. The spectral sensitivity functions which both authors obtained, are again similar to the green-receptor curves (Fig. 29). In principle, however, one cannot rule out the possibility that all three types of color receptors contribute to the motion detecting units by excitatory inputs, if one assumes that these are modulated by specific weighting factors. Bishop (1970), too, decided in favor of a system based on inputs from all types of receptors, but the stimuli he used were short light flashes. However, there are only weak responses of the motion detecting units to stationary lights. Furthermore, latencies instead of spike frequencies were measured in that investigation. Some support for the assumption that all types of color receptors are involved in motion detection, might be drawn from the spectral sensitivity function of the bee's retina (Goldsmith, 1960), because this function demonstrates that even the absolute sensitivity of the bee's retina is markedly dominated by the green receptors. Within the limits of error, it coincides with the spectral sensitivity functions of both, the motion detecting neurons and the optomotor turning responses of the bee.

This argument, however, is weakened by the following considerations. ERG measurements are markedly influenced by higher order neurons, e.g.

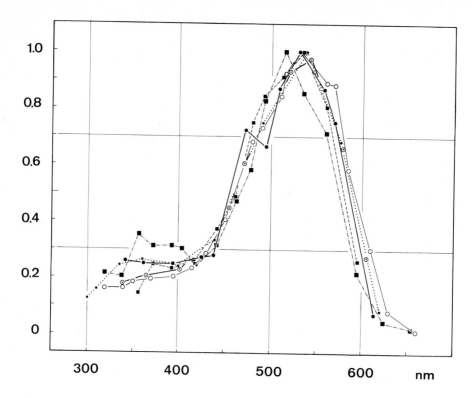

Fig. 29. Spectral sensitivity of the wide-field motion detecting system in the worker bee. ● (heavy solid line) optomotor flight behavior (Kaiser, 1975, Fig.16.9), ⊙ (medium solid line) optomotor walking behavior (Kaiser, 1974, Fig. 2), ■ extra-cellular recordings from motion detecting units in the optic lobe; tract between lobula and protocerebrum (dashed line: Kaiser, 1972, Fig. 2, Kaiser, 1975, Fig. 16.6; dashed and dotted line: Menzel, 1973, Fig. 4), · (dotted line) ERG, compound response (Goldsmith, 1960, Fig. 7 and 9), o (light solid line) intracel-lular recording from green sensitive retinular cells (Autrum and von Zwehl, 1964, Fig. 24)

monopolar cells, and hence are not conclusive in that context. Expe-cially in the ERG recordings performed in bees, compound responses and not isolated receptor components were used for calculating spec-tral sensitivities. Goldsmith (1960) was fully aware of the restric-tions caused by possible interactions between receptors on the inter-pretation of the spectral sensitivities he had obtained. On the other hand it seems to be beyond all doubt that ERG measurements are marked-ly influenced by the exact recording site within the retina. In any case, where spectral sensitivities have been deduced from ERG record-ings in species of Hymenoptera others than bees, UV sensitivities are well pronounced (wasps and ants; Menzel, 1971; Roth and Menzel, 1972; Nowak and Wehner, unpubl.; see Fig. 30). Furthermore, the spectral sensitivity function of the motion detecting units cannot be selective-ly adapted by illumination with green light of $\lambda = 532$ nm (Menzel, 1973), so that at least the UV receptors do not contribute to the responses by an excitatory input. Another possibility seems more like-ly. As at least one type of monopolar cell receives inhibitory input from the UV receptors (Menzel, 1973), these monopolar cells are candi-dates for the neurons that trigger optomotor responses. If they also

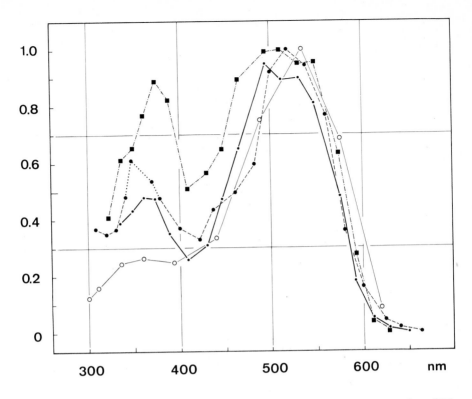

Fig. 30. Spectral sensitivities of the eyes of different Hymenopteran species (ERG measurements after dark adaptation). o (<u>light solid line</u>) worker bee, <u>Apis melli-fera</u> (Goldsmith, 1960, Figs. 7 and 9, ERG compound response), ■ (<u>dashed and dotted line</u>) wasp, <u>Paravespula germanica</u> (Menzel, 1971, Fig. 5, ERG plateau response), · (<u>heavy solid line</u>) wood ant, <u>Formica polyctena</u> (Roth and Menzel, 1972, Fig. 3, ERG compound response), ● (<u>dashed line</u>) desert ant, <u>Cataglyphis bicolor</u> (Nowak and Wehner, unpubl., ERG compound response; the <u>dotted</u> part of the curve between 341 and 378 nm needs further investigation)

contribute to the compound ERG response, the low UV sensitivity in the spectral sensitivity function recorded by Goldsmith (1960) could be explained. In any way, the only excitatory input and the major input at all that the wide-field motion detecting system receives, is from the green receptors.

2.7.4.3 Detection, Localization and Recognition of Objects

In flying or fast running foragers, as the highly developed Hymen-opteran species are, target detection means analyzing a specific vis-ual stimulus against a moving background. Rapidly moving objects, how-ever, cause the image to become blurred and distorted, and hence those objects are difficult to recognize during high-speed locomotion. As the distorting effect induced by the insect's own movement, is the smaller the less the angular position of the object deviates from the direction of locomotion, the insect should try to stabilize the pic-ture of the object in the frontal part of its visual field. This is exactly what a flying bee does during pattern recognition (Wehner and Flatt, in prep.). Furthermore, flies perform saccadic and stabilization movements of the head and hence the eyes (Land, 1975), just as it is done by humans during optokinetic nystagmus or voluntary visual scanning.

In Hymenoptera, nothing at all is known about motion sensitive neurons involved in the detection and identification of objects. At present, we can only outline the conditions such a motion sensitive system shall meet. Let us start with the assumption that, in general, objects first have to be detected in the lateral visual field and then have to be identified in the frontal visual field. The system detecting an object should be fed by small-field motion sensitive neurons that have not to be directionally sensitive by necessity. Electrophysiological measurements made by several authors in the medulla, lobula complex and the protocerebrum, mostly of Diptera and Orthoptera, provide an abundance of candidates for those units (for summaries see Mimura, 1975; Collett and King, 1975; Kien, 1975).

An important feature of the appropriate neurons is that they respond to small movements in the visual field, but not to large-field movements caused by the insect's translatory and rotatory components of movement. One example of such a novelty or "jittery" neuron that has been worked out most elaborately, is the descending movement detector (DMD) in locusts and crickets (Rowell, 1971; O'Shea and Rowell, 1975). This neuron is highly sensitive to abrupt movements of small dark objects or to a rapid decrease of illumination over small areas of the visual field. Although no related behavioral responses have been observed in locusts, it is most likely that neurons of that type are decisive constituents of any target detecting system. It would be interesting to know whether neurons that detect objects in the lateral field of view, are already involved in the identification of the object. It may well be that small-field motion detecting neurons only become attracted by specific parameters of object. Most probably, however, the main task of these neurons is to detect and to locate an object in the visual field and hence to induce the fixation process. At least in bees and ants, the neurons that subsequently come into play when the object has been stabilized in the frontal visual field have to be part of a system that is sensitive to a variety of parameters, e.g. color and spatial configuration of the stimulus. As to the physiological mechanisms, until now one can only state that the responses of monopolar cells within the lamina provide the higher order neurons with a high acuity, contrast efficient and color sensitive topological representation of single points in the environment (Menzel, 1974; Laughlin, 1975c). In search of strategies for target detection, one should expect to find small-field movement and color detectors in the medulla of foraging Hymenoptera.

2.7.4.4 Conclusion

Basic functions of the visual system of Hymenoptera are mediated by separate receptor systems. Because of receptor specializations, the color coding system is insensitive to polarized light, and the E-vector detecting system is insensitive to the hue of color. In bees, it is only the long twisted UV receptor that contributes to both, color and E-vector detection. Whereas the E-vector detecting system, which determines the direction of course in long-distance navigation, receives its only input from the UV receptors, stabilization of course by optomotor control, i.e. by wide-field motion detecting neurons, is predominantly or even exclusively done by the green receptors. Hence, only one type of color receptor contributes to E-vector navigation and predominantly to course control, but all three types of color receptors are involved in detection and recognition of objects

that are of biological significance for at least the higher developed
Hymenopteran species. Most likely, it is this type of behavior, which
we admire so much in a foraging bee, wasp or ant, that is mediated by
small-field motion detecting neurons. It seems reasonable to assume
that just these interneurons have been particularly shaped by natural
selection in recent evolutionary times.

Acknowledgments. The experimental work has been supported by Swiss
National Science Foundation Grant No. 3.814.72 and the Sandoz Foundation. I am especially grateful to Dr. W.A. Ribi (Australian National
University, Canberra), who allowed me to cite his unpublished material
(incorporated into Figs. 21, 23, 28 and Table 1), and to my graduate
students dipl. biol. E. Meyer and F. Räber for important measurements
and reconstructions. Furthermore, I would like to thank Prof. Dr. G.D.
Bernard (Yale Medical School, New Haven, Conn.) for helpful discussions
and Prof. Dr. R. Menzel (Technical School, Darmstadt, Germany) for
providing me with the main results of his unpublished Procion yellow
studies on the bee's retina-lamina projection (Fig. 15 of the paper
of Menzel and Blakers, submitted to J. Comp. Physiol.).

2.7.5 References

Arnett, D.W.: Receptive field organization of units in the first optic ganglion
of Diptera. Science 173, 929-931 (1971)
Arnett, D.W.: Spatial and temporal integration properties of units in first optic
ganglion of Dipterans. J. Neurophysiol. 35, 429-444 (1972)
Autrum, H., Kolb, G.: Spektrale Empfindlichkeit einzelner Sehzellen der Aeschniden.
Z. Vergl. Physiol. 60, 450-477 (1968)
Autrum, H., Zettler, F., Järvilehto, M.: Postsynaptic potentials from a single
monopolar neuron of the ganglion opticum I of the blowfly Calliphora.
Z. Vergl. Physiol. 70, 414-424 (1970)
Autrum, H., Zwehl, V. von: Zur spektralen Empfindlichkeit einzelner Sehzellen der
Drohne (Apis mellifica ♂). Z. Vergl. Physiol. 46, 8-12 (1962)
Autrum, H., Zwehl, V. von: Ein Grünrezeptor im Drohnenauge (Apis mellifica ♂).
Naturwissenschaften 50, 698 (1963)
Autrum, H., Zwehl, V. von: Die spektrale Empfindlichkeit einzelner Sehzellen des
Bienenauges. Z. Vergl. Physiol. 48, 357-384 (1964)
Baumann, F.: Slow and spike potentials recorded from retinula cells of the honey
bee drone in response to light. J. Gen. Physiol. 52, 855-876 (1968)
Baumann, F.: Electrophysiological properties of the honey bee retina. In: The
Compound Eye and Vision of Insects. Horridge, G.A. (ed.). Oxford: Clarendon 1975,
pp. 53-74
Baumgartner, H.: Der Formensinn und die Sehschärfe der Bienen.
Z. Vergl. Physiol. 7, 56-143 (1928)
Beier, W., Menzel, R.: Untersuchungen über den Farbensinn der deutschen Wespe
(Paravespula germanica, Hymenoptera, Vespidae): Verhaltensbiologischer Nach-
weis des Farbensehens. Zool. Jb. Physiol. 76, 411-454 (1972)
Berger, E.: Untersuchungen über den Bau des Gehirns und der Retina der Arthropoden.
Arb. Zool. Inst. Wien, Bd. 1 (1878)
Bernard, G.D.: Physiological optics of the fused rhabdom. In: Photoreceptor Optics.
Snyder, A.W., Menzel, R. (eds.). Berlin-Heidelberg-New York: Springer 1975,
pp. 78-97
Bernard, G.D., Wehner, R.: Dichroism, birefringence and structural twist in polar-
ized light detectors of insects. Biol. Bull. 149, 421 (1975)
Bernstein, S., Finn, C.: Ant compound eye: size-related ommatidium differences
within a single wood ant nest. Experientia 27, 708-710 (1971)
Bishop, L.G.: The spectral sensitivity of motion-detector units recorded in the
optic lobe of the honeybee (Apis mellifera). Z. Vergl. Physiol. 70, 374-381
(1970)

Boschek, C.B.: On the fine structure of the peripheral retina and lamina ganglio-
   naris of the fly, Musca domestica. Z. Zellforsch. 118, 369-409 (1971)
Braitenberg, V.: Patterns of projection in the visual system of the fly.
   I. Retina-lamina projections. Exp. Brain Res. 3, 271-298 (1967)
Braitenberg, V., Debbage, P.: A regular net of reciprocal synapses in the visual
   system of the fly, Musca domestica. J. Comp. Physiol. 90, 25-31 (1974)
Brunnert, A., Wehner, R.: Fine structure of light and dark adapted eyes of desert
   ants, Cataglyphis bicolor. J. Morph. 140, 15-30 (1973)
Burkhardt, D., De La Motte, I.: Electrophysiological studies on the eyes of
   Diptera, Mecoptera and Hymenoptera. In: Information Processing in the Visual
   Systems of Arthropods. Wehner, R. (ed.). Berlin-Heidelberg-New York: Springer
   1972, pp. 147-153
Burton, P.R., Stockhammer, K.A.: Electron microscopic studies of the compound eye
   of the toadbug, Gelastocoris oculatus. J. Morph. 127, 233-258 (1969)
Butler, R.: The identification and mapping of spectral cell types in the retina
   of Periplaneta americana. Z. Vergl. Physiol. 72, 67-80 (1971)
Cajal, S.R., Sanchez, D.: Contribucion al conocimiento de los centros nerviosos
   de los insectos. Trab. Lab. Invest. Biol. Univ. Madrid 13, 1-168 (1915)
Collett, T., King, A.J.: Vision during flight. In: The Compound Eye and Vision
   of Insects. Horridge, G.A. (ed.). Oxford: Clarendon 1975, pp. 437-466
Corneli, W.: Von dem Aufbau des Sehorgans der Blattwespenlarven und der Entwicklung
   des Netzauges. Zool. Jb. Anat. Ontogen. Tiere 46, 573-608 (1924)
Daumer, K.: Reizmetrische Untersuchungen des Farbensehens der Bienen.
   Z. Vergl. Physiol. 38, 413-478 (1956)
Duelli, P., Wehner, R.: The spectral sensitivity of polarized light orientation
   in Cataglyphis bicolor (Formicidae, Hymenoptera). J. Comp. Physiol. 86, 37-53
   (1973)
Eguchi, E.: Fine structure and spectral sensitivities of retinular cells in the
   dorsal sector of compound eyes in the dragonfly Aeschna.
   Z. Vergl. Physiol. 71, 201-218 (1971)
Eheim, W.P., Wehner, R.: Die Sehfelder der zentralen Ommatidien in den Appositions-
   augen von Apis mellifica und Cataglyphis bicolor (Apidae, Formicidae; Hymen-
   optera). Kybernetik 10, 168-179 (1972)
Fisher, S.K., Boycott, B.B.: Synaptic connexions made by horizontal cells within
   the outer plexiform layer of the retina of the cat and the rabbit.
   Proc. Roy. Soc. London 186B, 317-331 (1974)
Gemperlein, R., Smola, U.: Übertragungseigenschaften der Sehzelle der Schmeissfliege
   Calliphora erythrocephala. 1. Abhängigkeit vom Ruhepotential.
   J. Comp. Physiol. 78, 30-52 (1972)
Gogala, M.: Die spektrale Empfindlichkeit der Doppelaugen von Ascalaphus macaronius
   (Neuroptera, Asclaphidae). Z. Vergl. Physiol. 57, 232-243 (1967)
Gogala, M., Hamdorf, K., Schwemer, J.: UV-Sehfarbstoff bei Insekten.
   Z. Vergl. Physiol. 70, 410-413 (1970)
Goldsmith, T.H.: The nature of retinal action potential and the spectral sensitiv-
   ities of ultraviolet and green receptor systems of the compound eye of the
   worker honeybee. J. Gen. Physiol. 43, 775-799 (1960)
Goldsmith, T.H.: Fine structure of the retinae in the compound eye of the honeybee.
   J. Cell Biol. 14, 489-494 (1962)
Goldsmith, T.H.: Photoreceptor processes: some problems and perspectives.
   J. Exp. Zool. (in press)
Grenacher, H.: Untersuchungen über das Sehorgan der Arthropoden insbesondere der
   Spinnen, Insekten und Crustaceen. Göttingen: von Vandenhoeck and Ruprecht 1879
Gribakin, F.G.: Ultrastructural organization of the photoreceptor cells of the
   compound eye of the honeybee, Apis mellifera (russ.). Zh. evolutsionnoi biochemii
   i fiziologii 3, 66-72 (1967a)
Gribakin, F.G.: The types of photoreceptor cells of the compound eye of the honeybee
   worker as revealed by electron microscopy (russ.). Tsitologia 9, 1276-1279 (1967b)
Gribakin, F.G.: Types of photoreceptor cells in the compound eye of the worker
   honeybee relative to their spectral sensitivities (russ.). Tsitologia 11,
   308-313 (1969a)
Gribakin, F.G.: Cellular basis of colour vision in the honeybee.
   Nature 223, 639-641 (1969b)

Gribakin, F.G.: The distribution of the long wave photoreceptors in the compound
  eye of the honeybee as revealed by selective osmic staining.
  Vision Res. 12, 1225-1230 (1972)
Gribakin, F.G.: Functional morphology of the compound eye of the bee. In: The Com-
  pound Eye and Vision of Insects. Horridge, G.A. (ed.). Oxford: Clarendon 1975,
  pp. 154-176
Grundler, O.J.: Elektronenmikroskopische Untersuchungen am Auge von Apis mellifera.
  Zulassungsarbeit zur wiss. Prüfung für das Lehramt an Gymnasien, Universität
  Würzburg (1972)
Grundler, O.J.: Morphologische Untersuchungen am Bienenauge nach Bestrahlung mit
  Licht verschiedener Wellenlänge. Cytobiology 7, 105-110 (1973)
Grundler, O.J.: Elektronenmikroskopische Untersuchungen am Auge der Honigbiene
  (Apis mellifera). I. Untersuchungen zur Morphologie und Anordnung der neun
  Retinulazellen in Ommatidien verschiedener Augenbereiche und zur Perzeption
  linear polarisierten Lichtes. Cytobiology 9, 203-220 (1974)
Grundler, O.J.: Elektronenmikroskopische Untersuchungen am Auge der Honigbiene
  (Apis mellifera). II. Untersuchungen zur Reaktion der Feinstrukturen des Rhabdoms
  auf verschiedene experimentelle Einflüsse, vor allem Reizlicht unterschiedlicher
  Wellenlänge. Microscopica Acta 77, 241-258 (1975)
Helversen, O. von: Zur spektralen Unterschiedsempfindlichkeit der Honigbiene.
  J. Comp. Physiol. 80, 439-472 (1972)
Helversen, O. von, Edrich, W.: Der Polarisationsempfänger im Bienenauge: ein Ultra-
  violettrezeptor. J. Comp. Physiol. 94, 33-47 (1974)
Herrling, P.L.: Measurements on the arrangement of ommatidial structures in the
  retina of Cataglyphis bicolor (Formicidae, Hymenoptera). In: Information Pro-
  cessing in the Visual Systems of Arthropods. Wehner, R. (ed.). Berlin-Heidelberg-
  New York: Springer 1972, pp. 49-53
Herrling, P.L.: Topographische Untersuchungen zur funktionellen Anatomie der Retina
  von Cataglyphis bicolor (Formicidae, Hymenoptera). Dissertation Universität
  Zürich (1975)
Hickson, S.J.: The eye and optic tract of insects. Quart. J. Micr. Sci. N. Ser. 25,
  215-251 (1885)
Horridge, G.A.: The retina of the locust. Wenner-Gren Center Int. Symp. Ser. 7,
  513-541 (1966)
Horridge, G.A.: Unit studies on the retina of dragonflies. Z. Vergl. Physiol. 62,
  1-37 (1969)
Horridge, G.A., Meinertzhagen, I.A.: The exact neural projection of the visual
  fields upon the first and second ganglia of the insect eye.
  Z. Vergl. Physiol. 66, 369-378 (1970)
Horridge, G.A., Mimura, F.R.S.: Fly photoreceptors. I. Physical separation of two
  visual pigments in Calliphora retinula cells 1-6. Proc. R. Soc. London 190B,
  211-224 (1975)
Horridge, G.A., Mimura, F.R.S., Tsukahara, Y.: Fly photoreceptors. II. Spectral
  and polarized light sensitivity in the drone fly Eristalis.
  Proc. R. Soc. London 190B, 225-237 (1975)
Janzen, D.H.: Euglossine bees as long-distance pollinators of tropical plants.
  Science 171, 203-205 (1971)
Kaiser, W.: A preliminary report on the analysis of the optomotor system of the
  honey bee. Single unit recordings during stimulation with spectral lights.
  In: Information Processing in the Visual Systems of Arthropods. Wehner, R. (ed.).
  Berlin-Heidelberg-New York: Springer 1972, pp. 167-172
Kaiser, W.: The spectral sensitivity of the honeybee's optomotor walking response.
  J. Comp. Physiol. 90, 405-408 (1974)
Kaiser, W.: The relationship between visual movement detection and colour vision
  in insects. In: The Compound Eye and Vision of Insects. Horridge, G.A. (ed.).
  Oxford: Clarendon 1975, pp. 359-377
Kaiser, W., Bishop, L.G.: Directionally selective motion detecting units in the
  optic lobe of the honeybee. Z. Vergl. Physiol. 67, 403-413 (1970)
Kaiser, W., Liske, E.: A preliminary report on the analysis of the optomotor system
  of the bee. Behavioural studies with spectral lights. In: Information Processing
  in the Visual Systems of Arthropods. Wehner, R. (ed.).
  Berlin-Heidelberg-New York: Springer 1972, pp. 163-165

Kaiser, W., Liske, E.: Die optomotorischen Reaktionen von fixiert fliegenden Bienen bei Reizung mit Spektrallichtern. J. Comp. Physiol. 89, 391-408 (1974)

Kenyon, F.C.: The optic lobes of the bee's brain in the light of recent neurological methods. Am. Naturalist 31, 365-376 (1897)

Kien, J.: Motion detection in locusts and grasshoppers. In: The Compound Eye and Vision of Insects. Horridge, G.A. (ed.). Oxford: Clarendon 1975, pp. 410-422

Kiepenheuer, P.: Farbunterscheidungsvermögen bei der roten Waldameise Formica polyctena. Z. Vergl. Physiol. 57, 409-411 (1968)

Kirschfeld, K.: Die Projektion der optischen Umwelt auf das Raster der Rhabdomere im Komplexauge von Musca. Exp. Brain Res. 3, 248-270 (1967)

Kirschfeld, K.: Die notwendige Anzahl von Rezeptoren zur Bestimmung der Richtung des elektrischen Vektors linear polarisierten Lichtes. Z. Naturforsch. 276, 578-579 (1972)

Kirschfeld, K.: Vision of polarized light. Int. Biophys. Congr., Moscow 4, 289-296 (1973a)

Kirschfeld, K.: Optomotorische Reaktionen der Biene auf bewegte "Polarisations-Muster". Z. Naturforsch. 28C, 329-338 (1973b)

Kirschfeld, K., Snyder, A.W.: Waveguide mode effects, birefringence and dichroism in fly photoreceptors. In: Photoreceptor Optics. Snyder, A.W., Menzel, R. (eds.). Berlin-Heidelberg-New York: Springer 1975, pp. 55-77

Kolb, G., Autrum, H.: Selektive Adaptation und Pigmentwanderung in den Sehzellen des Bienenauges. J. Comp. Physiol. 94, 1-6 (1974)

Kolb, H.: The connections between horizontal cells and photoreceptors in the retina of the cat: Electron microscopy of Golgi preparations. J. Comp. Neurol. 155, 1-14 (1974)

Land, M.F.: Head movements and fly vision. In: The Compound Eye and Vision of Insects. Horridge, G.A. (ed.). Oxford: Clarendon 1975, pp. 469-489

Laughlin, S.B.: Neural integration in the first optic neuropile of dragonflies. I. Signal amplification in dark-adapted second-order neurons. J. Comp. Physiol. 84, 335-355 (1973)

Laughlin, S.B.: Neural integration in the first optic neuropile of dragonflies. II. Receptor signal interactions in the lamina. J. Comp. Physiol. 92, 357-375 (1974a)

Laughlin, S.B.: Neural integration in the first optic neuropile of dragonflies. III. The transfer of angular information. J. Comp. Physiol. 92, 377-396 (1974b)

Laughlin, S.B.: Resistance change associated with the response of insect monopolar neurons. Z. Naturforsch. 29C, 449-450 (1974c)

Laughlin, S.B.: The function of the lamina ganglionaris. In: The Compound Eye and Vision of Insects. Horridge, G.A. (ed.). Oxford: Clarendon 1975a, pp. 341-358

Laughlin, S.B.: Receptor function in the apposition eye - an electrophysiological approach. In: Photoreceptor Optics. Snyder, A.W., Menzel, R. (eds.). Berlin-Heidelberg-New York: Springer 1975b, pp. 479-498

Laughlin, S.B.: Receptor and interneuron light-adaptation in the dragonfly visual system. Z. Naturforsch. 30C, 306-308 (1975c)

Laughlin, S.B., Horridge, G.A.: Angular sensitivity of the retinula cells of dark-adapted worker bee. Z. Vergl. Physiol. 74, 329-335 (1971)

Laughlin, S.B., Menzel, R., Snyder, A.W.: Membranes, dichroism and receptor sensitivity. In: Photoreceptor Optics. Snyder, A.W., Menzel, R. (eds.). Berlin-Heidelberg-New York: Springer 1975, pp. 237-259

Mazokhin-Porshnyakov, G.A., Trenn, W.: Electrophysiological study of eye in ants (russ.). Zool. J. 51, 1007-1017 (1972)

Meinertzhagen, I.A.: The first and second neural projections of the insect eye. Ph. D. Thesis, Univ. St. Andrews (1971)

Meinertzhagen, I.A.: Development of the compound eye and optic lobe of insects. In: Developmental Neurobiology of Arthropods. Young, D. (ed.). Cambridge Univ. 1973, pp. 51-104

Melamed, J., Trujillo-Cenoz, O.: The fine structure of the central cells in the ommatidia of dipterans. J. Ultrastruc. Res. 21, 313-334 (1968)

Menzel, R.: Untersuchungen zum Erlernen von Spektralfarben durch die Honigbiene (Apis mellifica). Z. Vergl. Physiol. 56, 22-62 (1967)

Menzel, R.: Über den Farbensinn von Paravespula germanica (Hymenoptera): ERG und selektive Adaptation. Z. Vergl. Physiol. 75, 86-104 (1971)

Menzel, R.: Feinstruktur des Komplexauges der roten Waldameise Formica polyctena
    (Hymenoptera, Formicidae). Z. Zellforsch. 127, 356-373 (1972)
Menzel, R.: Spectral response of moving detecting and "sustaining" fibres in
    the optic lobe of the bee. J. Comp. Physiol. 82, 135-150 (1973)
Menzel, R.: Spectral sensitivity of monopolar cells in the bee lamina.
    J. Comp. Physiol. 93, 337-346 (1974)
Menzel, R.: Colour receptors in insects. In: The Compound Eye and Vision of Insects.
    Horridge, G.A. (ed.). Berlin-Heidelberg-New York: Springer 1975a, pp. 121-153
Menzel, R.: Polarization sensitivity in insect eyes with fused rhabdoms. In: Photo-
    receptor Optics. Snyder, A.W., Menzel, R. (eds.).
    Berlin-Heidelberg-New York: Springer 1975b, pp. 372-387
Menzel, R., Blakers, M.: Functional organization of an insect ommatidium with
    fused rhabdom. Cytobiology 11, 279-298 (1975)
Menzel, R., Blakers, M.: Colour receptors in the bee's eye - morphology and spectral
    sensitivity. J. Comp. Physiol. (in press)
Menzel, R., Knaut, R.: Pigment movement during light and chromatic adaptation in the
    retinula cells of Formica polyctena (Hymenoptera, Formicidae).
    J. Comp. Physiol. 86, 125-138 (1973)
Menzel, R., Lange, G.: Änderung der Feinstruktur im Komplexauge von Formica
    polyctena bei der Helladaptation. Z. Naturforsch. 26B, 357-359 (1971)
Menzel, R., Snyder, A.W.: Polarized light detection in the bee, Apis mellifera.
    J. Comp. Physiol. 88, 247-270 (1974)
Menzel, R., Wehner, R.: Augenstrukturen bei verschieden großen Arbeiterinnen von
    Cataglyphis bicolor (Formicidae, Hymenoptera). J. Vergl. Physiol. 68, 446-449
    (1970)
Mimura, K.: Units of the optic lobe, especially movement perception units of Diptera.
    In: The Compound Eye and Vision of Insects. Horridge, G.A. (ed.).
    Oxford: Clarendon 1975, pp. 423-436
Mote, M.I.: Focal recording of responses evoked by light in the lamina ganglionaris
    of the fly Sarcophaga bullata. J. Exp. Zool. 175, 149-158 (1970a)
Mote, M.I.: Electrical correlates of neural superposition in the eye of the fly
    Sarcophaga bullata. J. Exp. Zool. 175, 159-168 (1970b)
Mote, M.I., Goldsmith, T.H.: Spectral sensitivities of color receptors in the
    compound eye of the cockroach Periplaneta. J. Exp. Zool. 173, 137-146 (1970)
Mote, M.I., Goldsmith, T.H.: Compound eyes: localization of two color receptors
    in the same ommatidium. Science 171, 1254-1255 (1971)
Naka, K.I., Eguchi, E.: Spike potentials recorded from the insect photoreceptor.
    J. Gen. Physiol. 45, 663-680 (1962)
Ninomiya, N., Tominaga, Y., Kuwabara, M.: The fine structure of the compound eye
    of a damsel-fly. Z. Zellforsch. 98, 17-32 (1969)
O'Shea, M., Rowell, C.H.F.: A spike-transmitting electrical synapse between visual
    interneurons in the locust movement detector system.
    J. Comp. Physiol. 97, 143-158 (1975)
Patten, W.: Studies on the eyes of arthropods. I. Development of the eyes of Vespa,
    with observations on the ocelli of some insects. J. Morph. 1, 193-226 (1887)
Peckham, G.W., Peckham, E.G.: On the instincts and habits of solitary wasps.
    Wisconsin Geol. Nat. Hist. Survey, Bull. Sci. Ser. 1, 2, 1-148 (1898)
Perrelet, A.: The fine structure of the retina of the honey bee drone: an electron-
    microscopical study. Z. Zellforsch. 108, 530-562 (1970)
Perrelet, A., Baumann, F.: Presence of three small retinula cells in the ommatidium
    of the honey bee drone eye. J. Microscop. 8, 497-502 (1969)
Phillips, E.F.: Structure and development of the compound eye of the honey bee.
    Proc. Acad. Nat. Sci. Philadelphia 57, 123-157 (1905)
Portillo, J. del: Beziehungen zwischen den Öffnungswinkeln der Ommatidien, Krümmung
    und Gestalt der Insektenaugen und ihrer funktionellen Aufgabe.
    Z. Vergl. Physiol. 23, 100-145 (1936)
Ratliff, F., Hartline, H.K., Miller, W.H.: Spatial and temporal aspects of retinal
    inhibitory interaction. J. Opt. Soc. Am. 53, 110-120 (1963)
Rehbronn, W.: Gleichzeitige intrazelluläre Doppelableitungen aus dem Komplexauge
    von Calliphora erythrocephala. Z. Vergl. Physiol. 76, 285-301 (1972)
Ribi, W.A.: Neurons in the first synaptic region of the bee, Apis mellifera.
    Cell Tiss. Res. 148, 277-286 (1974)

Ribi, W.A.: The neurons of the first optic ganglion of the bee, Apis mellifera.
    Adv. Anat. Embryol. Cell Biol. 50 (4), 1-43 (1975a)
Ribi, W.A.: Golgi studies of the first optic ganglion of the ant, Cataglyphis
    bicolor. Cell Tiss. Res. 160, 207-217 (1975b)
Röhlich, P., Török, L.: Fine structure of the compound eye of Daphnia in normal,
    dark- and strongly light-adapted state. In: Eye Structure. Rohen, J.W. (ed.).
    Stuttgart: Schattauer 1965, 2nd Symp., pp. 175-186
Roth, H., Menzel, R.: ERG of Formica polyctena and selective adaptation. In: Infor-
    mation Processing in the Visual Systems of Arthropods. Wehner, R. (ed.).
    Berlin-Heidelberg-New York: Springer 1972, pp. 177-181
Rowell, C.H.F.: The orthopteran descending movement detector (DMD) neurones: a
    characterization and review. Z. Vergl. Physiol. 73, 167-194 (1971)
Schinz, R.H.: Structural specialization in the dorsal retina of the bee, Apis
    mellifera. Cell Tiss. Res. 162, 23-34 (1975)
Scholes, J.: The electrical responses of the retinal receptors and the lamina in
    the visual system of the fly Musca. Kybernetik 6, 149-162 (1969)
Shaw, S.R.: Coupling between receptors in the eye of the drone honeybee.
    J. Gen. Physiol. 50, 2480-2481 (1967)
Shaw, S.R.: Interreceptor coupling in ommatidia of drone honeybee and locust
    compound eyes. Vis. Res. 9, 999-1029 (1969)
Skrzipek, K.H., Skrzipek, H.: Die Morphologie der Bienenretina, Apis mellifica,
    in elektronenmikroskopischer und lichtmikroskopischer Sicht.
    Z. Zellforsch. 119, 552-576 (1971)
Skrzipek, K.H., Skrzipek, H.: Die Anordnung der Ommatidien in der Retina der
    Biene (Apis mellifica). Z. Zellforsch. 139, 567-582 (1973)
Skrzipek, K.H., Skrzipek, H.: The ninth retinula cell in the ommatidium of the
    worker honey bee (Apis mellifica). Z. Zellforsch. 147, 589-593 (1974)
Smola, U., Gemperlein, R.: Übertragungseigenschaften der Sehzelle der Schmeiss-
    fliege Calliphora erythrocephala. 2. Die Abhängigkeit vom Ableitort: Retina-
    Lamina ganglionaris. J. Comp. Physiol. 79, 363-392 (1972)
Snyder, A.W.: Optical properties of invertebrate photoreceptors. In: The Compound
    Eye and Vision of Insects. Horridge, G.A. (ed.). Oxford: Clarendon 1975,
    pp. 179-235
Snyder, A.W., Laughlin, S.B.: Dichroism and absorption by photoreceptors.
    J. Comp. Physiol. 100, 101-116 (1975)
Snyder, A.W., McIntyre, P.: Polarization sensitivity of twisted fused rhabdoms.
    In: Photoreceptor Optics. Snyder, A.W., Menzel, R. (eds.).
    Berlin-Heidelberg-New York: Springer 1975, pp. 388-391
Snyder, A.W., Menzel, R., Laughlin, S.B.: Structure and function of the fused
    rhabdom. J. Comp. Physiol. 87, 99-135 (1973)
Sommer, E.W., Wehner, R.: The retina-lamina projection in the visual system of
    the bee, Apis mellifera. Cell Tiss. Res. 163, 45-61 (1975)
Stavenga, D.G.: Waveguide modes and refractive index in photoreceptors of inverte-
    brates. Vis. Res. 15, 232-330 (1975)
Strausfeld, N.J.: Golgi studies on insects. Part II. The optic lobes of Diptera.
    Phil. Trans. R. Soc. London 258B, 135-223 (1970a)
Strausfeld, N.J.: Variations and invariants of cell arrangements in the nervous
    systems of insects. (A review of neuronal arrangements in the visual system
    and corpora pedunculata). Verh. Dtsch. Zool. Ges. 64, 97-108 (1970b)
Strausfeld, N.J.: The organization of the insect visual system (light microscopy).
    I. Projections and arrangements of neurons in the lamina ganglionaris of
    Diptera. Z. Zellforsch. 121, 377-441 (1971)
Strausfeld, N.J.: Atlas of an Insect Brain. Berlin-Heidelberg-New York: Springer 1975
Strausfeld, N.J., Blest, A.D.: Golgi studies on insects. Part I. The optic lobes
    of Lepidoptera. Phil. Trans. R. Soc. London 258B 81-134 (1970)
Strausfeld, N.J., Braitenberg, V.: The compound eye of the fly (Musca domestica):
    connections between the cartridges of the lamina ganglionaris.
    Z. Vergl. Physiol. 70, 95-104 (1970)
Swammerdam, J.: Biblia naturae sive historia insectorum. Boerhaave, H. (ed.).
    Leyden: I. Severinum, B. and P. Vander 1737
Trujillo-Cenoz, O.: Some aspects of the structural organization of the intermediate
    retina of dipterans. J. Ultrastruct. Res. 13, 1-33 (1965)

Trujillo-Cenoz, O.: The structural organization of the compound eye in insects.
In: Physiology of Photoreceptor Organs. Handbook of Sensory Physiology.
Fuortes, M.G.F. (ed.). Berlin: Springer 1972, Vol. VII/2, pp. 5-62

Trujillo-Cenoz, O., Melamed, J.: Spatial distribution of photoreceptor cells in
the ommatidia of Periplaneta americana. J. Ultrastruct. Res. 34, 397-400 (1971)

Tuurala, O., Lehtinen, A.: Über die Einwirkung von Licht und Dunkel auf die Feinstruktur der Lichtsinneszellen der Assel Oniscus asellus. 2. Microvilli und
multivesikuläre Körper nach starker Belichtung. Ann. Acad. Sci. fenn. A,
IV Biologica 177, 1-8 (1971)

Varela, F.G.: Fine structure of the visual system of the honeybee (Apis mellifera).
II. The lamina. J. Ultrastruct. Res. 31, 178-194 (1970)

Varela, F.G., Porter, K.R.: Fine structure of the visual system of the honeybee
(Apis mellifera). I. The retina. J. Ultrastruct. Res. 29, 236-259 (1969)

Varela, F.G., Wiitanen, W.: The optics of the compound eye of the honeybee
(Apis mellifera). J. Gen. Physiol. 55, 336-358 (1970)

Viallanes, H.: Etudes histologiques sur les centres nerveux et les organes des
sens animaux articulés - quatrième mémoire - cerveau de la guêpe (Vespa cabro
et Vespa vulgaris). Ann. Sci. Natur. 7e. 2, 5-100 (1887)

Viallanes, H.: Contribution à l'histologie du système nerveux des invertebrés.
La lame ganglionnaire de la langouste. Ann. Sci. Natur. 7e. 13, 385-398 (1892)

Vowles, D.M.: The orientation of ants. II. Orientation to light, gravity and
polarized light. J. Exp. Biol. 31, 356-375 (1954)

Waddington, C.H., Perry, M.M.: Inter-retinular fibres in the eyes of Drosophila.
J. Insect Physiol. 9, 475-478 (1963)

Wehner, R., Bernard, G.D., Geiger, E.: Twisted and non-twisted rhabdoms and their
significance for polarization detection in the bee. J. Comp. Physiol. 104,
225-245 (1975)

Wehner, R., Brunnert, A., Herrling, P.L., Klein, R.: Periphere Adaptation und
zentralnervöse Umstimmung im optischen System von Cataglyphis bicolor
(Formicidae, Hymenoptera). Rev. Suisse Zool. 79, 197-228 (1972)

Wehner, R., Toggweiler, F.: Verhaltensphysiologischer Nachweis des Farbensehens
bei Cataglyphis bicolor (Formicidae, Hymenoptera). J. Comp. Physiol. 77,
239-255 (1972)

White, R.H.: The effect of light and light deprivation upon the ultrastructure
of the larval mosquito eye. II. The rhabdom. J. Exp. Zool. 166, 405-426 (1967)

Wolken, J.J.: Photoprocesses, Photoreceptors and Evolution. New York-San Francisco-
London: Academic Press 1975

Zettler, F., Järvilehto, M.: Decrement-free conduction of graded potentials along
the axon of a monopolar neuron. Z. Vergl. Physiol. 75, 402-421 (1971)

Zettler, F., Järvilehto, M.: Lateral inhibition in an insect eye. Z. Vergl.
Physiol. 76, 233-244 (1972)

Zettler, F., Järvilehto, M.: Active and passive axonal progagation of non-spike
signals in the retina of Calliphora. J. Comp. Physiol. 85, 89-104 (1973).

2.8 Neuronal Architecture and Function in the Ocellar System of the Locust

J. A. Patterson

2.8.1 Introduction

A full understanding of the peripheral visual process must depend on
a knowledge of both the physiological properties of the cells involved
and the functional connectivity of those cells. As this symposium shows,
considerable effort has been devoted to the accomplishment of this task
for organs such as the retina of vertebrates and cephalopods and the
retina, lamina and medulla of the compound eye of arthropods. Many of
these organs contain some hundreds of thousands or millions of neural
elements and are capable of achieving a highly complex analysis of the
visual information furnished by the photoreceptors. Such complexity
would not be so amenable to experimental analysis were it not for the
fact that the visual system, in these cases, is composed of an array
formed by the repetition of groups of cells. There are relatively few
cells in each group and each cell may be placed in one or another of
a small number of classes of cell, each with distinctive anatomical
and physiological properties. The array of cell groups is perhaps most
clearly seen in the ommatidia and optic cartridges of the insect compound eye.

An alternative approach is to study a system, the analytical capability
of which is more limited and in which the total number of cells involved
is much smaller. It is still frequently possible to classify cells into
distinctive types, and in addition and more importantly, it is possible
to identify some cells as unique individuals, enabling the same cells
to be studied in a large number of preparations. One such system is
found in the so-called simple eyes or dorsal ocelli which are possessed
by some adult insects. The early literature concerning the anatomy,
physiology and possible roles of behaviour, of ocelli has been reviewed
by Goodman (1970). The present paper reviews some of the more recent
studies of ocelli, in some cases comparing these with similar studies
of the insect compound eye and emphasis is placed on the relationships
between structure and function which have been revealed, particularly
in the locust ocellar system.

2.8.2 The Structure of the Ocellus

The head of the locust, Schistocerca gregaria, bears three simple photoreceptor organs, the dorsal ocelli. The single median ocellus is situated in the anterior mid-line on the frons, while each of the two lateral ocelli lies at the antero-dorsal medial edge of a compound eye
(see Fig 1A). Each ocellus is a cup-shaped organ 0.5 mm in diameter
and 0.5-1.0 mm in depth, bounded distally by a convex cuticular lens
and connected to the brain by an ocellar nerve about 1.5 mm in length.
Fig. 1B is a diagrammatic, longitudinal section through the median
ocellus and a part of the ocellar nerve.

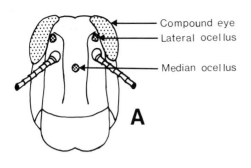
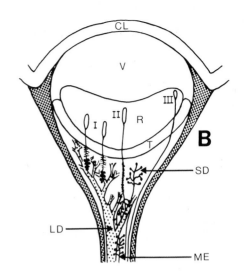

Fig. 1A and B. A) Diagram of the head of the locust, Schistocerca gregaria, viewed anteriorally, showing the relative positions of the compound eyes and the median and lateral ocelli. B) Diagrammatic longitudinal section through the median ocellus of S. gregaria, showing some of the principal features of ocellar organisation. CL: corneal lens, V: vitreous cell layer, R: receptor cell layer, T: tapetum, LD: large field dendrite of a giant neuron of the ocellar nerve, SD: small field dendrite of a medium-sized neuron of the ocellar nerve, ME: a mossy efferent neuron synapsing with the giant axons of the ocellar nerve. I, II and III refer to the three classes of receptor cell described in the text. Material for this figure is taken from Goodman (1974)

Beneath the cuticular lens and the corneageal cells responsible for its formation, are several layers of vitreous cells; beneath these lie the 600-800 photoreceptive cells. In some insect ocelli, such as those of the dragonfly (Ruck and Edwards, 1964), the receptor cells are arranged as a single, orderly layer, but in the locust ocellus the position of individual receptor cells in the receptor cell region is variable so that no layering is evident (Goodman, 1970). In the locust ocellus, as in other ocelli and many compound eyes, the rhab-domere regions of adjacent receptor cells are fused to form a rhab-dom. The receptor and higher order cells of the ocellus do not show the highly ordered spatial arrangement seen in the ommatidia and op-tic cartridges of the insect compound eye and the variety of rhabdom configurations seen in cross sections of ocelli reflects, to a certain extent this less ordered structure. In the locust, between two and seven cells can contribute to any individual rhabdom complex (Goodman, 1970), while in the dragonfly, although the majority of rhabdoms are three-limbed and formed from the rhabdomeres of three receptor cells, a few may be made up from a greater number of cells (Chappell and Dowling, 1972). In contrast, the rhabdoms of the worker honeybee ocel-lus are apparently uniformly formed by paired retinular cells producing rhabdoms which are rectangular in section (Toh and Kuwabara, 1974). Cooter (1975) reports that in the cockroach, Periplaneta americana, five receptor cells regularly contribute to the formation of each rhabdom, and that cells reduce to visual cell axons at differing depths in the receptor cell region producing a variety of arrangements of rhabdomeres and receptor cell axons in cross sections.

Proximal to the rhabdomere bearing region in the locust, the receptor
cell narrows to form an axon. Bundles, each formed from a number of
axons surrounded by a common glial sheath, pass through the tapetal
layer which underlies the receptor cell region and through a basement
membrane to enter the synaptic plexus region which occupies the proximal part of the cup and the distal part of the ocellar nerve (Goodman,
1970, 1974). Once the synaptic region is reached, the bundles together
with the glia, lose their grouped organisation and the majority of
axons form synaptic contacts with the distal ends of the neurons of the
ocellar nerve. It is in the synaptic plexus that the difference between the orderly arrangement of the optic cartridge of the compound
eye and the equivalent structures in the ocellus is most striking.

Goodman (1974) describes three classes of receptor cell terminal in
the locust median ocellus (see Fig. 1B). Class I terminals have a
'bulbous irregular club-like ending' and possess brush-like arborisations at a higher level of the axon. These axons tend to end more distally in the synaptic plexus than do the Class II axons which possess
stubby collaterals and end in a tree of fine branches, each bearing
a small bulbous termination. Class III axons are mainly situated around
the rim of the synaptic plexus and retain a glial sheath so that they
bypass the main synaptic plexus area and pass down the ocellar nerve,
possibly terminating in the brain (Goodman, 1974).

The Class III receptor cell axons presumably form a proportion of the
50-60 small (< 1.0 μm) diameter axons seen in cross sections of the
ocellar nerve. The nerve also contains around 20 medium sized fibres,
with diameters in the range 1.0-1.5 μm and six giant and one slightly
smaller second-order neurons with diameters between 10 and 15 μm. As
the giant fibres pass distally into the ocellar cup they branch repeatedly to form large field dendritic trees which contribute to the
synaptic plexus. The medium sized axons of the ocellar nerve also
branch distally and form small field dendritic trees (Goodman, 1974).
The synaptic connections between receptor cell axons and the larger
branches of the giant second-order cells are particularly well known
by virtue of the distinctive appearance of both these cell types in
the electron microscope. The organisation of the synaptic plexus has
been described for the locust (Goodman, 1970, 1974), dragonfly
(Dowling and Chappell, 1972), worker honeybee (Toh and Kuwabara, 1974)
and fleshfly (Toh and Kuwabara, 1975) ocellus.

In the locust ocellus, Goodman (1974) finds a variety of types of synapse and of arrangements of pre- and postsynaptic cells. Conventional
synapses where there is little structural specialisation other than
a thickening of the pre- and postsynaptic membrane, and a concentration of vesicles around the presynaptic membrane, are only found in
limited numbers in the ocellus, notably in association with giant
axon bulbs in the proximal region of the cup. Peg-type synapses are
also found, forming reciprocal connections between adjacent receptor
cell axons and between receptor cell axons and giant axon dendrites.
The most abundant synapse type in the locust ocellus is the ribbon
type synapse where an electron dense ribbon is found in association
with the presynaptic membrane in an arrangement which resembles that
found in the bipolar terminals of the inner plexiform layer of the
vertebrate eye (Dowling and Boycott, 1966). Essentially similar structures – termed button synapses – are found in the dragonfly ocellus
(Dowling and Chappell, 1972). In the locust these synapses are found
in the axon branches and terminal bulbs of both Class I and Class II
receptor cells, in the dendrites of giant second-order neurons and in
their axons at least as far as the brain, and in the dendrites of the
small field second-order neurons of the ocellar cup. Many of the rib-
bon-type synapses are dyads where the presynaptic ribbon opposes two
postsynaptic cells. The commonest form of dyad is one in which the

presynaptic cell is a receptor cell and the postsynaptic components
are either a large branch of a giant axon, a small second-order dendrite or another receptor cell axon. The most frequent arrangement
is for the receptor cell to be presynaptic to a branch of a giant axon
and to a small dendrite. Both large and small second-order dendrites
form synaptic connections with the receptor cell terminals and with
adjacent second-order cells, and ribbon type synapses may form reciprocal synapses between receptor cells in addition to those formed by
peg-type reciprocal synapses.

The important points are that giant axon branches and smaller dendrites
are postsynaptic to the receptor cells and that the synaptic arrangement is such as to allow feedback to occur between receptor cells, between second-order cells and from second-order cells back onto receptor cells. A similar arrangement is seen in the dragonfly ocellus and
has formed the basis for both theoretical and experimental analysis
of the transfer of signal parameters from receptor to second-order
cells (Dowling and Chappell, 1972; Chappell and Klingman, 1974). In
their studies of the ocellus of the worker honeybee, Toh and Kuwabara
(1974) find a rather different arrangement. In the bee there are two
types of fibres in the ocellar nerve, termed thick and thin. The thick
fibres are postsynaptic to the receptor cells as are the thin fibres.
While the thick fibres appear afferent in character, the thin fibres
are frequently presynaptic to receptor cell axons, to thick and also
to other thin fibres, suggesting that the thin fibres are efferent in
function. Efferent axons have been described as forming mossy type
endings in the distal region of the bee ocellar plexus (Kenyon, 1896).
Similar structures are seen in the distal part of locust ocellar nerve
(Goodman, 1974). There is both anatomical (Cajal, 1918) and physiological (Rosser, 1974) evidence to suggest the presence of efferent axons
in the dragonfly ocellar nerve. The so-called centrifugal fibres, which
are a feature of the neural arrangement seen in the optic cartridge of
the lamina ganglionaris of insect compound eyes, such as that of the
fly Calliphora, (Boschek, 1972) are also presumably efferent in function.

The structural organisation of the ocellus suggests that the functional
capabilities of the system are limited. The failure of the corneal dioptrics to create an image of the visual environment which falls onto the
receptor cell region, or indeed within the ocellar cup, coupled with
the high degree of convergence between receptor cells and the giant
second-order neurons, would seem to preclude form vision. Similarly
this convergence, while presumably enhancing the sensitivity of responses of the second-order cells, would militate against colour vision by
the ocellus even though the receptor cells of the dragonfly have been
shown by Chappell and Devoe (1975), to have the ability for colour vision. It is possible that colour information could be preserved to a
more central site by the dragonfly equivalent of the Class III receptor
cells of the locust ocellus, if these are present. The randomness with
which the rhabdomeres of the receptor cells are orientated in the receptor cell region makes the detection of the plane of polarised light
also unlikely. All this does not necessarily reduce the possible function of the ocellus to the detection of the absolute level or of changes
in the level of illumination falling on the photoreceptors. Some directional sensitivity may be afforded to second-order cells by the restric-
tion of dendritic fields to certain regions of the ocellar cup. This is
certainly possible for the small field, medium diameter second-order
cells (Goodman, 1974) and there are indications that this may also apply to some extent to the large field giant second-order cells. Again
the Class III receptor cells could also preserve directional information to a more central site. Rosser (1972) has demonstrated the ability
of the corneal dioptrics of the dragonfly ocellus to impose crude direc-

tional information on the receptor cell region, and Zenkin and Pigarev
(1971) have obtained responses which are direction dependent from afferent neurons of a dragonfly ocellus when the ocellus is stimulated
by the movements of a bar of wide visual angle.

2.8.3 The Electrophysiology of the Ocellus and Ocellar Nerve

2.8.3.1 Extracellular Studies

Goodman (1970) has reviewed those studies of the extracellular consequences of the illumination of the ocellus which have been performed
by a number of authors working with the ocelli and ocellar nerves of
a variety of insects. A major criticism of these studies is that they
can only provide information about events such as the ocellar electroretinogram (ERG) which are the result of the summated activity of a
large number of cells and where more than a single type of cell is
contributing to the summation. Nevertheless, by comparing the results
obtained primarily from the ocelli of two insects, the dragonfly and
cockroach, combined with the observation of the effects of K^+ ions
and TTX and the effects of severing the ocellar nerve, Ruck was able
to produce an elegant series of papers which provided a convincing
model of the events occurring in the various major classes of cell
found in the ocellus and ocellar nerve (Ruck, 1961a,b,c, 1966). The
central part of this analysis is the description of the four components which summate to form the ERG and action potentials recorded
from the ocellar cup and the ocellar nerve. The components are as
follows:

Component 1. The depolarising response of the receptor cells to illumination, recorded extracellularly as a cornea negative wave and originating in the distal ends of the receptor cells.

Component 2. A transient cornea positive wave which follows immediately after illumination of the ocellus and is recorded in some ocelli.
Ruck suggests that this may be the initial and synchronous discharge
of action potentials in the receptor cell axons, the transient nature
of the wave resulting from the increasing asynchrony of discharges as
time after illumination increases.

Component 3. The activity in the receptor cells evokes component 3,
the hyperpolarisation of the second-order cells of the ocellus and
ocellar nerve which is recorded extracellularly as a cornea positive
wave.

Component 4. The action potentials of the neurons in the ocellar nerve
comprise component 4 of the ocellar ERG. In extracellular recordings,
the larger membrane area of individual, large diameter axons means that
the action potentials of these cells are more readily recorded than
those of smaller diameter cells. The ocellar nerve of the dragonfly
produces a continuous discharge of action potentials in the dark which
can be recorded extracellularly and which is inhibited by illumination
of the ocellus (Ruck, 1961b). A similar dark discharge has been recorded from the ocellar nerve of Locusta (Höyle, 1955), but not from the
Schistocerca ocellus (Patterson and Goodman, 1975). In addition an excitation of the ocellar second-order neurons occurs at the end of a
period of illumination, seen as a burst of action potentials in the
dragonfly, or as the production of one or two 'off-spikes' in the
Locusta (Burtt and Catton, 1956) and Schistocerca (Patterson and Goodman, 1975) ocellus. A dark discharge inhibited by illumination and excited by the onset of darkness or dimming is also seen in the ocellus

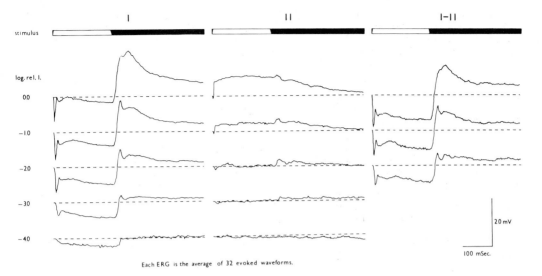

stimulus

log. rel. I.

00

−10

−20

−30

−40

20 mV

100 mSec.

Each ERG is the average of 32 evoked waveforms.

Fig. 2. Componental analysis of the averaged ERG waveforms obtained from a median ocellus of S. gregaria in response to light flashes of differing intensities. Each waveform is the average of 32 evoked potentials obtained from an ocellus which had stabilised to repetitive stimulation at the log relative intensity indicated at the left-hand side of the rows. Column I shows the averaged normal ERG obtained at the beginning of the experiment. Column II shows the averaged ERG obtained after cutting the ocellar nerve. Column I-II shows the waveform obtained when that shown in Column II was subtracted from the corresponding waveform in Column I. Upward deflection indicates increasing negativity of the recording electrode. (From Patterson and Goodman, 1975)

of the fly, Calliphora (Metschl, 1963). In the adult cabbage looper moth, however, Eaton (1975) records an excitatory spike discharge from the second-order ocellar neurons during illumination of the ocellus.

Components 3 and 4 of the ERG can be removed from corneal recordings by severing the ocellar nerve, since the cell bodies of the second-order cells lie proximal to the ocellus (see below) and cutting the nerve will, therefore, bring about the depolarisation of the distal processes of these cells which lie in the ocellar cup. This technique coupled with the signal averaging and waveform subtraction facilities of an on-line computer has been used by Patterson and Goodman (1975) to repeat Ruck's extracellular analysis using the locust ocellus. Some of the results of this work are shown in Fig. 2. Column I of Fig. 2 shows the averaged ERGs obtained from the intact median ocellus of a locust from an electrode which was advanced into the ocellus through a hole made in the cornea. Column II shows the responses obtained about an hour after the electrode had been withdrawn, the ocellar nerve severed and the electrode replaced to approximately the same position. The waveforms obtained at the three highest light intensities in column I had been stored in the computer during this time and from these waveforms the respective waveforms of column II were subtracted to give the waveforms shown in column I-II. Column II, therefore, shows the contribution to the total ERG made by the receptor cells, predominantly component 1, the depolarising response of the

340

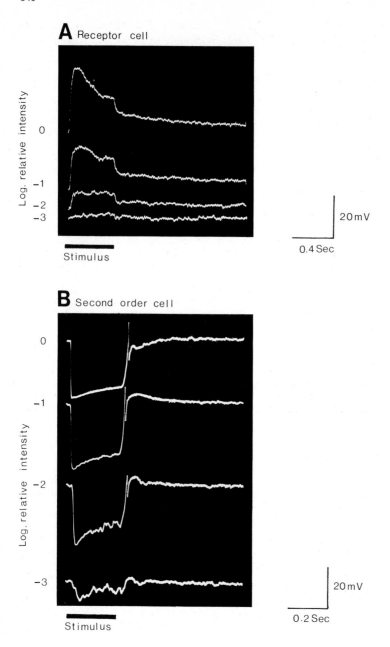

Fig. 3A and B. A) Intracellular recordings of the depolarisation produced in a receptor cell of an ocellus of <u>S. gregaria</u> in response to flashes of white light of differing intensity. Records are of the steady state response produced by repetitive flashes of 0.5 s duration and 2.5 s separation. B) Intracellular recordings of the hyperpolarisation produced in a giant second-order cell of an ocellus of <u>S. gregaria</u> in response to flashes of white light of differing intensity. In this case records are of the steady state response produced by repetitive flashes of 0.25 s duration and 1.25 s separation so that the events occurring at light off may be more clearly seen

receptor cells to illumination. In common with the general properties of many receptor cells the amplitude of component 1 appears to halve with each logarithmic decrease in the stimulus light intensity and at log relative intensities of -2.0, -3.0 and -4.0 the receptor cells do not contribute significantly to the overall ERG shown in column I. A small transient positive deflection precedes the receptor potential, shown in column II, but as Patterson and Goodman (1975) point out, this may be a residual response of the second-order neurons rather than component 2 of Ruck's analysis. Column I-II should correspond to components 3 and 4 of Ruck's analysis. In the Schistocerca ocellus the only action potentials (component 4) recorded are the small number of 'off-spikes' which occur after a period of illumination. 'Off-spikes' were not, however, present in the preparation illustrated in Fig. 2 so that in this case the waveforms shown in column I-II correspond to component 3 alone.

These records show that the amplitude of the responses of the second-order cells does not decrease with decreasing intensity of illumination in the same way as do those of the receptor cells, so that at low light intensities the ERG is composed primarily of the responses of second-order cells. In addition the hyperpolarising nature of component 3 suggests that the majority of axons in the locust ocellar nerve are hyperpolarised by illumination of the photoreceptors since, in terms of membrane area and, therefore, contribution to the extracellular potential, the medium and small diameter axons of the ocellar nerve contribute more to the ERG than do the giant axons.

Although extracellular studies can provide a considerable amount of useful information about ocellar function, it is with intracellular techniques that records are obtained which allow more reliable statements to be made about the way that individual cells respond to differing stimulus intensities and about the nature of the transfer of the receptor cell signal across the receptor cell second-order cell synaptic system.

2.8.3.2 Intracellular Studies

Studies of the intracellular events following illumination of the ocellus have only been performed on two insect species, namely the dragonfly (Chappell and Dowling, 1972; Dowling and Chappell, 1972; Chappell and Klingman, 1974; Chappell and DeVoe, 1975) and the locust, Schistocerca, (Patterson and Goodman, 1974a). In these studies the activity in both receptor and second-order cells is recorded. There have been several studies of the intracellular responses of the receptor cells of the insect compound eye, in a number of species (for a recent review see Goldsmith and Bernard, 1974), but there are only a few thorough studies of both receptor and second-order cells. These studies are those of Zettler and Järvilehto (1971, 1972a, 1973) and Järvilehto and Zettler (1971, 1973) on the compound eye and lamina of the fly Calliphora and Laughlin (1973, 1974a,b,c, 1975) on the compound eye and lamina of the dragonfly.

The waveform recorded from individual receptors when the photoreceptors are illuminated varies with the preparation studied. In the locust ocellus (see Fig. 3A and Patterson and Goodman, 1974a), and the locust compound eye (Shaw, 1968), the receptor cell response is a relatively slow wave which may rise to an initial peak at high light intensities before producing a plateau depolarisation which lasts throughout the period of illumination, followed by a relatively slow return to the resting potential at the end of the stimulus. A similar waveform is recorded from the receptor cells of the compound eye of the fly Calliphora by Zettler and Järvilehto (1973), Järvilehto and Zettler (1973) and those of the dragonfly compound eye by Laughlin (1973, 1974a,c). In the receptor cell of the dragonfly ocellus there is a spike-like transient, which can be removed by the application of TTX, associated with the initial phase of the cell's response to illumination (Chappell and Dowling, 1972), and a similar spike-like potential, again sensitive to TTX has been recorded, in the receptor cell of the compound eye of the drone honeybee by Baumann (1968). Similar differences in detail are seen in the types of response recorded from second-order visual cells. The majority of insect second-order visual cells studied intracellularly to date show a hyperpolarisation when the photoreceptors are illuminated. After an initial transient hyperpolarisation the cell may return, especially at high light intensities, to a plateau value which is close to the resting membrane potential of the cell. This happens in the dragonfly ocellus (Chappell and Dowling, 1972), the large monopolar cells of the lamina of the dragonfly (Laughlin, 1973, 1974b,c) and in the lamina monopolar cells of the fly, Calliphora (Järvilehto and Zettler, 1971, 1973; Zettler and Järvilehto, 1972b). At the end of the period of illumination there is a rapid return to the resting potential with usually a depolarising overshoot which may persist for some time. Associated with this overshoot there is frequently a spike-like potential which is quite marked in the giant second-order cells of the locust ocellus (see Fig. 3B), and which in this case is almost certainly the attenuated form of one of the 'off-spikes' which can be recorded extracellularly from the ocellar nerve (Patterson and Goodman, 1974a). A similar but much less marked event is seen in the dragonfly ocellus second-order cells (Chappell and Dowling, 1972), and also in the lamina monopolar cells of the fly. In the latter, action potentials are not normally recorded and information is probably transmitted by a form of electronic conduction (Zettler and Järvilehto, 1973, 1973).

Much more striking than these differences in the detail of the responses produced by different receptor or second-order cells are those general differences between the responses of receptor as compared to second-order cells which are common to all the photoreceptive organs referred to above - differences which reflect the nature of the first synapse of the insect visual system. The most notable of these, other than the reversal of the polarity of the visual signal, is the difference in the position of what Laughlin (1973) refers to as the 'dynamic range' of receptor as compared to second-order cells. The dynamic range is that range of light intensities where the magnitude of either the peak or plateau response of the cell is linearly related to the logarithm of the light intensity. This is bounded by a plateau at higher light intensities, signifying a saturation of the cell's response, and a toe at lower light intensities, where the threshold for a response is approached. It is commonly found that the dynamic range of second-order cells is confined to a lower range of light intensities than that of the receptor cells, and that the responses of second-order cells saturate at lower light intensities than do those of receptor cells. These features as seen in the locust ocellus are shown in Fig. 4.

These differences in the positions of the dynamic range for the responses of receptor and second-order cells are consistent with the notion that a process of signal amplification occurs at the receptor/second-order cell synapse. The methods which have been employed to obtain a numerical assessment of this amplification vary. Järvilehto and Zettler (1972) compared the responses of receptor cells and lamina monopolar neurons in the fly by a method which compared the amplitudes of the potentials recorded from cells in response to sinusoidally modulated light intensities and found values for the gain varying from x3 to x8 depending on the frequency of modulation used. Laughlin (1973)

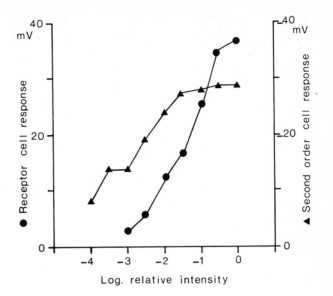

Fig. 4. V-log I curves showing the peak response amplitude plotted as a function
of log relative light intensity for a receptor cell
(o) and a second-order
cell (▲) of the ocellus of
S. gregaria. These data
have been replotted from
those given in Patterson
and Goodman (1974a)

compared the amplitudes, both peak and plateau, of the responses of
the same classes of cells (receptor and large monopolars) of the dragonfly compound eye to square wave light stimuli. Laughlin obtained
values for the amplification in the dark adapted eye of x12 and x14
with this method. It is also possible to estimate any amplification
by comparing the rate of polarisation, the slope steepness, of the
initial response to illumination of the receptor and second-order
cells. Although our experiments were not specifically designed to measure the amplification at the receptor cell/giant second-order axon
synapse in the locust ocellus (Patterson and Goodman, 1974a), we were
able to obtain some estimates of the gain using methods based on a
comparison of slope steepnesses and on the amplitudes of responses
obtained with square wave stimuli. A comparison of the amplitudes of
the responses gave values between x1.8 and x4.3 and a comparison of
the slope steepnesses of the 'on' effect gave values between x4.7 and
x7.8 depending on the intensity of the light stimulus.

These values for the amplification produced in the locust ocellus do
not suggest that the gain produced in this organ is greatly different
from that found in the compound eye and lamina of the fly or dragonfly. This is perhaps surprising when one considers that the degree of
convergence between the 800 receptor cells and the six wide-field dendritic trees of the giant axons of the ocellus could be one to two
orders of magnitude greater than that occurring between the six receptor cell axons and the two large monopolar cells of the fly (Boschek,
1972) or the dragonfly (Laughlin, 1974c) optic cartridge. It is possible that the reason for the greater convergence seen in the ocellus
is simply that the much larger membrane area of the ocellar giant
second-order cells requires a much larger number of synaptic inputs
from receptor cells in order to achieve a given response than does
the smaller membrane area of the lamina monopolar neurons of the compound eye. This explanation could be supported by the results obtained
from the locust ocellus. Alternatively, one may speculate that it will
be at low light intensities that the number of synaptic inputs which
the postsynaptic cell receives from receptor cells will be most significant in determining to what extent the weak receptor cell signal is

amplified by the synaptic system. Laughlin (1973) has shown that the gain, produced at the receptor cell/large monopolar cell synaptic system in the dragonfly compound eye, rises to a maximum as the stimulating light intensity decreases, before falling in value as the cells involved fail to respond at the lowest light intensities used. In the locust ocellus also, the highest values for the gain are found at lower light intensities (Patterson and Goodman, 1974a). It is possible, therefore, that it is at low light intensities that the apparently much greater convergence seen in the ocellus may be reflected in higher numerical values for any estimate of the amplification of the system than those which have been obtained for the compound eye. Although a direct experimental comparison between the amplification produced in the ocellus and the compound eye of the same insect has yet to be performed, it is possible to compare the records obtained from the dragonfly ocellus by Chappell and Dowling (1972) with those obtained from the dragonfly compound eye by Laughlin (1973). In the dragonfly compound eye the light intensity at which both the receptor and large monopolar cells fail to respond is about the same, and both cell types respond over a range of five log units of intensity. The dynamic range of the second-order cell responses is of course situated to the left of that of the receptor cells as is the case in the dragonfly ocellus (Chappell and Dowling, 1972), and the locust ocellus (see Fig. 4). Chappell and Dowling's (1972) records of the responses of the dragonfly ocellus show that the receptor cells respond over a range of 4-5 log units of light intensity, while the second-order cells respond over a range of 6-7 log units. It appears, therefore, that the second-order cells of the ocellus respond to illumination at an intensity which is about a log unit below the threshold for the production of a recordable signal in the receptor cells. Chappell and Dowling (1972) have not used the amplitudes of the cellular responses which they obtained to calculate the amplification occurring in the dragonfly ocellus but one might expect that, in a situation where potentials are recorded from second-order cells although the same low light intensity fails to elicit recordable potentials from receptor cells, calculation would reveal very high values for the gain. Clearly, further experiments are required to discern the functional significance of the greater convergence seen in the ocellus as compared with the compound eye.

Another feature of the receptor cell/second-order cell system is the enhancement of transient phenomena which is in part revealed by examining the slope steepness of 'on' and 'off' effects, since most second-order cells show a more rapid onset of response than do receptor cells. Chappell and Dowling (1972) and Dowling and Chappell (1972) have remarked on this enhancement of transients, which is also seen in the dragonfly ocellus, and suggest that the presence of feedback and lateral synapses in the ocellus is responsible for the production of the transient responses of second-order ocellar neurons. These are especially pronounced in preparations in which light flashes are presented while an adapting light source is shone onto the ocellus. In a brief report, Chappell and Klingman (1974) describe the feedback of second-order onto receptor cells in the dragonfly ocellus and find that the input transmitter, producing the hyperpolarisation of the second-order cells, is apparently acetylcholine and that the feedback transmitter is GABA. The results of experiments involving the selective blocking of input and feedback transmitters enabled the authors to produce a model of the events occurring during the feedback process. One of the consequences of this model is that the feedback transmitter is released in the dark to increase the sensitivity of the receptors to dim illumination, rather than to reduce the input when the intensity of illumination is high.

A

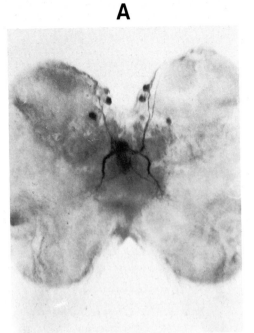

B

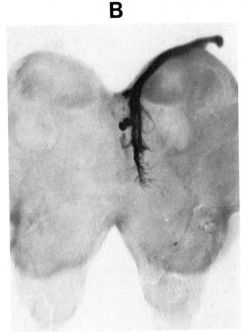

Fig. 5A-C. A) Photograph of the posterior aspect of a whole mount of the brain of S. gregaria in which the neurons of the median ocellar nerve have been visualised by Cobalt sulphide precipitation. B) Photograph of the posterior aspect of a whole mount of the brain of S. gregaria in which the neurons of the right lateral ocellar nerve have been visualised by Cobalt sulphide precipitation. C) A summary diagram of the pathways of ocellar neurons within the brain of S. gregaria, viewed from the postero-dorsal surface. The neurons of the median and right lateral ocellus only are shown. Higher order neurons are indicated by dotted lines for the left side of the brain only. l.o., m.o. and r.o.: left, median and right ocellar nerves, m.o.f.: medium diameter neuron of the right lateral ocellar nerve; a. and p.: antero-lateral and postero-dorsal dendritic complexes. The individual named neurons are described in the text. (From Goodman et al., 1975)

C

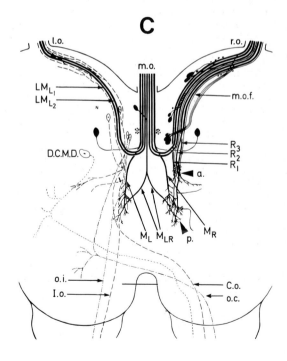

2.8.4 The Central Projections of Ocellar Neurons

The visualisation of neurons which have previously been filled with Cobalt ions by diffusion, axonal transport or by the application of iontophoresis, can be achieved by the precipitation within the filled cells of the black precipitate of Cobaltous sulphide (Pitman et al., 1972). This technique has allowed a fairly rapid description of the central projections of the giant and some of the medium sized neurons of the ocellar nerves of the locust (Goodman, 1974; Goodman et al., 1975 for S. gregaria; Goodman, C., 1974 for Schistocerca vaga). The results of Cobalt studies when coupled with electron microscope studies of ultrastructure and light microscope studies of Golgi-impregnated material have resulted in a description of not only the central projections of second-order neurons but has allowed the tentative identification of higher-order neurons and of possible sites of synaptic interaction within the ocellar system (Goodman et al., 1975). Information about the central projections of locust ocellar neurons has recently been obtained by Williams (1975) as part of an ultrastructural study of the pars intercerebralis and Cooter (1975) has used Cobalt filling and electron microscopy to study the ocellar system of the cockroach, P. americana, an insect which possesses only two ocelli, both lateral in position.

The anatomical arrangement of the central projections of the neurons of the ocelli of the locust are shown in Fig. 5. Different accounts of the arrangement of ocellar neurons have employed different nomenclature. The nomenclature used in this account is that of Patterson and Goodman (1974b) and Goodman et al. (1975).

2.8.4.1 The Projections of the Neurons of the Median Ocellar Nerve

The axons of the neurons of the median ocellar nerve can be divided into three main groups. The first group consists of the four neurons which do not form dendritic arborisations within the brain, but pass to one of the lateral ocelli where they branch in the distal ocellar nerve and form large field arborisations in the synaptic plexus of the ocellar cup. Two of these four axons LM_{L1}, LM_{L2} pass to the left lateral ocellus while the other two LM_{R1}, LM_{R2} pass to the right. The four neurons are collectively referred to as the lateromedial axons. These axons also have large field dendritic trees in the synaptic plexus of the median ocellus.

The second group of median ocellar axons is composed of axons M_L and M_R. These axons on entering the brain pass respectively to the left and right postero-lateral regions of the protocerebral lobes of the brain where they give rise to complex terminal arborisations. These axons together with the lateromedials, pass under the protocerebral bridge as they enter the brain. The third type of axon is that seen in M_{LR}, an axon which passes over the protocerebral bridge and then divides sending a branch to both left and right protocerebral lobes, where each branch forms a dendritic tree which lies in the same region as, although at a different level to, those formed by M_L and M_R. There is some indication that a second cell may show the same arrangement as M_{LR1}, or that M_{LR} may branch to form duplicate structures, although this point has yet to be clarified.

2.8.4.2 Projections of the Neurons of the Lateral Ocellar Nerve

These projections may be divided into three groups. The first of these is made up by the two lateromedial axons which pass from each lateral to the median ocellus and which have been described above. The second group are the two axons in each lateral nerve, L_1, L_2 in the left and R_1, R_2 in the right, which pass to the ipsilateral postero-lateral region of the protocerebral lobe and form terminal arborisations in the same region (but not the same plane) as axons M_L, M_R and M_{LR} of the median ocellar nerve. The third group of lateral ocellar axons are those which follow the same path as L_1, L_2 or R_1, R_2 but which terminate in arborisations at a more dorsal position in the protocerebrum. This group is made up of one giant or large diameter axon from each lateral ocellus, L_3 or R_3, together with a number of the medium diameter lateral ocellar axons.

2.8.4.3 Sites of Synaptic Interaction, and Possible Third-Order Neurons

There are two areas on each side of the brain where ocellar neurons form dendritic arborisations and where integration of ocellar information may be expected to take place. These have been described as the postero-dorsal dendritic complex and the antero-lateral dendritic complex (Goodman et al., 1975). The former is composed of, on each side of the brain, the dendritic arborisations of either L_1, L_2 and M_L or R_1 R_2 and M_R together in each case with the appropriate branch of M_{LR}. The latter is made up on each side of the brain from either the dendritic branches of L_3 or R_3 together with the terminal arborisations of these medium sized fibres of each lateral ocellar nerve which terminate in the same region. The presence of axo-axonic synapses in the ocellar nerves and ocellar tracts of both the median and the lateral ocelli, indicates that these may also be important sites for the integration of input from the ocelli.

Cobalt iontophoresis from the circumesophageal connectives back into the brain has revealed a number of neurons which may be postsynaptic to the central projections of ocellar neurons, (Goodman et al., 1975, and see Fig. 5c). These neurons pass close to both the postero-dorsal protocerebral and the antero-lateral dendritic complexes and in some cases pass into the lateral ocellar tracts where they are, apparently, closely associated with ocellar axons descending into the brain. Four of these neurons pass from each connective back into the brain. Two axons stay in the ipsilateral side of the brain while two cross to the contralateral side of the brain at the level of the deutocerebrum. All the axons pass close to either the ipsilateral or contralateral dendritic complexes and one axon on each side appears to pass into the ipsilateral or the contralateral ocellar tract.

2.8.5. The Processing of Ocellar Information

On the basis of some simple generalisations about the anatomical arrangement of the second-order and of possible third-order neurons of the ocellar pathway Patterson and Goodman (1974b) have been able to make some predictions about the categories of responses to ocellar stimulation which it may be possible to record in third- or higher-order neurons of the ocellar system. These generalisations are as follows:

1. Second-order neurons from the lateral ocelli are confined to the ipsilateral side of the brain.

2. Second-order neurons of the median ocellus project to both sides of the brain.

3. Each lateromedial axon passes from the median to one lateral ocellus.

4. Second-order ocellar axons terminate within the brain.

5. Of the possible third-order neurons of the ocellar system, some pass from the circumoesophageal connective to the ipsilateral side of the brain to associate with second-order neurons, while others pass to the contralateral side before so doing.

Assuming no further crossing over or integration of ocellar input one may expect the following types of ocellar driven unit to occur in the circumeosophageal connectives and ventral nerve cord. The origin of these units is explained in Fig. 6.

A. Units driven by the ipsilateral ocellus alone, by the route indicated in Fig. 6A.

B. Units driven by the ipsilateral and median ocellus, but not by the contralateral (Fig. 6C or D).

C. Units driven by the median ocellus alone (Fig. 6B).

D. Units driven by the median and the contralateral ocellus but not by the ipsilateral (Fig. 6C or D).

E. Units driven by the contralateral ocellus alone (Fig. 6A).

Similarly the following units are not expected to occur, since no direct pathway, involving a single synapse, is evident in the generalised anatomy.

G. Units driven by all three ocelli.

H. Units driven by both the ipsilateral and contralateral ocellus, but not by the median.

The validity of this classification was tested (Patterson and Goodman, 1974b) by examining the types of ocellar stimulation which elicited responses in one readily recordable class of ocellar driven unit, the ocellar 'off' units which can be found in the ventral nerve cord of the locust from the supraoesophageal ganglion as far as the metathoracic ganglion (Goodman, 1970). Ocellar 'off' units respond to dimming or darkening of the ocellus by producing one to three spikes (Burtt and Catton, 1954) with a short latency and without adapting to repetitive stimuli (Goodman, 1970), factors which made 'off' units both easy to recognise and the obvious choice for a unit with which to test the above classification. Similar units have been recorded in the cockroach cord by Cooter (1973).

In recording from the ventral nerve cord between the pro- and mesothoracic ganglia we found that the only consistently occurring classes of ocellar driven unit of the 'off' type were those of types B, C and D above (see Fig. 7). Units of types G and H were not encountered, as our predictions from the anatomy suggested. More importantly, it appears that units of types A and E are also absent from the ventral

Fig. 6A-D. Some possible connections between second- and higher-order ocellar neurons in S. gregaria, together with the types of ocellar driven higher order neurons which the connections may produce. Only the possible connections occurring in the left half of the brain are shown. Individual second-order neurons which may contribute to any given pathway are named, but the higher order neurons have been generalised so that only a single neuron is shown passing into the left (ipsilateral) circumoesophageal connective and only a single neuron crossing over in the brain and passing into the right (contralateral) connective. For discussion, see text

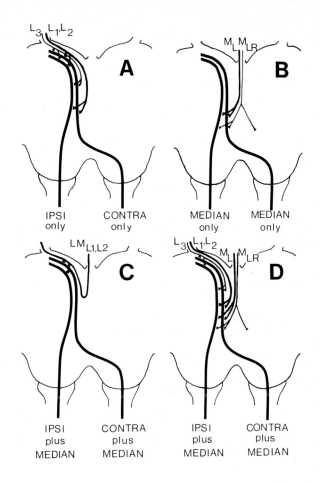

nerve cord of the prothorax, implying that visual information from the ipsilateral of the contralateral ocellus is pooled with that from the median to produce either ipsilateral plus median or contralateral plus median driven units. Information from the median ocellus appears, however, to have a separate pathway so that, if required, the input from each ocellus could to some extent be reconstituted from a simple subtraction of the median information from that provided by units driven by the median plus a lateral ocellus. This process would be satisfactory if the general illumination levels at each ocellus were the required product of the integration, but it is difficult to see how any directional information provided by the wide field ocellar dendrites of the giant second neurons, which are presumably involved in this pathway, could survive the convergence and later reintegration of the information involved in this process. The response patterns of those ocellar 'off' units which fire tonically in the dark and which are inhibited by the ocellus or ocelli which generate 'off' spikes in the unit, also fall into the same classes; namely, units of types B, C and D, as do the 'off' units studied with repetitive stimulation of the ocelli (Patterson and Goodman, 1974b). There is, however, the added complication that illumination of the ocellus or ocelli which do not inhibit the tonic activity, or produce 'off' spikes, tends to produce an excitation of the tonic activity rather than having no effect.

350

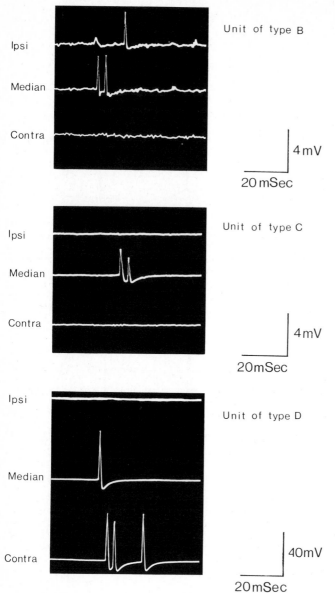

Fig. 7. Recordings obtained from ocellar 'off' units in the ventral nerve cord between the pro- and mesothoracic ganglia of S. gregaria. Each group of traces shows the response of a unit to a flash of white light presented to either the ipsilateral, median or contralateral ocellus. The three types of unit shown here (types B, C and D) are the only ones regularly recorded under these experimental conditions. The data are rearranged from Patterson and Goodman (1974b)

Our generalisations as to the anatomy of the second-order neurons of the ocelli do not provide a simple pathway for this type of response but Patterson and Goodman (1974b) have suggested some possible mechanisms. These include interaction between the giant second-order ocellar axons at the axo-axonic feedback synapses which have been observed in the ocellar nerves and tracts, or the action of efferent ocellar axons, or more simply that the responses are the consequence of stray illumination entering the compound eye adjacent to an ocellus, the illumination of which may produce an excitation of the unit. A further possibility is that the response is mediated by some of the small diameter neurons of the ocellar nerve, the central projections of which have not been revealed by cobalt studies.

2.8.6 Conclusion

Perhaps this discussion of the current state of knowledge about some aspects of the structure and function of the insect ocellus can be most usefully concluded, not by simply underlining the similarities and differences between the ocellus and the retina of vertebrates and cephalopods or the compound eye of arthropods, but by attempting to point to those areas where studies of the ocellus are most likely to yield information which may compliment that obtained from the study of more complex visual systems.

The study of signal amplification in particular, and of signal transfer in general, at the first synapse in the visual system could perhaps be more fully understood as a result of comparative studies of events occurring at the receptor cell second-order cell synaptic system in both the compound eye and ocellus of a single insect species. If the amplification of the receptor cell signal is dependent, at least in part on the number and arrangement of synapses between the receptor and any given postsynaptic cell, as would seem empirically to be the case, then sufficient differences would appear to exist in the synaptic arrangement of these two organs for some significant differences to be seen in the function of the two systems.

The most notable feature of the locust ocellar system is that the three ocelli together contain the distal processes of some thirteen, large diameter, readily identifiable, and anatomically unique second-order neurons which terminate in the brain or in another ocellus. The fact that these cells are relatively easy to record from at their distal ends and that stimuli can be applied to each of three spatially separated visual organs, should make the insect ocellar system a valuable preparation for the study of a number of aspects of the peripheral visual process. Such a preparation can be used to study the feedback processes between specific cells, the integration of the visual input from the three photoreceptor organs and the transfer of information from second-order to higher-order neurons in the system. The generalised anatomy of the arrangement of these second-order cells has already been used to make some useful predictions about the properties of the higher-order neurons, and the explanation of such discrepancies between the predicted and observed behaviour of higher-order neurons as exist or may be found will undoubtedly reveal other facts relevant to the study of the integration of visual information.

Finally, the distinctive central projections of the second-order ocellar neurons would seem to be ideally suited to studies of the development and plasticity of both the anatomy and physiology of the peripheral visual system.

2.8.7 References

Baumann, F.: Slow and spike potentials recorded from retinula cells of the honeybee drone in response to light. J. Gen. Physiol. 52, 855-875 (1968)

Boschek, C.B.: Synaptology in the lamina ganglionaris in the fly. In: Information
Processing in the Visual Systems of Arthropods. Wehner, R. (ed.).
Berlin-Heidelberg-New York: Springer 1972, pp. 17-22

Burtt, E.T., Catton, W.T.: Visual perception of movement in the locust.
J. Physiol. (London) 125, 566-580 (1954)

Burtt, E.T., Catton, W.T.: Electrical response to visual stimulation in the optic
lobes of the locust and certain other insects. J. Physiol. (London) 133, 68-88
(1956)

Cajal, S.R.: Observaciones sobre la estructura de los ocelos y vias nerviosus
ocelares de algunos insectos. Trab. Lab. Invest. Biol. Univ. Madrid 16, 109-139
(1918)

Chappell, R.L., DeVoe, R.D.: Action spectra and chromatic mechanisms of cells in
the median ocelli of dragonflies. J. Gen. Physiol. 65, 399-419 (1975)

Chappell, R.L., Dowling, J.E.: Neural organisation of the median ocellus of the
dragonfly. I. Intracellular electrical activity. J. Gen. Physiol. 60, 121-147
(1972)

Chappell, R.L., Klingman, A.D.: Synaptic feedback in the retina. A model from
pharmacology of the dragonfly ocellus. Proceedings of the Intern. Union Physiol.
Sci. New Delhi: 1974, Vol. XI

Cooter, R.J.: Visual and multimodal interneurones in the ventral nerve cord of the
cockroach, Periplaneta americana. J. Exp. Biol. 59, 675-696 (1973)

Cooter, R.J.: Ocellus and ocellar nerves of Periplaneta americana L. (Orthoptera:
Dictyoptera). Intern. J. Insect Morphol. Embryol. 4(3), 273-288 (1975)

Dowling, J.E., Boycott, B.B.: Organisation of the primate retina: Electron microscopy. Proc. R. Soc. London 166B, 80-111 (1966)

Dowling, J.E., Chappell, R.L.: Neural organisation of the median ocellus of the
dragonfly. II. Synaptic structure. J. Gen. Physiol. 60, 148-165 (1972)

Eaton, J.L.: Electroretinogram components of the ocellus of the adult cabbage moth
Trichoplusia ni. J. Insect. Physiol. 21, 1511-1515 (1975)

Goldsmith, T.H., Bernard, G.D.: The visual system of insects. In: The Physiology
of Insecta. Rockstein, M. (ed.). New York-London: Academic 1974, Vol. II,
pp. 165-272

Goodman, C.: Anatomy of locust ocellar interneurons: Constancy and variability.
J. Comp. Physiol. 95, 185-202 (1974)

Goodman, L.J.: The structure and function of the insect dorsal ocellus. In: Advances
in Insect Physiology. Treherne, J.E., Beament, J.W.L., Wigglesworth, V.B. (eds.).
London-New York: Academic 1970, Vol. VII, pp. 97-195

Goodman, L.J.: The neural organisation and physiology of the insect dorsal ocellus.
In: Compound Eye and Vision of Insects. Horridge, G.A. (ed.).
Oxford: Clarendon 1974, pp. 515-548

Goodman, L.J., Patterson, J.A., Mobbs, P.G.: The projection of ocellar neurons
within the brain of the locust, Schistocerca gregaria.
Cell. Tiss. Res. 157, 467-492 (1975)

Hoyle, G.: Functioning of the insect ocellar nerve. J. Exp. Biol. 32, 397-407 (1955)

Järvilehto, M., Zettler, F.: Localised intracellular potentials from pre- and postsynaptic components in the external plexiform layer of an insect retina.
Z. Vergl. Physiol. 75, 422-440 (1971)

Järvilehto, M., Zettler, F.: Electrophysiological-histological studies on some
functional properties of visual cells and second-order neurons of an insect
lamina. Z. Zellforsch. 136, 291-306 (1973)

Kenyon, F.C.: The brain of the bee. A preliminary contribution to the morphology
of the nervous system of the Arthropoda. J. Comp. Neurol. 6, 133-210 (1896)

Laughlin, S.B.: Neural integration in the first optic neuropile of dragonflies.
I. Signal amplification in dark-adapted second-order neurons.
J. Comp. Physiol. 84, 355 (1973)

Laughlin, S.B.: Neural integration in the first optic neuropile of dragonflies.
II. Receptor signal interactions in the lamina. J. Comp. Physiol. 92, 357-375
(1974a)

Laughlin, S.B.: Neural integration in the first optic neuropile of dragonflies. III. The transfer of angular information. J. Comp. Physiol. 92, 377-396 (1974b)

Laughlin, S.B.: The function of the lamina ganglionaris. In: The Compound Eye and Vision of Insects. Horridge, G.A. (ed.). Oxford: Clarendon 1974c, pp. 341-358

Laughlin, S.B.: Receptor and interneuron light-adaptation in the dragonfly visual system. Z. Naturforsch. 30C, 306-308 (1975)

Metschl, N.: Electrophysiologische Untersuchungen an den Ocellen von Calliphora. Z. Vergl. Physiol. 62, 382-394 (1963)

Patterson, J.A., Goodman, L.J.: Intracellular responses of receptor cells and second-order cells in the ocelli of the desert locust. Schistocerca gregaria. J. Comp. Physiol. 95, 237-250 (1974a)

Patterson, J.A., Goodman, L.J.: Relationships between ocellar units in the ventral nerve cord and ocellar pathways in the brain of Schistocerca gregaria. J. Comp. Physiol. 95, 251-262 (1974b)

Patterson, J.A., Goodman, L.J.: Componental analysis of the ocellar electroretinogram of the locust, Schistocerca gregaria. J. Insect Physiol. 21, 287-298 (1975)

Pitman, R.M., Tweedle, C.D., Cohen, M.J.: Branching of central neurons: Intracellular cobalt injection for light and electron microscopy. Science 176, 412-414 (1972)

Rosser, B.L.: Electrophysiological studies on the ocelli of Aeschna cyanea and their implications for ocellar functions. Ph.D. thesis. Univ. Bristol 1972

Rosser, B.L.: A study of afferent pathways of the dragonfly larval ocellus from extracellularly recorded spike discharges. J. Exp. Biol. 60, 135-160 (1974)

Ruck, P.: Electrophysiology of the insect dorsal ocellus. I. Origin of the components of the electroretinogram. J. Gen. Physiol. 44, 605-627 (1961a)

Ruck, P.: Electrophysiology of the insect dorsal ocellus. II. Mechanisms of generation and inhibition of impulses in the ocellar nerve of dragonflies. J. Gen. Physiol. 44, 629-639 (1961b)

Ruck, P.: Electrophysiology of the insect dorsal ocellus. III. Responses to flickering light of the dragonfly ocellus. J. Gen. Physiol. 44, 641-657 (1961c)

Ruck, P.: Extracellular aspects of receptor excitation in the dorsal ocellus. In: The Functional Organisation of the Compound Eye. Bernhard, C.G. (ed.). Oxford: Pergamon 1966, pp. 195-206

Ruck, P., Edwards, G.A.: The structure of the insect dorsal ocellus. I. General organisation of the ocellus in dragonflies. J. Morphol. 115, 1-26 (1964)

Shaw, S.R.: Organisation of the locust retina. Symp. Zool. Soc. London 23, 135-163 (1968)

Toh, Y., Kuwabara, M.: Fine structure of the dorsal ocellus of the worker honeybee. J. Morphol. 143, 285-306 (1974)

Toh, Y., Kuwabara, M.: Synaptic organisation of the fleshfly ocellus. J. Neurocytol. 4, 271-287 (1975)

Williams, J.L.D.: Anatomical studies of the insect central nervous system: A ground-plan of the midbrain and an introduction to the central complex in the locust, Schistocerca gregaria. (Orthoptera). J. Zool. London 176, 67-86 (1975)

Zenkin, G.M., Pigarev, I.N.: Optically determined activity in the cervical nerve chain of the dragonfly. Biofizika 16, 299-306 (1971)

Zettler, F., Järvilehto, M.: Decrement-free conduction of graded potentials along the axon of a monopolar neuron. Z. Vergl. Physiol. 75, 402-421 (1971)

Zettler, F., Järvilehto, M.: Lateral inhibition in an insect eye. Z. Vergl. Physiol. 76, 233-244 (1972a)

Zettler, F., Järvilehto, M.: Intraaxonal visual responses from visual cells and second-order neurons in an insect retina. In: Information Processing in the Visual Systems of Arthropods. Wehner, R. (ed.). Berlin-Heidelberg-New York: Springer 1972b, pp. 217-222

Zettler, F., Järvilehto, M.: Active and passive axonal propagation of non-spike signals in the retina of Calliphora. J. Comp. Physiol. 85, 89-104 (1973)

2.9 The Resolution of Lens and Compound Eyes

K. KIRSCHFELD

2.9.1 Introduction

Two distinctly different types of eyes have been highly developed in evolution: lens eyes (= camera eyes) in vertebrates, some molluscs and arachnids and compound eyes in arthropods. Based on his comparative studies of the optical properties of compound and lens eyes, Exner (1891) concluded that both types of eyes are optimally adapted for different functions: lens eyes with their high angular resolution seem to more useful for pattern recognition, whereas the compound eyes, with their poor resolution, are thought to be specialized for movement perception. This view is still generally accepted (see the textbooks of Scheer, 1969, Kaestner, 1972). Furthermore, the small facet diameters of the ommatidia in compound eyes seem to cause a poor absolute sensitivity (Exner, 1891; Barlow, 1952; Kirschfeld, 1966; Prosser and Brown, 1969; Snyder et al., 1973). Some insects are said, however, to have higher temporal resolution than humans (Autrum, 1948).

Irrespective of the mentioned disadvantages of compound eyes - poor resolution and sensitivity - many more individual animals as well as animal species are equipped with compound rather than with lens eyes, since even the number of known insect species ($\sim 10^6$) is at least 10 times larger than that of vertebrates (Weber and Weidner, 1974). Though primitive lens eyes (the ocelli) are also common to many insects, these must not be as useful as compound eyes since evolution has clearly favored the latter.

If it is true that both types of eyes are adaptations for different functions, we expect that the world as seen through a compound eye looks different from the world observed by a lens eye. Information on the optical environment available from both types of eyes should be different at the receptor level.

We will consider in this chapter if this is really the case. In order to illustrate the situation we will answer the questions of what would a compound eye look like if it had the optical performance of a human eye, and how a lens eye with the performance of the compound eye of a fly would need to be constructed.

2.9.2 Subjective Resolution

There is no doubt that the absolute resolution of compound eyes is far inferior to that of lens eyes. The angular distance between stripes of a striped pattern that is just able to induce an optomotor turning response must be larger than approximately 2° in the fly (Eckert, 1973), whereas under optimal conditions the "minimum separabile" in man is $0.6 - 1.8 \times 10^{-2}$ degrees (Buddenbrock, 1952).

These variations in performance do not necessarily reflect differences between the various principles according to which lens and compound eyes are realized. It might be due rather to the fact that the eye of such a small animal as a fly is just much smaller than a human eye. It may be more germane in terms of function to compare acuity relative to eye size or, biologically relevant as well, to animal size instead of absolute resolution, since, as we will see below, physical dimensions of an animal's eye place severe restraints upon its performance.

Angular resolution as determined by physiological methods apparently is dependent on the quality of the dioptrics of an eye ("optical resolution") as well as on the angular separation $\Delta\varphi$ of the receptors ("anatomical resolution"). The resolution of the whole visual system has been determined with physiological methods. Test objects have been striped patterns or two point sources, the critical distances of which have been determined. These numbers, here called "physiologically resolution ε", have been measured for many animal species. They characterize the performance of eyes sufficiently well for our purpose and will be used for comparison, even if they do not give such a precise description as the modulation transfer or linespread functions which are known only for a few species.

Fig. 1 relates experimentally determined values of anatomical ($\Delta\varphi$) and physiological (ε) resolution to body height, H, for several species of animals.

We find in the first order a simple interrelationship between resolution ($\Delta\varphi$ or ε) and body height H:

$$\Delta\varphi \approx \varepsilon = k\frac{1}{H} \text{ [deg]}, \tag{1}$$

where k is a factor of proportionality. For most of the animals listed in Fig. 1, k is between 0.2 and 3 deg x cm.

Whereas $\Delta\varphi$ and ε in degrees are measures for an absolute spatial resolution, we may use k in deg x cm as a measure for "subjective resolution", the resolution being the better the smaller k. If two animals have the same subjective resolution, this means that for the same "subjective distance" of an object the same number of points per object area are scanned or resolved, where subjective distance is measured not in units of centimeters but in units of body height. For example, if we (H ≈ 2 m) look at a fly in a distance of 5 m we resolve this fly into the same number of points as a fly (H ≈ 2 mm) looking at another fly from a distance of 5 mm.

Eq. (1) is only a first-order approximation of the data of Fig. 1. There are, in fact, interesting deviations from this relationship. For instance, of all the larger animals, birds have the smallest value of k, that is the highest subjective resolution. The bat Myotis and the jumping spider Metaphidippus represent two extreme cases of low and high "subjective resolution". These facts will be considered again when we have developed a concept that allows an interpretation of the data on the basis of the performances of idealized lens and compound eyes.

The data suggest that smaller animals are adequately endowed even with a smaller absolute resolution because they have sufficient "subjective resolution". This is reasonable because small animals are concerned with objects in closer proximity than are large animals. At these shorter distances, a small animal can then resolve the same objects as well as can a larger animal at a greater distance. On the other hand,

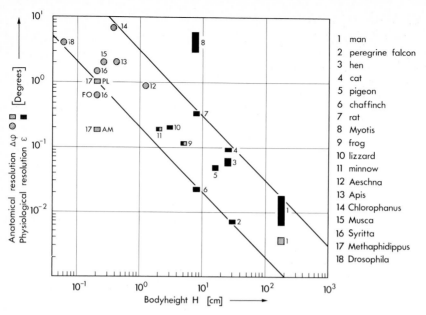

Fig. 1. Anatomical ($\Delta\varphi$) and physiological (ε) resolution as a function of body height H of different animals. Circles: data from animals with compound eyes, rectangles: data from animals with lens eyes. The numbers indicate species and source of the data (1, 2, 3, 5, 6, 8, 9, 11, Buddenbrock, 1952; 4, 7, 10, Penzlin, 1970; 12, del Portillo, 1936; 13, Kirschfeld, 1973; 14, Hassenstein, 1951; 15, Kirschfeld and Franceschini, 1968; 16, Collett and Land, 1975; 17, Land, 1969; 18, Götz, 1965). The body height H (center of eyes above ground) has been estimated on the basis of the size of the animals according to Garms (1969) as far as vertebrates are concerned, or it was measured directly in the insects. For the jumping spider Metaphidippus resolution of antero-median AM eyes as well as of postero-lateral eyes PL is indicated. For the hoverfly Syritta the resolution in the front region of the male is indicated by FO ("fovea"). The two lines of slope - 45° indicate k = 0.2 and 3.0 deg x cm respectively

it means that within the limits of the scatter of the points in Fig. 1, the whole visual environment is poorer in detail for smaller animals.

From the data it appears that "subjective resolution" for all animals varies approximately over one order of magnitude while the range of body height spans three orders of magnitude, irrespective of whether the animal uses a lens or a compound eye. This analysis, therefore, suggests that, contrary to popular belief, the practical resolution of compound eyes is comparable to, rather than worse than, that of lens eyes. The difference in absolute spatial resolution seems to be due not so much to the fact that these animals have compound rather than lens eyes, but to the fact that these animals are so small. This begs the question as to why only relatively small animals are equipped with compound eyes, whereas all known large animals have lens eyes. In order to interpret these facts we will consider the factors that limit the angular resolution in lens as well as in compound eyes.

Fig. 2a and b. Graphical representation
of Eq. (2). Double arrows: lenses,
dotted: intensity distributions in the
focal planes of the lenses. a) The angular size of Airy's disk depends only on
the diameter of a lens, irrespective of
its focal length. b) The absolute size
of Airy's disk depends only on the ratio
A/f, irrespective of the focal length of
a lens

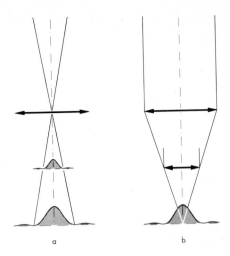

a   b

2.9.3 Physical Parameters Limiting Angular Resolution

2.9.3.1 The Diffraction Limit

An absolute limit to resolution of any eye is set by its optical resolution; information that is not transmitted by the dioptrics is irre-
trievably lost and cannot be restored by neural mechanisms. An upper
limit of optical resolution is easily estimated if the dioptric is made
by a lens system. In this case optical resolution is limited by Fraunhofer diffraction of light.

It has been shown that at small pupil sizes optical resolution in the
human eye is basically diffraction limited (Campbell and Gubisch, 1966).
The dioptric systems of the ommatidia of compound eyes are also lenses
or lens systems, which in the cases that have been analyzed in some
detail are also diffraction limited (e.g. Kirschfeld and Franceschini,
1968; Franceschini and Kirschfeld, 1971). It appears justified, therefore, to introduce diffraction as limiting the optical resolution of
lens and compound eyes. From this point of view all eyes considered
here are "lens eyes". Hence it would be more precise to call the "lens
eye" a single lens eye (or single camera eye) and the compound eye a
"multiple lens eye". Nevertheless, we will use the common terminology.

In the focal plane of a perfect lens system with circular aperture of
diameter A the Fraunhofer diffraction image of a pointlike object at
infinite distance may be characterized by the radius Δr of Airy's
disk, that is the central zone of the diffraction image, included within the first diffraction minimum. We have the equations

$$\Delta r = 1{,}22 \frac{\lambda}{A} \quad [\text{rad}] \tag{2a}$$

$$r = \Delta r f = 1{,}22 \frac{\lambda}{A} f \quad [\text{mm}], \tag{2b}$$

where λ is the wavelength of light in the image medium, f is the focal
length. Small Δr means better optical resolution. Δr, in angular units,
depends only upon the pupil diameter and is independent of the focal
length, whereas r, measured in units of length, depends only upon the
ration f/A, that is the "f-number" of the system (Fig. 2).

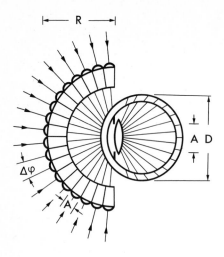

Fig. 3. Schematic diagram of a lens and a compound eye. $\Delta\varphi$: divergence angle between neighboring receptors (anatomical resolution). A: diameter of the pupil or facet lens, respectively, D: diameter of the lens eye, R: radius of the compound eye

2.9.3.2 The Maximal Lens Diameter of an Eye

Diffraction limited optical resolution is determined by the aperture of the lens. If we characterize the size of an eye by its diameter D (Fig. 3), we find that the diameter A of its lens is

$$A = c_1 \times D ,\qquad(3)$$

where c_1 ($\leq$ 1) characterizes the relative size of the lens.

Compound eyes differ basically in design from lens eyes in that the ommatidial aperture can never be made as large as the diameter of the whole eye, since there must be space to accommodate many separate ommatidia (Mallock, 1922; Barlow, 1952; de Vries, 1956; Götz, 1965). In the compound eye the equation

$$A \approx \Delta\varphi R\qquad(4)$$

must be applied, where R is the radius of the eye and $\Delta\varphi$ the angle between neighboring ommatidia, as can be derived directly from Fig. 3.

2.9.3.3 The Anatomical Resolution of an Eye

The overall angular resolution of any eye depends not only upon its optical resolution but also upon the angular separation $\Delta\varphi$ of the receptors. Intuitively one might expect that there could be a constant, optimal ratio for all eyes between optical and anatomical resolution: anatomical resolution might be matched to the optical one in order to just transmit the information available from the dioptric system. Higher anatomical resolution would serve no purpose while lower resolution would sacrifice some of the qualities of the dioptrics. The latter statement is true only if the eye makes use of the information available to the receptors at one and the same time, i.e. if there is no temporal scanning. This is assumed for the moment; temporal scanning will be considered later.

The Rayleigh resolution-criterion, usually applied to telescopes, was often used when discussing the acuity of eyes. More recently such questions have been discussed using a formalism based on information theory, which appears to be less arbitrary than the older approaches (e.g. Barlow, 1965; Westheimer, 1972a).

It is well known that a lens of finite size can transmit spatial frequencies only to an upper limit f_0 and $f_0 = A/\lambda$ [lines/radian] with incoherent illumination. Shannon (1949) has shown that all the information available in such a band-limited function is obtained if the values of the function are known at sampling intervals $\Delta\varphi_s$. The size of $\Delta\varphi_s$ is related to the highest spectral (= Fourier) component f_0 of the function by

$$\Delta\varphi_s = \frac{1}{2f_0} .$$ (5)

Thus for our optical system

$$\Delta\varphi_s = \frac{\lambda}{2A} \text{ [rad] .}$$ (6)

$\Delta\varphi_s$ gives the angular separation of independently acting receptors necessary in order to transmit all the information available from the dioptric system[1]. Introducing Δr from Eq. (2a) instead of A gives

$$\Delta\varphi_s = \frac{\Delta r}{2,44} = g\Delta r .$$ (7)

Fig. 4a is a graphical representation of this equation. It shows that approximately 5 receptors must scan the diameter of Airy's disk or approximately 20 receptors its area according to Eq. (7), in order to transmit the angular information available.

2.9.4 Eyesize and Resolution

2.9.4.1 Lens Eyes

For the sake of simplicity let us consider first so-called "isometric" animals: animals which are exact scale models of each other. Then the diameter D of the eye will be proportional to the body height H. The relation between optical resolution $1/\Delta r$ and body height H, becomes according to Eqs. (2a) and (3)

$$\frac{1}{\Delta r} = \frac{C_1 D}{1.22 \ \lambda} = \frac{C_2 H}{1.22 \ \lambda} \text{ [rad}^{-1}]$$ (8)

where C_1 and C_2 are constants of proportionality. This means, that larger body height and the accompanying larger eyesize yield better (diffraction limited) optical resolution for any given wavelength of light, since $\Delta r \times H$ for isometric animals is constant.

These elementary considerations rationalize the experimental finding that subjective resolution is in first approximation constant for all animals with lens eyes. It is not surprising, however, that in reality there is variation of resolution for a given H, since the assumptions inherent in Eq. (8) are not always valid. For instance, eye size and pupil diameter may either exceed the average or fall below it in some

---

[1] In this paper we use only the simplest concepts from sampling theory and do not concern ourselves with the effects of aperture shapes and sampling lattices. Such refinements do not change the general conclusions presented here.

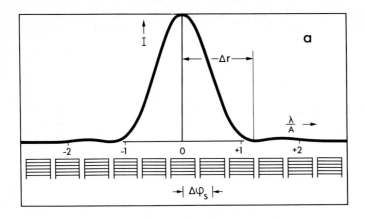

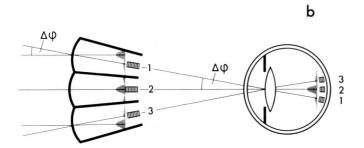

Fig. 4a and b. a) Array of receptors with angular separation $\Delta\varphi_s = \Delta r/2.44$ in the focal plane of a lens eye. According to Shannon's sampling theorem receptors with this separation could transmit all the information available from the lens as far as angular resolution is concerned. The Airy pattern is shown to illustrate its size relative to the receptors. b) In the compound eye (classical apposition type) we have not one Airy pattern projected from a point source onto a retina with many receptors as in the lens eye. Instead in several ommatidia Airy patterns from the point source are projected, each covering the single rhabdom by a different angle. The relative position of the receptors 1 to 3 to the Airy pattern is basically the same, however, in the two types of eyes. Therefore, Eq. (7) can also apply to the compound eye. In this schematic diagram $\Delta\varphi$ is greater than $\Delta\varphi_s$

species. In addition resolution may not always be diffraction limited as assumed in Eq. (2). Some of the variations of the subjective resolution shown in Fig. 1 reflect special adaptations. That birds (with the exception of the domesticated hen) usually have a better resolution than mammals of similar size is explained by their relatively large eyeballs (see Walls, 1967, Fig. 70), and might have to do with the fact that rapidly moving animals need better eyes. The antero-median (AM) eyes of the jumping spider Metaphidippus with their exceptionally high subjective resolution ($k \approx 0.05$ deg x cm, Fig. 1) is also explained by their large relative eye size with an aperture of one sixth and a focal length of one quarter of the body height (Land, 1969). The resolution of the smaller posterolateral (PL) eyes is within the range of other animals. The below average subjective resolution of the bat Myotis ($k \approx 30$ deg x cm) apparently has to do with the bats reliance on echolocation, which works even in the dark and is, therefore, of greater functional value for such an animal,

active at night, than is vision. Nevertheless, these exceptions need
not invalidate the general rule that appears to hold for the majority
of species.

2.9.4.2 Compound Eyes

By combining Eq. (2a), which describes the eyes as diffraction limited,
with Eq. (4), which introduces the geometrical arrangement of the ommatidia in compound eyes, and Eq. (7), which claims an optimal matching
between optical and anatomical resolution (comp. Fig. 4b) we obtain

$$\frac{1}{\Delta r} = \left[\frac{gR}{1.22\ \lambda}\right]^{1/2} \left[rad^{-1}\right] . \tag{9}$$

That is: in contrast to the diffraction limited lens eye, where the
resolution is proportional to eye size (Eq. (8)), we find in the diffraction limited compound eye that the resolution increases only with
the square root of the size of the eye (Mallock, 1894; Barlow, 1952;
de Vries, 1956; Kuiper and Leutscher-Hazelhoff, 1965) or, if eye size
and body height are proportional, to the square root of body height H.

If we look in Fig. 1 at the subjective resolution of animals with compound eyes we again find that it is better in flying insects than e.g.
in the usually slowly moving snout beetle Chlorophanus. The highest
subjective resolution of the selected animals with compound eyes is
found in the hover fly Syritta. The male has a specialized foveal
region FO in the front of the eye with increased angular resolution
($k \approx 0.12$). The high resolution in Syritta is explained by the fact
that it has eyes which are exceptionally large compared to its body
size.

2.9.5 Minimal Size of a Lens Eye

Diffraction and, hence, the absolute size of the aperture A of the
lens poses an absolute limit to the angular resolution of any eye.
A further question is what determines the minimal size of an eye,
given an absolute optical resolution and, hence, aperture size. The
aperture size defines a resolution $1/\Delta r$. Since we match the anatomical
resolution to the optical one according to Eq. (7) the angular distance
of the receptors must remain constant if we reduce an eye in size,
keeping the aperture constant. Therefore the receptor diameter must
be reduced. The small eye and the large one superimposed in Fig. 5,
therefore, show basically the same performance; angular resolution
as well as absolute sensitivity are principally the same. The latter
is due to the fact that the mean number of light quanta q from an extended optical environment available per receptor and time unit is
given by

$$q \sim B \frac{A^2}{f^2} \delta^2 , \tag{10}$$

where B is the mean brightness, f the focal length of the dioptric
system and δ the diameter of the receptors (Rodieck, 1973; Kirschfeld,
1974). Since $A \times \delta/f$ is constant for the small and the large eye as
well this is also the case for q.

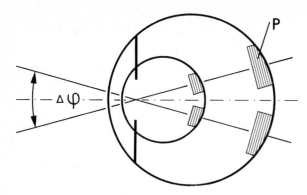

Fig. 5. A diffraction limited lens eye with given pupil diameter theoretically can be reduced in size without losses in angular resolution $\Delta\varphi$ or in absolute sensitivity as long as the diameter of the receptors is reduced proportional to the focal length of the system. A lower limit is reached, however, if the diameter of the receptors becomes so small that they are no longer able to work independently of each other due to optical cross talk. This appears to be at receptor diameters and distances of approx. 1-2 μm, given the realized refraction indices of receptors and surrounding media. P: Photoreceptors

There are, obviously, several factors that pose a limit to reduction of the eye size. For example, the f-number f/A of the lens must be reduced more and more with reducing the size of the eye which might introduce difficulties in its construction as far as aberrations are concerned (cf. Westheimer, 1972b). Second, there is clearly a lower limit to the photoreceptor diameter set by the size of nuclei, mitochondria, etc. A basic limit, however, seems to be posed by the wave properties of light.

By acting as light guides, outer segments and rhabdomeres are able to increase the total amount of light absorbed by increasing the optical path length. If light guides are too small, however, most of the light will travel along the outside of the light guide and, therefore, cannot be absorbed by the visual pigment inside. That is, the "optical diameter" is larger than the anatomical one. Furthermore, problems of optical cross-talk become serious in small waveguides (Snyder, 1975). This means that the receptors are no longer able to act independently of each other, a condition necessary if the sampling theorem (Eq. (5)) is applied. Lastly, the acceptance angle of the waveguide (directivity) becomes smaller as their size is reduced, which prevents the receptors from making use of a small f-number of the dioptrics. This reduces angular resolution as well as absolute sensitivity. The actual size to which receptors may be reduced depends basically upon the waveguide parameter

$$V = \frac{\pi\delta}{\lambda_o} \sqrt{n_1^2 - n_2^2} \qquad (11)$$

where n_1, n_2 are the refractive indices of the waveguide and its surround respectively, λ_o is the wavelength of light in vacuum and δ the diameter of the waveguide, that is the outer segment or rhabdomere. The problems mentioned become serious if V becomes smaller than 2 to 3. Since the difference between n_1 and n_2 cannot be increased beyond some limit with the substances (lipoproteins and water respectively) available to the receptor cells, δ cannot be reduced beyond some limit.

This limit is, given realistic differences $n_1 - n_2$ of 0.02 to 0.04, in the order of 1 to 2 μm. Thus, the minimal "grainsize" for a biological retina cannot be reduced below this value.

From this point of view the high concentration of membranes in the photopigment containing structures of outer segments and rhabdomeres may not only provide a means to increase the photopigment concentration within these structures thus increasing the number of photons absorbed, but also to <u>increase their refractive index so that the grain of the retina may be as fine as possible</u>.

Though there is a lower limit of photoreceptor-diameter, determined by waveguide properties, receptors could be larger than this limit. However, optimum resolution could not be achieved in this case without also making the eye larger in order to maintain the matching between optical and anatomical resolution. It seems reasonable, therefore, that the eyesize with a given absolute aperture A has been reduced by evolution just to the point where the minimal receptor diameter that allows the receptors to act independently is reached. This can be realized as we have seen without any loss of angular resolution (Fig. 5) as well as absolute sensitivity.

These general arguments show that a more or less constant receptor diameter for all eye sizes is a functionally adequate adaptation. They explain why the absolute size of retinal elements only varies within narrow limits however large or small a lens eye may be (Walls, 1967). For instance the diameter δ of cones in the human fovea is in the order of 1-2 μm. Surprisingly this is also just the diameter of the rhabdomeres in the compound eye of the fly (receptors 1 to 6: 2 μm, receptors 7 and 8: 1 μm, Boschek, 1971). Every individual fly ommatidium with a retinula composed of 8 receptors with 7 rhabdomeres, all acting as independent waveguides (Kirschfeld, 1967), may be considered in this context as a lens eye, which explains the convergence. In compound eyes of the classical apposition type which have only one rhabdom per ommatidium acting as one single waveguide it is of course not the cross-talk between receptors that is limiting. Here the relative increase of the "optical diameter" compared with the anatomical one and the consequences for absorption as well as angular resolution (Pask and Snyder, 1975) are the main limiting factors.

Realizing that there exists a lower limit to the size of receptors, determined by the difference of the refractive indices between receptors and surrounding medium, one is able to calculate an optimal f-number of a diffraction limited lens eye, which is still useful with respect to the angular resolution of the system. Considering the case of an optimal matching between the "graininess" of the retina (anatomical resolution) and the optical resolution, one finds from Eq. (7) with the angular separation δ/f of the receptors $\approx \Delta\varphi_s$

$$\Delta r \approx 2.44 \, \frac{\delta}{f} \tag{12}$$

combined with Eq. (2a) one arrives at

$$\frac{f}{A} \approx \frac{2\delta}{\lambda} \tag{13}$$

which determines the smallest useful f-number. Introducing $\lambda = 0.37$ μm (wavelength λ_o of light = 0.5 μm, refractive index n = 1.34) and δ = 1-2 μm yields f/A in the range of 5 to 10. <u>Smaller f-numbers are of no use to any eye with respect to angular resolution</u>. This holds as long as temporal scanning does not come into play. Smaller f-numbers might be advantageous, however, by increasing the absolute sensitivity.

2.9.6 Realized Lens and Compound Eyes

2.9.6.1 Human Eye

Optical measurements on human eyes have shown that diffraction is a limiting factor for optical resolution only at small pupil sizes. If the pupil dilates, other aberrations of the lens become more and more limiting. The best optical resolution occurs at a pupil diameter A of 2.4 mm (Campbell and Gubisch, 1966). According to Eq. (2a) the diameter of Airy's disk at this pupil size is 1.30 min of arc ($\lambda = 0.37$ µm).

The diameter of a cone in the human fovea is 1 to 2 µm (Buddenbrock, 1952) which corresponds to 0.2 to 0.4 min of arc, a value considerably less than the diameter of Airy's disk. This cone diameter corresponds nicely, however, to the angle $\Delta\varphi_s$ given by Eq. (7) to 0.27 min of arc for the optimal sampling interval. If signals of individual cones in the fovea act independently of one another, and if their signals are processed independently with respect to spatial resolution, the anatomical resolution of the human eye is sufficient to transmit all the spatial frequencies that pass the dioptric system. Optical and anatomical resolution are fitted according to Shannon's sampling theorem.

2.9.6.2 Fly's Eye

The radius at the front region of the eye of a Musca female is approximately 600 µm. By means of Eqs. (2a), (7) and (9) we calculate a value of $\Delta\varphi_s = 1.0°$ and of A = 11 µm (g = 1/2.44; $\lambda = 0.37$ µm). Actual measured values are $\Delta\varphi = 2.3°$ and A = 24 µ. This means that the anatomical resolution $\Delta\varphi$ is worse than it could be theoretically for a purely diffraction limited compound eye of the same size approximately by a factor of two. The deviation, though greater than the limit of experimental error, is not so large as to merit detailed discussion here.

2.9.7 Comparison of the Angular Resolution of Lens and Compound Eyes

Using the principle of the preceding sections we can determine how a compound eye might look if it where to have the same angular resolution as a human eye, and also the minimal size of a lens eye, with angular resolution equivalent to that of a fly's eye.

As we have seen, the human lens eye has its best resolution at a pupil size of 2.4 mm. Since its diameter is 24 mm, C_1 becomes 0.1. Fig. 6 shows the relationship between resolution of lens and compound eyes and their size (Eqs. (8) and (9)). It can be seen easily that a lens eye of D = 24 mm corresponds to a compound eye of radius R = 31 m ($\lambda = 0.37$ µm).

This means that a diffraction limited compound eye with the same angular resolution must be enormous (Fig. 7a). However, the calculation is somewhat erroneous for we know that the optimum resolution of the human eye is only confined to a rather small foveal region, decreasing rapidly with increasing angular distance α from the fovea (Fig. 8). This means that we can reduce the size of the "equivalent" compound eye by reducing A and R with increasing angular distance from the "fovea". The result is a compound eye of elongated shape with a long axis of 31 m (Fig. 7b). Its size can be reduced, however, still further, without loss of resolution. Since the radius is drastically

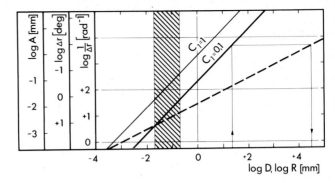

Fig. 6. Graphical representation of Eqs. (8) and (9). Ordinate: Pupil diameter $\underline{A}$, size of Airy's disk Δr and resolution $1/\Delta r$. Thick continuous line: ordinate parameters of the lens eye as a function of its diameter $\underline{D}$. The ratio C_1 was chosen to 0.1 which represents the case valid for the human eye at optical resolution (pupil size 2.4 mm). Thin continuous line: $c_1 = 1$. Interrupted line: ordinate parameters for the compound eye as a function of its radius $\underline{R}$. Hatched area: size of photoreceptors (lengths 20 µm to 200 µm). Arrows: indicate that a human eye of diameter $D = 24$ mm (log D = 1.38) with $c_1 = 0.1$ corresponds to a compound eye of radius $R = 3.1 \times 10^4$ cm (log R = 4.49)

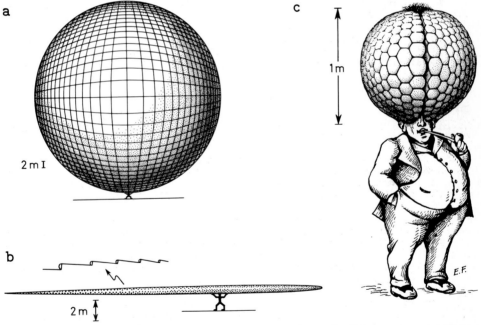

Fig. 7a-c. Human being equipped with compound eyes instead of lens eyes. (a) Compound eye with the same resolution as a diffraction limited lens eye with A = 2.4 mm. (b) The decrease in angular resolution with distance from the fovea has been taken into account. (c) Compound eye with the overall resolution of a human eye and of minimally possible size. The minimal surface necessary for all facets has been calculated and the size of the hemispheres has been determined so that the surface just allows space for all of the facets. Facet size not to scale, instead of the 100 facets per eye 10^6 should have been drawn

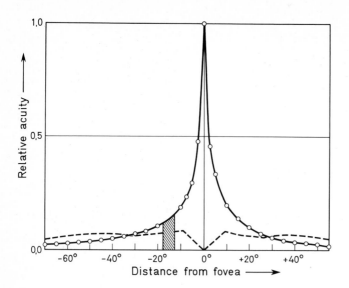

Fig. 8. Relative angular resolution of the human eye at different angular distances from the fovea as determined psychophysically by Wertheim. (From Rein Schneider, 1956). Circles: photopic system. Interrupted line: scotopic system. Hatched: blind spot

changing with angular distance of the fovea, most of the eye surface is not used for the entrance of light (inset Fig. 7b). If we calculate the integral only over the surface of all facets, we determine the surface the eye must have available at the minimum for the entrance of light into the ommatidia.

At angular distance α from the fovea the facets of size A (α) form a ring of radius R (α) sinα and thus contribute to the corneal surface an area $\Delta 0$ given by $\Delta 0 (\alpha) = 2\pi A (\alpha) R (\alpha) \sin\alpha$. $1/\Delta\varphi = \pi R/180 A$ gives the number of ommatidia rings per degree. Integrating over $\Delta 0/\Delta\varphi$ from 0 to 90° finally results in the total surface 0. It comes out to be 1.7×10^6 mm². A hemisphere with equivalent surface has a radius R of 0.52 m (Fig. 7c). And this is the minimal size that a diffraction limited compound eye with human angular resolution must have. It is impossible to reduce its size further without loss in angular resolution.

The size of a lens eye equivalent to that of a fly's compound eye might be estimated as follows: the diameter A of the lens must be 24 µm in order to have the same absolute optical resolution as found in a fly's ommatidium. This aperture has to be combined with a minimal focal length of 50 µm which is necessary to match the absolute size of Airy's disk to the 1 to 2 µm diameter of the rhabdomeres in the same way as in the fly's eye. The lens and the vitreous body alone therefore would provide a lens eye of diameter 50 µm. However, for such a small eye the size of the receptors must be considered. If we want to have the same total absorption within the photoreceptors of the hypothetical fly eye, we need receptors of a length eqivalent to those in the real Musca compound eye that is of 200 µm. As seen in Fig. 9, it is now the size of the receptors, which determines the actual diameter of the equivalent lens eye, and this yields a total diameter of some 500 µm. However, the lens eye shown in Fig. 9 would be considerably less efficient than the real compound eye of the fly. High resolution combined with high light gathering power could be reached only in a small foveal area since aberrations of the lens in extra foveal regions would probably become serious. Furthermore, the actual light gathering power in the Musca compound eye is, due to a special arrangement of receptors in the ommatidia, higher by a factor

Fig. 9. Comparison of a lens and compound
eye, both equivalent in angular resolution
to that of a fly's compound eye. Lens eye
vitreousbody and ommatidial dioptric-sys-
tem (clear) are drawn at the left- and
right-hand side of the receptors (stippled)
of 200 µm length respectively. Lens and
facets have the same diameter of 24 µm in
order to give the same optical resolution
as in the fly's eye. It can be seen that
the size of the compound eye is somewhat
bigger than the lens eye, in both cases
the size is basically determined by the
receptors, however. The compound eye has
the advantage that angular resolution and
light-gathering power do not decrease with
increasing angular distance of the "fovea"
of the eye, as it is usually the case in
lens eyes due to aberrations of the lens

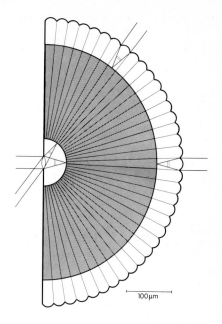

100 µm

of 7 compared to the light gathering power defined by the area of the
single facet (Kirschfeld, 1967). This means that the estimate of a
diameter D of some 500 µ for the equivalent lens eye is too low. And
even such there is no considerable difference in size.

2.9.8 Temporal Scanning

In contrast to the poor absolute resolution of compound eyes some insects, especially flies, have a rather high temporal resolution
(Autrum, 1948). The frequency-response curve of their photoreceptors
falls down to 10% of the maximal value at 150 cps (Zettler, 1969).
Intracellular photoreceptor recordings of warmblooded animals are not
available for comparison. Flicker-threshold response-curves determined
psychophysically in man in bright-adaptation show a decay of sensitivity to 10% of the maximum at 65 cps (Kelly, 1961). This indicates that
fly-eyes are by a factor of 2 to 3 faster than the human visual system.
One might ask if this higher temporal resolution can be used to increase angular resolution by means of temporal scanning.

In the human, temporal scanning in principle is not expected to increase angular resolution. The fitting between optical and anatomical
resolution according to the sampling theorem (9.6, this chapter) ensures that all the information passing the dioptrics in principle
might be transmitted by the foveal receptors already in stationary
conditions. This is in accordance with the finding that angular resolution determined under stabilized image conditions is the same as
that determined in nonstabilized conditions (e.g. Westheimer, 1972b).
It explains also why the f-number of approximately 10 in the human
for optimal resolution just fits the minimal f-number as calculated
above.

As we have seen there is some mismatch between optical and anatomical resolution in the fly. Even if not considerable this leaves the possibility of improvement of angular resolution by temporal scanning. Resolution then depends not only on the number of points that are scanned simultaneously by the eye, but also upon the sequence of signals that are elicited in the receptors if eye and surround are shifted relative to each other. The facets of diameter 24 µm in the fly's eye yield a $\Delta\varphi_s$ (Eq. (6), $\lambda = 0.37$ µm) of $0.5°$. This is by a factor of 4 better than the actually measured divergence angle $\Delta\varphi$ between receptors which is approximately $2°$. Temporal scanning, therefore, in principle can improve the angular resolution up to $0.5°$. The high temporal resolution of the receptors could help the animal to carry out a short time scanning process.

In general, angular resolution may be improved by temporal scanning up to the limit set by the aperture size of the dioptric system. The ratio between optical resolution $1/\Delta r$ in man and fly is, therefore, given according to Eq. (2a) by $A_{man} = 2.4$ mm / $A_{fly} = 24$ µm = 100 and cannot be reduced by high temporal resolution combined with special neural mechanisms in the fly.

2.9.9 Discussion

The above considerations show that lens eyes are apparently the better solution for large animals. These animals need high absolute angular resolution and in this case lens eyes need considerably less space than compound eyes. However, in small animals with poor absolute (but still adequate subjective) resolution the space needed for lens and compound eyes becomes comparable, especially if the eyes are so small that the size of the receptors and not that of the dioptric systems become a limiting factor. This explains why lens and compound eyes fit the relation between bodysize and resolution (Fig. 1) to the same degree, since compound eyes are realized only in small animals.

For small animals, compound eyes seem to be rather advantageous as far as size of visual field and distribution of angular resolution over the visual field are concerned. A single lens for the whole visual field gives high resolution only in a foveal region, since for practical lenses increasing angular distance from the fovea is usually accompanied by greater aberrations. Also it might well be that small animals with a poor absolute resolution could not tolerate a resolution still considerably worse at angles off the "fovea": this might bring predators out of the range of visibility, for example.

What the brain does with the information available from the receptors depends upon the logic of its wiring diagram. If we compare the numbers of neurons in those parts of the brain which seem to be directly related to vision we find 5×10^5 in the fly (neurons of the retina and the three optic ganglia; Campos-Ortega, pers. comm.) and 7×10^8 in man (retina, geniculate body and cortical areas 17 to 19; Blinkov and Glezer, 1968). This difference of a factor of approximately 10^3 is, indeed, considerable. If we relate the absolute numbers of neurons to the number of points resolved by the eyes (4×10^5 in man; Steinbuch, 1965, and 6×10^3 in the fly; Braitenberg, 1967) we arrive at 2×10^3 neurons per point discriminated in man and at 100 neurons per ommatidium in the fly. There is still an obvious difference of a factor of 20. But this difference is not so high as to lead us to expect from the outset different ways of processing of visual information in vertebrates and in insects.

Exner's (1891) statement that lens eyes seem to be more useful for pattern recognition whereas compound eyes are specialized for movement perception is no longer supported by the knowledge we have today of the optical properties of the two types of eyes. The image projected by a lens eye at the receptor level is not so different from that in a compound eye, as far as resolution is concerned, to support such a conclusion. And the belief that compound eyes have a poor absolute sensitivity as mentioned in the introduction, is also not justified (Rodieck, 1973; Kirschfeld, 1974).

Without any doubt there are differences between lens and compound eyes. Discrimination of E-Vector orientations of linearly polarized light seems to be a domain of animals with compound eyes. But what the differences are cannot be derived in general terms from the optical performance of the eyes or from the gross anatomy. And with increasing understanding of how visual systems work, it might transpire that the differences between the visual systems of animals with lens and compound eyes are smaller than was initially believed.

Acknowledgments. It is a pleasure to acknowledge the valuable comments and criticisms of D.G.M. Beersma, N. Franceschini, S. Laughlin, P. Liebman, C. Pask, A.W. Snyder and D. Stavenga, all helping to clarify this presentation. I thank also E. Freiberg and M. Heusel for preparing the figures.

2.9.10 References

Autrum, J.: Über Energie- und Zeitgrenzen der Sinnesempfindungen. Naturwissenschaften 12, 361 (1948)

Barlow, H.B.: The size of ommatidia in apposition eyes. J. Exp. Biol. 29, 667 (1952)

Barlow, H.B.: Visual resolution and the diffraction limit. Science 149, 553 (1965)

Blinkov, S.M., Glezer, I.J.: The human brain in figures and tables, a quantitative handbook. New York: Plenum 1968

Boschek, C.B.: On the fine structure of the peripheral retina and lamina ganglionaris of the fly, Musca domestica. Z. Zellforsch. 118, 369-409 (1971)

Braitenberg, V.: Patterns of projection in the visual system of the fly. I. Retina-lamina projection. Exp. Brain Res. 3, 271-298 (1967)

Buddenbrock, W. von: Vergleichende Physiologie I. Basel: Birkhäuser 1952

Campbell, F.W., Gubisch, R.W.: Optical quality of the human eye. J. Physiol. 186, 558-578 (1966)

Collett, T.S., Land, M.F.: Visual control of flight behaviour in the hoverfly, Syritta pipiens L. J. Comp. Physiol. 99, 1-66 (1975)

Eckert, H.: Optomotorische Untersuchungen am visuellen System der Stubenfliege Musca domestica L. Kybernetik 14, 1-23 (1973)

Exner, S.: Die Physiologie der facettierten Augen von Krebsen und Insecten. Leipzig-Wien: Franz Deuticke 1891

Franceschini, N., Kirschfeld, K.: Etude optique in vivo des éléments photorécepteurs dans l'oeil composé de Drosophila. Kybernetik 8, 1-13 (1971)

Garms, H.: Pflanzen und Tiere Europas. Braunschweig: Westermann 1969

Götz, K.G.: Die optischen Übertragungseigenschaften der Komplexaugen von Drosophila. Kybernetik 2, 215-221 (1965)

Hassenstein, B.: Ommatidienraster und afferente Bewegungsintegration. Versuche an dem Rüsselkäfer Chlorophanus viridis. Z. Vergl. Physiol. 33, 301-326 (1951)

Kaestner, A.: Lehrbuch der Speziellen Zoologie, Band I Wirbellose, 3. Teil Insecta: A. Allgemeiner Teil. Stuttgart: Gustav-Fischer 1972

Kelly, D.H.: Visual responses to time-dependent stimuli. I. Amplitude sensitivity measurements. J. Opt. Soc. Am. 51, 422 (1961)

Kirschfeld, K.: Discrete and graded receptor potentials in the compound eye of the fly (Musca). The functional organization of the compound eye. Bernhard, C.G. (ed.). Oxford, Pergamon 1966

Kirschfeld, K.: Die Projektion der optischen Umwelt auf das Raster der Rhabdomere im Komplexauge von Musca. Exp. Brain Res. 3, 248-270 (1967)

Kirschfeld, K., Franceschini, N.: Optische Eigenschaften der Ommatidien im Komplexauge von Musca. Kybernetik 5, 47-52 (1968)

Kirschfeld, K.: Optomotorische Reaktionen der Biene auf bewegte "Polarisations-Muster". Z. Naturf. 28C, 329-338 (1973)

Kirschfeld, K.: The absolute sensitivity of lens and compound eyes. Z. Naturf. 29C, 592-596 (1974)

Kuiper, J.W., Leutscher-Hazelhoff, J.T.: Linear and nonlinear responses from the compound eye of Calliphora erythrocephala. Cold Spring Harb. Symp. Quant. Biol. 30, 319-428 (1965)

Land, M.F.: Structure of the retinae of the principal eyes of jumping spiders (Salticidae: Dendryphantinae) in relation to visual optics. J. Exp. Biol. 51, 443-470 (1969)

Mallock, A.: Insect sight and the defining power of composite eyes. Proc. R. Soc. London 55B, 85 (1894)

Mallock, A.: Divided composite eyes. Nature 110, 770-771 (1922)

Pask, C., Snyder, A.W.: Angular sensitivity of lens-photoreceptor systems. In: Photoreceptor Optics. Snyder, A.W., Menzel, R. (eds.). Berlin-Heidelberg-New York: Springer 1975

Penzlin, R.: Kurzes Lehrbuch der Tierphysiologie. Jena: Fischer 1970

Portillo, J. del: Beziehungen zwischen den Öffnungswinkeln der Ommatidien, Krümmung und Gestalt der Insektenaugen und ihrer funktionellen Aufgabe. Z. Vergl. Physiol. 23, 100-145 (1936)

Prosser, C.L., Brown, F.A.: Comparative Animal Physiology. Philadelphia-London: Saunders 1961 (reprinted 1969). 2nd ed.

Rein, H., Schneider, M.: Physiologie des Menschen. Berlin-Heidelberg-New York: Springer 1956

Rodieck, R.W.: The Vertebrate Retina: Principles of Structure and Function. San Francisco: W.H. Freeman 1973

Scheer, B.T.: Tierphysiologie. Stuttgart: Gustav-Fischer 1969

Shannon, Cl.E., Weaver, W.: The Mathematical Theory of Communication. Urbana: Univ. Illinois 1949

Snyder, A.W., Menzel, R., Laughlin, S.B.: Structure and function of the fused rhabdom. J. Comp. Physiol. 87, 99-135 (1973)

Snyder, A.W.: Photoreceptor optics - theoretical principles. In: Photoreceptor Optics. Snyder, A.W., Menzel, R. (eds.). Berlin-Heidelberg-New York: Springer 1975

Steinbuch, K.: Automat und Mensch. Berlin-Heidelberg-New York: Springer 1965

Vries, H. de: Physical aspects of sense organs. In: Progress in Biophysics and Biophysical Chem. Butler, J.A.V. (ed.). Oxford: Pergamon 1956, Vol. VI

Walls, G.L.: The Vertebrate Eye. New York-London: Hafner 1967

Weber, H., Weidner, H.: Grundriß der Insektenkunde. Stuttgart: Gustav-Fischer 1974

Westheimer, G.: Optical properties of vertebrate eyes. In: Handbook of Sensory Physiology. Fuortes, M.G.F. (ed.). Berlin: Springer 1972a, Vol. VII/2, pp. 449-482

Westheimer, G.: Visual acuity and spatial modulation thresholds. In: Handbook of Sensory Physiology. Jameson, D., Hurvich, L.M. (eds.). Berlin-Heidelberg-New York: Springer 1972b, Vol. VII/4

Zettler, F.: Die Abhängigkeit des Übertragungsverhaltens von Frequenz und Adaptationszustand; gemessen am einzelnen Lichtrezeptor von Calliphora Erythrocephala. Z. Vergl. Physiol. 64, 432-449 (1969).

3 MOLLUSCS

3.1 Ultrastructural Observations on the Cortex of the Optic Lobe of the Brain of Octopus and Eledone

N. M. CASE

3.1.1 Introduction

The nervous system of cephalopods, because of its relative simplicity compared to that of vertebrates, has long been favored for investigations relating brain structure to function. The octopus, rated by some as the most intelligent of all the invertebrates, is particularly suited to experimentally relating the brain structure to the animal's behavior.

A comprehensive study of the brain, the optic lobe, and the learning and memory system in cephalopods has been made by the combined efforts of a number of investigators (Boycott, 1954; Boycott and Young, 1955; Wells and Wells, 1957; Wells, 1959, 1966; Young, 1962a,b) to list but a few. By the light microscope, both normal and degenerating connecting pathways of the retina and optic lobe of Octopus vulgaris have been studied in detail by Young (1962a,b). This study and others by Young (1960, 1961, 1964a,b, 1965a,b) have broadened the foundation for understanding the neural basis of the animal's learning and memory systems.

With this focus of importance, Dilly et al. (1963) extended the morphologic studies of the optic lobe to the ultrastructural level with the electron microscope. Young (1971) produced a monumental monograph covering the anatomy of the nervous system of O. Vulgaris to that time. Further ultrastructural findings about the optic lobe cortex of O. vulgaris and Eledone were reported by Case et al. (1972).

The purpose of this presentation is to review the overall ultrastructural organization of the optic lobe cortex, adding to it any recent and as yet unpublished observations.

3.1.2 Materials and Methods

Optic lobe material for this study has been obtained from Eledone, O. vulgaris and Octopus bimaculatus following ethyl carbamate (urethane) anaesthesia. In the early part of the work, using Eledone and O. vulgaris, the optic lobes were exposed and small pieces excised as rapidly as possible. These were immediately immersed in 1% osmium tetroxide buffered to pH 7.2 with veronal acetate. Additional osmium solution was injected with a micropipette having a tip bore of approximately 300 μm, in a pattern of punctures approximately 300 μm apart (Gray, 1970). Following the osmium injections, the tissues were rinsed in distilled water, dehydrated through a graded series of ethanol dilutions, and embedded in Araldite using epoxypropane as an intermediate step (Luft, 1961). During dehydration, some pieces were block-stained with 1% uranyl acetate in 70% ethanol for 2 h. Sections were further stained on the grid with lead citrate (Reynolds, 1963).

In more recent work done on O. bimaculatus, the animals were perfused through the dorsal aorta with a solution of 2.5% glutaraldehyde in 0.2 M phosphate buffer (Gueffroy, 1975) for 10-15 min with a flow rate of 30-50 ml/min. For the last Octopus prepared, 4% formaldehyde (made from paraformaldehyde) was added to the above perfusate. Following perfusion, the optic lobes were removed, sliced in half, and placed in a vial of the perfusion fluid for an additional 5-10 min. Pieces of tissue of a suitable size for embedding were sliced from the alde- hyde fixed material and postfixed for two h in 1% osmium tetroxide made up in 0.2 M phosphate buffer solution. Following the postfixation, the tissues were rinsed in distilled water, dehydrated through a graded series of acetone dilutions and embedded in Epon 812, using epoxypro- pane as an intermediate step. Sections were stained on the grid with saturated aqueous uranyl acetate (Sjöstrand, 1967), followed by lead citrate (Reynolds, 1963). If selection of the right area of the block to trim for sectioning proved difficult, the method of Gray (1961) was followed.

Serial sections were made on a Reichert Om U3 microtome using a Du Pont diamond knife. From these, serial photographic montages were construct- ed.

3.1.3 Observations and Discussion

When one looks at a light microscope section of the optic lobe of Octopus or Eledone, one sees a body of tissue organized with a cortex and a medulla. The cortex is formed of three layers (outer granular, plexiform and inner granular), and, though not analogous to, do some- what resemble the nuclear and plexiform layers of the vertebrate eye. Thus the cortex is sometimes referred to as the "deep retina" (Young, 1962b).

The medulla is composed of a neuropil in which are situated island clumps of neurons (possibly interconnected) which are somewhat ana- logous to the layer of ganglion cells of the vertebrate eye, in that some of them will receive the encoded and integrated visual messages from the optic lobe cortex and send them on to the visual interpretive centers via the optic tracts, of which their axons are one component (Young, 1962b).

3.1.3.1 The Optic Nerves

The optic nerves, arising in the retina, stream to the periphery of the optic lobe where they penetrate a layer of glial cells and course inward among the microneurons of the outer granular layer of the cor- tex. Here they are dark in appearance in contrast to adjacent fibers arising from the other neuronal cells, and contain mitochondria, micro- fibrils and microtubules (Figs. 2, 4). In O. bimaculatus they also

---

Fig. 1. A microneuron mn of the outer granular layer showing its trunk t passing ▷ down through the basement layer b into the top of an optic nerve terminal ot. gf glial folds

Fig. 2. A microneuron trunk t and optic axon on passing down through the outer granular layer. Glial fold investments gf are indicated

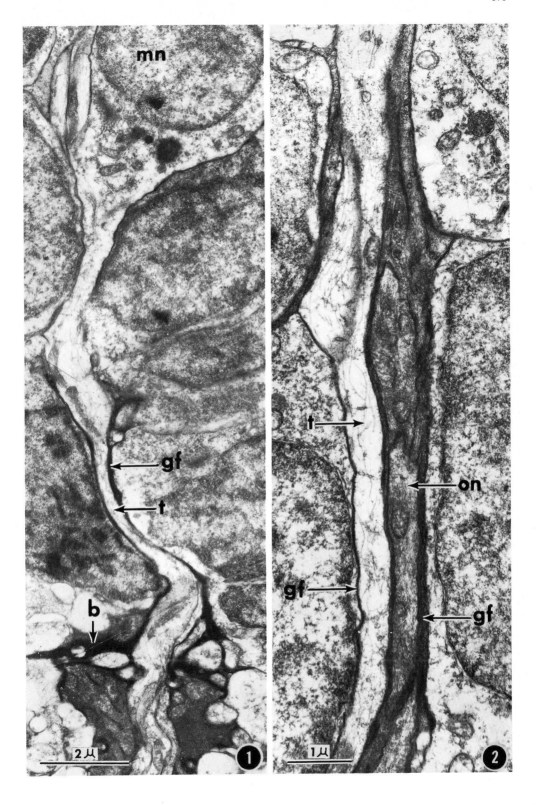

contained considerable amounts of glycogen granules, as do their terminal bags in the plexiform layer (Fig. 7). Except in O. bimaculatus,
in their passage through the outer granular layer, the fibers have an
investment of dark, granular appearing glial folds (Figs. 2, 3) which
seem to be continuous with a layer of neuroglial cells covering the
surface of the optic lobe, and, with a thick outer basement layer of
similar glial folds and sheets which separates the outer granular
layer from the plexiform layer (Fig. 3).

Though direct continuity has not been traced, the temptation is strong
to believe that much of the dark glial material of the outer basement
layer and that surrounding the nerve fibers passing through the outer
granular layer, derives from dark cells scattered among the microneurons in this layer. Similar cells were described by Gray (1969) in
the vertical lobe of Octopus brain, and later continuity between these
dark cells and dark glial laminae around nerve cells, similar to the
situation in the optic lobe, was traced (Case, unpublished). For this
reason it was suggested that the term basement "layer" is more appropriate than basement "membrane" (Case et al., 1972).

For about 4 μm just before penetrating the outer basement layer, the
slender optic nerve fibers frequently, but not always, enlarge, the
cytoplasm becomes lighter, mitochondria become more numerous and microtubules and microfibrils become more prominent (Fig. 3). Synaptic vesicles may also occasionally be seen in this preterminal varicosity of
the optic nerve axon.

Narrowing somewhat, the varicose part of the nerve passes through the
outer basement layer to enter the plexiform layer (Fig. 3) where it
expands into a terminal bag packed with synaptic vesicles, and contains a few mitochondria and microfilaments (Dilly et al., 1963). This
terminal presynaptic bag will be described in greater detail under the
section on the plexiform layer.

3.1.3.2 The Outer Granular Layer

This layer has been described at the light microscope level by Young
(1962b, 1971) as consisting of microneurons with only one inwardly
directed process, which was frequently seen to branch, but without
any one branch being identifiable as an axon. These cells are, therefore, generally referred to as "amacrine cells".

At the ultrastructural level, the cells of the outer granular layer
have a low density cytoplasm containing Golgi membranes, mitochondria,
centrioles and scattered ribosomes (Case et al., 1972). The cytoplasm
of the amacrine trunk ground substance has a similar density to the
cytoplasm of the perikaryon and contains microtubules, a few vesicles
and an occasional mitochondrion. Fig. 2 shows an amacrine trunk which
was traceable to its perikaryon, and an optic nerve fiber traced to
its terminal bag, adjacent to each other, with the optic nerve fiber
being darker and containing more mitochondria and tubules. Varicosities, characteristic of the optic nerve fibers before they enter the
optic lobe (Dilly et al., 1963), are not present as the nerve makes
its way through the outer granular layer; here ephaptic interaction
between fibers is prevented by an insulation of glial folds.

Considering the large number of microneurons in the outer granular
layer, it is surprising that the region where a trunk joins the perikaryon is seen so rarely; only once has the full extent of an amacrine
fiber from the perikaryon to and through the basement layer been seen
(Fig. 1).

3.1.3.3 The Plexiform Layer

Optic Nerve Terminations. Both the optic nerve fibers and the amacrine trunk fibers pass through the basement layer to enter the plexiform layer. In the plexiform layer the optic nerve fibers have been described by Young (1962b) as having three types of endings: retinal types 1, 2, 3. With the electron microscope those terminating in the first radial layer (retinal type 1) form terminal enlargements or bags containing synaptic vesicles. In O. vulgaris and Eledone these presynaptic bags are irregularly conical in shape, with the base of the cone next to the deep surface of the basement layer, and the taper of the cone penetrating into the plexiform layer. Many times an optic nerve could be traced in direct continuity through the basement layer with its terminal bag (Fig. 3), except in O. bimaculatus, where the fibers remain narrow and tortuous as they pass through the layer of astrocyte processes.

From the tips of some of the cones a long slender fiber has been described with the light microscope, which continues deeper into the plexiform zone (retinal type 2). Though the beginnings of such fibers have been seen at the tips of some presynaptic bags in electron micrographs, it has not been possible to trace them in the neuropil. It has been impossible in electron micrographs to identify with certainty the type 3 retinal fibers, those passing deep into the neuropil without forming a terminal bag.

Near the basement layer thin dark sheets of glial cytoplasm appear at times to invest at least partially the upper part of the presynaptic bags. It is presumed that most of this glial material is continuous with that of the basement layer. In O. bimaculatus from the southwest coast of California which have been studied, the basement layer of dark glial processes is absent. Though dark cells are present in the outer granular layer, dark glial folds are seldom seen to invest the optic nerve fibers and amacrine trunks as they pass among the outer granule cells. Instead, a thick, very irregular layer of cell processes separate the outer granular and plexiform layers (Fig. 4). These fibers are large and pale, contain much filamentous material, but have virtually no ribosomes or other cytoplasmic organelles. They thus fit the description of protoplasmic astrocytes (Gray, 1969). The cell bodies for these astrocytes have not been identified. There is also much extracellular space among the astrocyte processes in which varying amounts of a fibrous material termed "collagen" by Gray (1969) is seen. Profiles of similar glial processes can be found scattered deeper in the neuropil, a not unexpected finding since glial cells have been identified there (Gray, 1969).

Glial processes, usually thought of as supportive elements, serve to outline glio-vascular tunnels in the plexiform layer (Gray, 1969). In an area of such densely packed neuronal processes as the neuropil under consideration, might they not also function as an "insulation", minimizing "cross talk" between fibers?

At the ultrastructural level, there are some marked differences between the presynaptic terminals of the optic nerve fibers between O. vulgaris and Eledone, and O. bimaculatus. In O. vulgaris and Eledone, the presynaptic bags have a broad base approximately 5-6 μm in diameter and a tapering length of about 25-30 μm. Because of this shape they have been vulgarly termed "carrots". In a montage of a stretch of the junction between the outer granular and plexiform layers in Eledone (Case and Gray, 1971) four such "carrots" were seen to be several micrometers apart, with the surrounding neuropil reaching to the basement layer

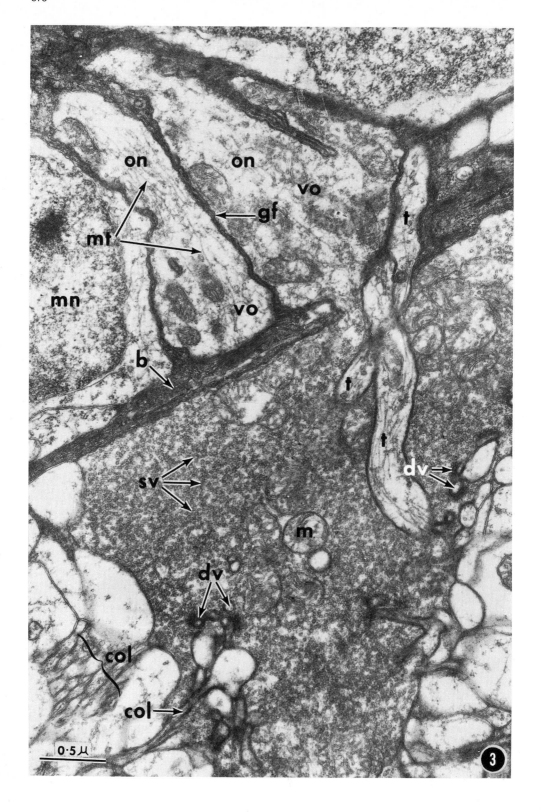

between them. None were seen to lie immediately adjacent to each other
so that their plasma membranes were apposed. This is in contrast to
the situation in O. bimaculatus where the presynaptic bags, though
about the same length as in O. vulgaris and Eledone, lack the carrot
shape, being very irregular and thin, with diameters of the order of
2-2.5 µm, with domed or slightly pointed tops where they press up into
the astrocyte fiber layer (Fig. 4). They also frequently lie close together throughout a significant part of their length, with their adjacent plasma membranes closely apposed for a considerable distance
(Fig. 4), though no synaptic contacts between them have been seen.

A third difference was in the population density of synaptic vesicles
within the bags. In O. vulgaris and Eledone the bags were uniformly
very densely packed with vesicles, giving them a dark appearance
(Fig. 3). On O. bimaculatus some bags were densely packed with vesicles, others had a very light population of vesicles, and still others
with a concentration between these two extremes (Figs. 4, 4 inset,
5, 7). This may possibly represent a functional difference among these
terminations, or it may be a reflection of the light conditions to
which the eye was exposed at the time of anaesthesia. It should be
noted that the O. vulgaris and Eledone used in the work at University
College London were in normal room daylight conditions before and during anaesthesia. The O. bimaculatus used at Loma Linda University were
in darkness to the time of anaesthesia, were in light only long enough
to transfer them to the anaesthetic bath, where, with the lid on the
container, they were again in darkness until anaesthesia was complete.
This had not been done with any experimental aim in mind, and further
animals should be sacrificed under strictly controlled light conditions
to determine if light or darkness does have any affect on the presynaptic bulb vesicles.

In O. bimaculatus, considerable amounts of glycogen granules were found
in the optic nerves as they appeared in the outer granular layer, and
in the terminal bags (Figs. 4, 7), a feature not observed in O. vulgaris and Eledone in such quantity.

Amacrine trunks in the neuropil. Dilly et al. (1963) described "enig-
matic tunnel fibers" passing through the presynaptic bags. These were
shown to be amacrine trunks (Case et al., 1972) penetrating the broad
tops of the presynaptic bags (Fig. 3) and coursing downward through
their substance for varying distances before emerging into the neuropil (Fig. 5). One tunnel fiber was observed to extend clear to the tip
of a carrot, where upon emerging, it expanded into a much larger, bulbous appearing profile (Fig. 6).

Though they do occur, tunnel fibers are not as common in the more narrow optic nerve terminal bags of O. bimaculatus, leading to the conclusion that possibly the majority of the amacrine trunks enter the
neuropil between the terminal bags, directly from the outer granular
layer.

Throughout the course of the amacrine trunk in the tunnel, the plasma
membranes of the trunk and of the presynaptic bag are closely apposed;

◁ Fig. 3. The border between the outer granular layer (above) and the plexiform layer
(below) in Eledone. Two optic axons form varicosities vo above the basement layer b,
one of which (right) narrows and passes through the basement layer, accompanied by
an amacrine trunk which then becomes a tunnel fiber t, to form the terminal bag ot
packed with synaptic vesicles sv. Horizontal dendritic collaterals col and dendritic
invaginations dv are indicated. mn nucleus of a microneuron, mt microtubule, on optic
nerve axon, gf glial folds (dark)

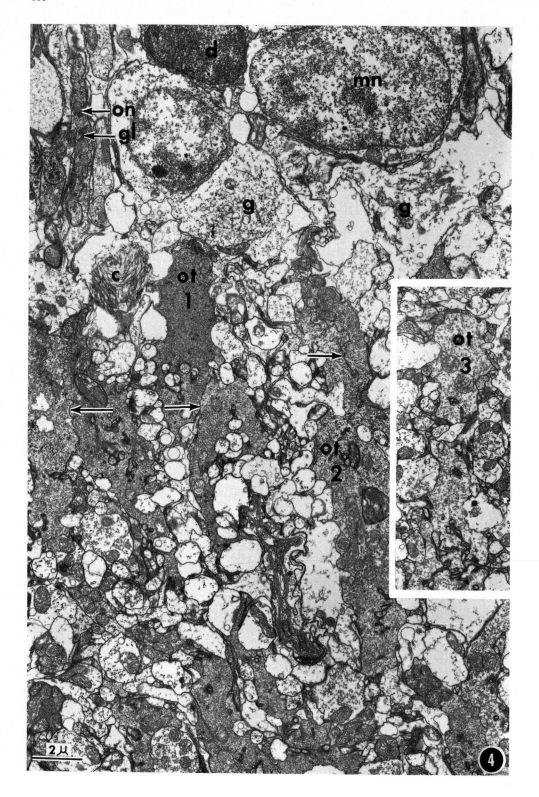

glial sheets surrounding the trunk above the basement layer do not
follow the trunk into the tunnel. Though careful search has been made,
no evidence of synaptic contacts within the tunnel have been found.

Although tunnel fibers have been seen emerging from presynaptic bags,
it has not been possible to follow them for any distance in the neuropil. It is presumed that they send collaterals to make synaptic connection with the presynaptic bags and probably with other fibers of
the neuropil. Young (1962b) showed that fibers of the neuropil include
centrifugal cell fibers from the inner granular layer, fibers from
cells of the medulla, which after entering the plexiform layer, return
to the medulla, and optic nerve fibers (retinal class 3) which penetrated deep and possibly through the plexiform layer without forming
a presynaptic bulb. There are also some amacrine trunks of the outer
granular layer which have directly entered the neuropil without tra-
versing a tunnel through a carrot (Dilly et al., 1963). Synaptic connections with all of these fibers including the last mentioned would
seem to assist in the classifying and encoding of impulses from the
visual receptor cells.

Synaptic Relations of the Presynaptic Bag. With the electron microscope, many small fibers can be seen making invaginations into the
presynaptic bags, some only denting the surface moderately, others
penetrating deeply into the bag (Fig. 3). Those penetrating into the
bag may branch and end in "grape-like" clusters of terminal expansions
which make synaptic-appearing contacts with the presynaptic bag
(Figs. 3 and 7; Case et al., 1972). The frequency of occurrence of
these apparent synapses within the bag appears to be the same regardless of the population density of vesicles in the bag.

Although tunnel fibers may branch, but are never seen to give rise to
collaterals within the synaptic bag, it would seem certain that the
small collaterals seen invaginating the bags must arise from the amacrine trunks after they have left the tunnels, or from amacrine trunks
which entered the neuropil directly through the basement layer. Other
collaterals of the amacrine trunks likely synapse with the other types
of fibers in the neuropil, making the microneurons of the outer granular layer the mediators and modifiers of visual impulses between the
optic nerve terminations and fibers of the neuropil arriving from, or
passing to, other parts of the optic lobes and brain.

The Deeper Layers of the Neuropil. With light microscopy, Young (1962a)
divided the plexiform layer into eight sublayers of fibers: four radial layers alternating with four tangential layers. The discussion so
far of the optic nerve terminal bags and collateral fibers from other
sources, involve only the outermost two of these eight layers; the
first radial layer and at least part of the first tangential layer.
The deeper six layers, as seen with the electron microscope, are an
intricate feltwork of interlacing fibers in which the distinction between radial and tangential layers is difficult to determine. Although
in most cases it it difficult to speculate about their origin, several
neuronal-type fibers can be described.

Fig. 4. The border between the outer granular layer (above) and the plexiform layer
(below), as seen in O. bimaculatus, separated by processes of glial cells g and
intercellular space with collagen c. Optic nerve terminals ot show differing populations of synaptic vesicles (1, 2, inset 3). Unlabeled arrows: close apposition
of plasma membranes of adjacent terminal bags. An optic nerve axon on containing
glycogen granules gl is indicated. mn nucleus of a microneuron, d dark cell nucleus

382

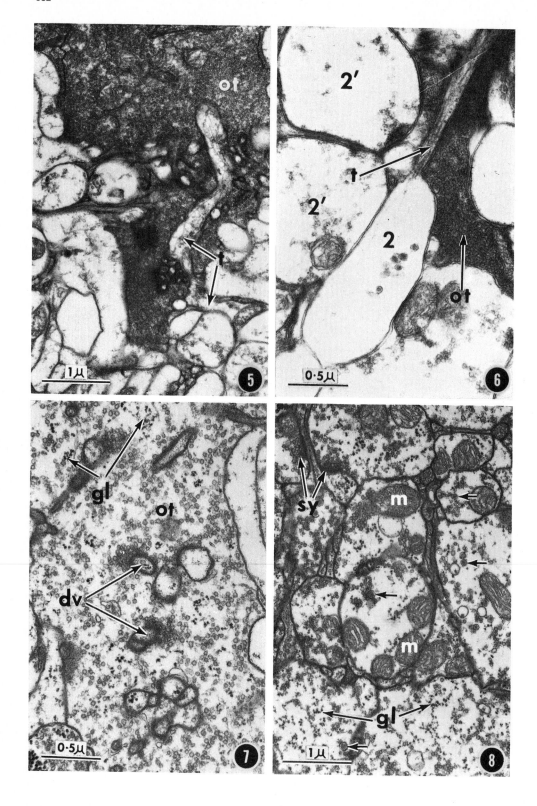

Fiber type 1 contains many hollow synaptic vesicles and numerous mitochondria. These may also contain occasional dense cored vesicles similar to those found in adrenergic synapses (Grillo and Palay, 1962;
Fig. 8). This fiber type varies in observed size from 0.4 μm to 1.5 μm
and sometimes show small collaterals or dendritic processes branching
from them. These fibers also show numerous synapses with collaterals
or dendritic processes (Fig. 8); on occasions two have been seen to
synapse with the same process.

Type 1 fibers occur in all layers of the plexiform layer and are the
most common fiber type seen. In the first radial zone they occur in
close apposition to the plasma membrane of presynaptic bags, but do
not show any synaptic contact with the bag. No clear evidence has been
found that this type of fiber sends collaterals to penetrate the presynaptic bags.

Though the type 1 fibers are here placed in one group, based on their
cytoplasmic content, because of their variability in size (by a factor
of at least 3) they may not necessarily have the same origin or termination. Some have been seen to synapse with each other.

Fiber type 2 averages smaller in size than type 1, usually show very
few mitochondria, and have almost no vesicles, though they may contain
some flocculent-appearing material (Figs. 6, 10, 11). In general they
have a "watery-appearing" cytoplasm somewhat similar to processes from
astrocytes in the neuropil.

Synapses with fiber type 1 being presynaptic to type 2 occur (Fig. 10).
Fibers of type 2 have been seen to give off small collaterals which
may be the very small spine-like fibers with which type 1 fibers are
frequently seen to synapse.

Type 2 processes, though they can occur deep in the neuropil, are most
common in the first radial and first tangential layers where many are
in close contact with the periphery of presynaptic bags, occasionally
forming a synapse with the bag (Figs. 9, 11). These are the fibers
which have been referred to previously as sending dendritic branches
to penetrate, branch and ramify within the bag. The terminal bulbs
form synapses with the bag membrane which has been pushed in and follows the dendritic branch, always separating the plasma membrane of
the dendrite from the cytoplasm of the bulb.

Most type 2 fibers, especially in the vicinity of the presynaptic bags
are believed to originate from the branches of amacrine trunks, either
penetrating directly through the neuropil, or after emerging from the
presynaptic bag tunnels, and all appear to be postsynaptic. More extensive serial sections will be needed (and are planned) to either
prove or disprove this belief.

◁ Fig. 5. A tunnel fiber t seen emerging from the side of an optic nerve terminal ot

Fig. 6. A tunnel fiber t seen emerging from the tip of an optic nerve terminal ot
and expanding into a bulbous profile of plexiform layer fiber type 2 appearance 2.
Two other fiber type 2 fibers are indicated 2'

Fig. 7. Terminations of dendritic invaginations dv showing synapses with the optic
nerve terminal ot. Glycogen granules are indicated gl

Fig. 8. Plexiform layer type 1 fibers, some showing occasional dense cored vesicles
(unlabeled arrows) as well as clear vesicles. Synapses with small dendritic processes are indicated sy. m mitochondrion

384

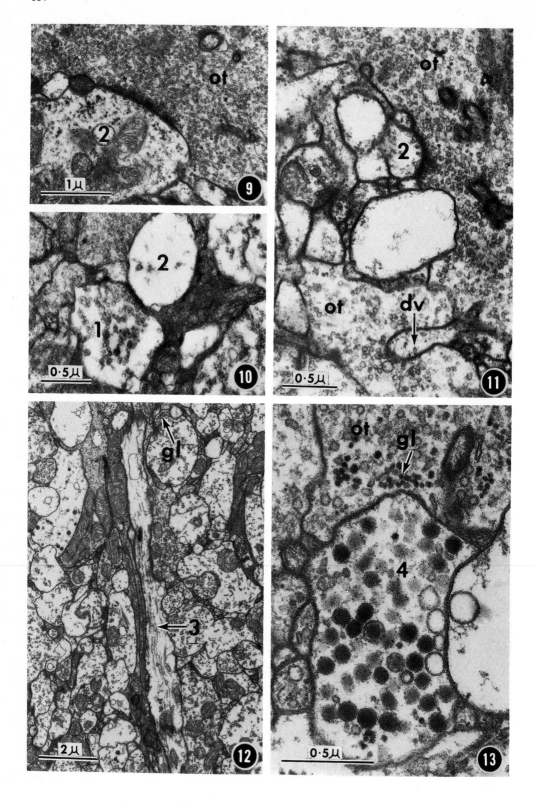

Fiber type 3 fibers are those that can be described as similar in size and morphology to the amacrine trunks above the basement layer or when seen as tunnel fibers, which have been previously described (Figs. 2, 3). Type 3 fibers have been seen oriented vertically, radially and/or nearly tangential in the first radial and first tangential layers; very occasionally as deep as the second and third radial and tangential layers (Fig. 12). Although these have not been traced entering the neuropil directly through the basement layer, they are believed to be amacrine trunks which have not been tunnel fibers. Since no fibers have been found in the plexiform layer which resemble retinal fibers before they enter the neuropil, some of these fibers described as type 3 with the electron microscope, may be the same as the type 3 retinal fiber of Young (1962b).

Fiber type 4 fibers are those which are packed with what appear to be dense, unit-membrane bound, secretion-type granules (Fig. 13) first noted by Dilly et al. (1963). For the most part these granules are completely packed with dense material. A few smaller granules similar to dense cored synaptic vesicles are usually mixed among the larger granules. No synapses between these fibers and other types have been found, nor have collaterals or dendritic branches been observed. So, whether the smaller dense cored granules are synaptic in function or only immature forms of the larger granules cannot be told.

The type 4 fibers are comparatively rare, but are found scattered throughout the plexiform layer, seemingly without preference for location. Some are closely apposed to, but show no synaptic contact with presynaptic bags, nor do they show any penetrating branches into the bags. This type of fiber has not been seen in either granular layer of the cortex and the location of the cell body is unknown. That these granules are a neurosecretion is likely, though its function is a problem yet to be solved.

In addition to the four types of fibers already described, much smaller fibers, collaterals and dendrites, undoubtedly originating chiefly from the first three types described, exist in abundance. Small collaterals oriented tangentially and directed toward the presynaptic bags (Fig. 3) help account for the first tangential layer (Case et al., 1972). In the deeper layers, collaterals and dendritic spines are observed to have frequent synaptic contact with other fibers, especially types 1 and 2.

Fiber type 5 fibers, scarcely meriting a number because they are so infrequently seen, can best be described as "giant" fibers (Fig. 14). These have been encountered chiefly near the inner granular layer, sometimes emerging from among the microneurons there as a narrow fiber which expands to a diameter of up to 3.5 µm in the plexiform layer. They usually pursue a very straight course; one could be measured for

◁ Fig. 9. Synapse between a plexiform layer type 2 fiber 2 and a terminal bag ot

Fig. 10. Synapse between plexiform layer type 1 1 and type 2 2 fibers

Fig. 11. Synapse between plexiform layer type 2 fiber 2 and a terminal bag ot. A deeply invaginating collateral fiber into a terminal bag is indicated dv

Fig. 12. Plexiform layer type 3 fiber 3 in the central zone of the neuropil. A small dendritic-like process packed with glycogen is indicated gl

Fig. 13. Plexiform layer type 4 fiber 4 closely apposed to an optic nerve terminal ot containing some glycogen gl

386

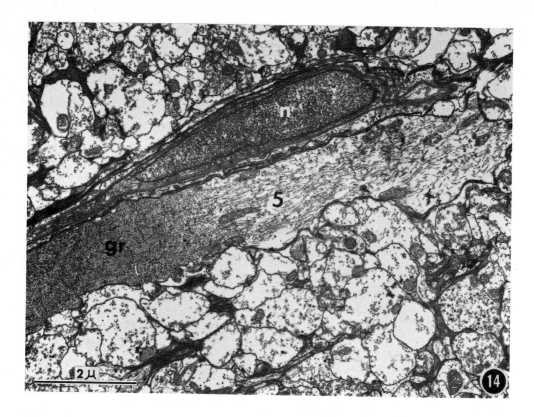

Fig. 14. A small segment of a plexiform layer type 5 fiber _5_. Finely granular material _gr_ occupied about 1/4 of the length of the 70 μm visible length of this giant fiber. Parts of the fiber were approximately 1/4 wider than seen here. Above the fiber is the nucleus _n_ of an endothelial cell of a blood vessel paralleling the fiber

a distance of 70 μm through the neuropil. Only once was one of these giant fibers seen to give off a collateral, and that very near the inner granular layer. Usually a blood vessel closely paralleled the giant fiber (Fig. 14).

The cytoplasm of these giant fibers contains many microfibrils, a few mitochondria and small light colored granules in addition to glycogen granules; these are, however, confined to localized segments where they are highly concentrated for a distance of several microns (Fig. 14). Vesicles have not been observed in these giant fibers.

Nonneural fibers of the plexiform layer belong to astrocytes which are associated with the vascular system in that area (Gray, 1969). Their fibers are similar to the type 2 fibers described above.

Glycogen granules occur to some degree in all of the fibers of the neuropil of O. bimaculatus, and show in most of the figures, with the greatest amounts appearing in fiber type 1, though some small fibers of dendritic spine dimensions were seen to be very densely packed with the granules, possibly in these instances advertising that their origin was from type 1 fibers.

As yet, a study of the inner granular layer and the medulla of the optic lobe has not been pursued. With the electron microscope large bundles of dark fibers have been seen among the microneurons and emerging into the neuropil, but the location of their cell bodies of origin is not known.

In conclusion it can be said, that, though a picture of the types and distribution of various nervous elements of the optic lobe of Octopus and Eledone is partially known, having been assembled from both light and electron microscopic observation, the origins and terminations of most of the fibers of the neuropil as seen with the electron microscope, have not been traced. Until this is done, a functional interpretation of the synaptic contacts seen between the various elements cannot be made. Such an interpretation is necessary before the relationship of the retinal and other inputs to the optic lobe and the memory system located there can be fully understood.

Acknowledgments. Appreciation is expressed to Dr. Robert Schultz for advice and for assistance with perfusions; to Mr. Mark Akland for supplying for study, electron micrographs taken in and near the inner granular layer, and for also assisting with the perfusions; to Mrs. Betty Chancellor and Mrs. Ruby Wheeler for technical assistance; to Mr. Michael Schultz for photographic assistance and to Mrs. Diane Burishkin for typing the manuscript.

3.1.4. References

Boycott, B.B.: Learning in Octopus vulgaris and other cephalopods. Pbbl. Staz. Zool. Napoli 25, 67-93 (1954)

Boycott, B.B., Young, J.Z.: A memory system in Octopus vulgaris Lamarck. Proc. R. Soc. London 143B, 449-480 (1955)

Case, N.M., Gray, E.G.: Nerve fibers in the optic lobe cortex of Octopus. 29th Ann. Proc. Electron Microscopy Soc. Am. Baton Rouge: Claitor's 1971, pp. 238-239

Case, N.M., Gray, E.G., Young, J.Z.: Ultrastructure and synaptic relations in the optic lobe of the brain of Eledone and Octopus. J. Ultrastructural Res. 39, 115-123 (1972)

Dilly, P.N., Gray, E.G., Young, J.Z.: Electron microscopy of optic nerves and optic lobes of Octopus and Eledone. Proc. R. Soc. London 158B, 446-456 (1963)

Gray, E.G.: Accurate localization in ultrathin sections by direct observations of the block face for trimming. Stain Tech. 36, 42-44 (1961)

Gray, E.G.: Electron microscopy of the glio-vascular organization of the brain of Octopus. Phil. Trans. R. Soc. London 255B, 13-32 (1969)

Gray, E.G.: The fine structure of the vertical lobe of Octopus brain. Phil. Trans. R. Soc. London 258B, 379-394 (1970)

Grillo, M.A., Palay, S.L.: Granule-containing vesicles in the autonomic nervous system. Electron Microscopy 2. Breese, S.S. (ed.). New York: Academic 1962

Gueffroy, D.E. (ed.): A Guide for the Preparation and Use of Buffers in Biological Systems. La Jolla, Calif.: Calbiochem 1975

Luft, J.H.: Improvements in Epoxy resin embedding methods. J. Biophys. Biochem. Cytol. 9, 409-414 (1961)

Reynolds, E.S.: The use of lead citrate at high pH as an electron opaque stain in electron microscopy. J. Cell Biol. 17, 208-212 (1963)

Sjöstrand, F.S.: Electron microscopy of cells and tissues. Instrumentation and Techniques. New York: Academic 1967, Vol. I

Wells, M.J.: A touch-learning center in Octopus. J. Exp. Biol. 36, 501-511 (1959)

Wells, M.J.: Learning in the Octopus. Symp. Soc. Exp. Biol. 20, 477-507 (1966)

Wells, M.J., Wells, J.: The effect of lesions to the vertical and optic lobes on tactile discrimination in Octopus. J. Exp. Biol. 34, 378-393 (1957)

Young, J.Z.: The failure of discrimination learning following the removal of the vertical lobes in Octopus. Proc. R. Soc. London 153B, 18-46 (1960)

Young, J.Z.: Learning and discrimination in the Octopus. Biol. Rev. 36, 32-96 (1961)

Young, J.Z.: The retina of cephalopods and its degeneration after optic nerve section. Phil. Trans. R. Soc. London 245B, 1-18 (1962a)

Young, J.Z.: The optic lobes of Octopus vulgaris. Phil. Trans. R. Soc. London 245B, 19-58 (1962b)

Young, J.Z.: Some essentials of neuronal memory systems. Paired centers that regulate and address the signals of the results of action. Nature (London) 198, 626-630 (1963)

Young, J.Z.: Paired centers for control of attack by Octopus. Proc. R. Soc. London 159B, 565-588 (1964a)

Young, J.Z.: A Model of the Brain. Oxford: Clarendon 1964b

Young, J.Z.: The nervous pathways for poisoning, eating and learning in Octopus. J. Exp. Biol. 43, 581-593 (1965a)

Young, J.Z.: The organization of a memory system. Proc. R. Soc. London 163B, 285-320 (1965b)

Young, J.Z.: The anatomy of the nervous system of Octopus vulgaris. Oxford: Clarendon 1971.

3.2 The Question of Lateral Interactions in the Retinas of Cephalopods

G. D. LANGE, P. H. HARTLINE, and A. C. HURLEY

3.2.1 Introduction

The usual way to describe visual neurons is to map their receptive
fields and to assign them a name related to their response dynamics.
Therefore, for example, one speaks of a "phasic, on center, off surround unit". These properties are often then explained in terms of
lateral nervous interactions. Both the connectivity and the response
dynamics of the various interneurons are important. For instance, the
properties of a given vertebrate ganglion cell might be ascribed to
the sizes of dendritic trees and the extent and response dynamics of
relevant receptors, bipolars, horizontals and amacrines. Even in a
simple retina such as Limulus the receptive fields and response dynamics of both the receptors and eccentric cells (interneuron-gang-
lion cell analogs) determine the overall properties.

In cephalopods we have a unique situation. The only neurons intrinsic
to the retina are the receptor cells themselves. Receptive field and
response dynamic properties in these retinas are similar to those
found in vertebrates but the anatomy seems to be much simpler. The
properties usually ascribed to a variety of interneurons in other systems must reside in the receptors and in their connectivity. This
paper will be concerned with the evidence (anatomical, physiological)
for (or against) lateral interactions in the retinas of cephalopods.

3.2.2 Anatomical Basis for Lateral Interactions

The photoreceptors in squid and octopus have microvillus outer segments which face the incident light source. These segments contain the
visual pigment and are the primary photoreceptors. They are arranged
so that adjacent cells cooperate in the formation of receptor structures analagous to the rhabdoms of arthropods. Structural cells are
also found in the layer and both these cells and the receptor outer
segments contain migratory screening pigments. The inner segment layer
contains the cell bodies of the receptors as well as glial elements.
Behind this layer is the plexiform layer containing the axons emerging from the receptor cell somas, collaterals from these axons, and
the axons of centrifugal neurons from the optic lobes. Electron microscopic work (Cohen, 1973) reveals that there are synapses among the
collaterals, between collaterals and receptor processes and between
the centrifugals and the receptor processes.

The pathways for lateral interactions could be by direct coupling of
the receptors and of course via the synapses of the plexiform layer.
One other possibility for lateral interactions involves interneurons
in the optic lobe with a return path along the centrifugal axons. In
fact, we have been able to record responses from efferents when light
is flashed on the retina. However, since we will restrict ourselves

to phenomena recordable in the excised eye, this pathway will not be extensively discussed.

3.2.3 Physiological Evidence on Lateral Interactions

3.2.3.1 Receptive Fields of Single Receptors

As we have previously said, the usual starting point for a discussion of lateral interactions is the explanation of a particular form of receptive field. In the case of squid optic nerve units this evidence is not compelling. We have mapped receptive fields for single units by flashing a small lamp and marking the angle at which responses can be obtained. The main finding is that the receptive field is more or less circular. There are two general size classes with diameters 1-4° and >15°. There is no evidence for any inhibitory field when the experiment is done at low light levels. Therefore, it can be seen that the receptive field data do not force us to a conclusion that lateral interactions exist. Totally excitatory fields can always be explained regardless of their size by light scatter or by assuming a large acceptance angle in the receptors.

3.2.3.2 Spatial Extent of ERG

Tasaki et al. (1963) obtained indirect evidence for lateral interactions in their studies of the octopus ERG. They recorded two different responses respectively from the surface or deep in the excised retina. These two responses were of opposite polarity which is consistent with a record of a single response from opposite sides of a

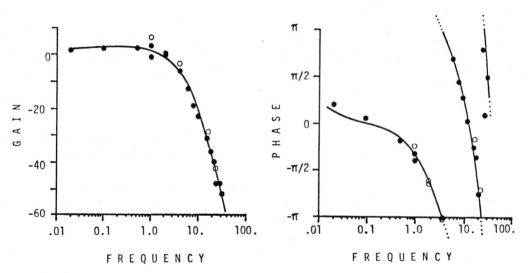

Fig. 1. Frequency response characteristic of the ERG of Loligo opalescens. Stimuli were a series of sinusoidally modulated light intensities at various frequencies. Gain is in arbitrary logarithmic amplitude units expressed as dB. Phase is in radians. Both are plotted against the logarithm of the frequency. The solid curves result from a trial-and-error fitting procedure in which a high-pass filter with corner frequency at 0.01 Hz (corresponding to our amplifier's filter) and low-pass filters with corner frequencies at 2.5, 4.5, 6.0 Hz were placed in series with a 32 ms delay

sheet dipole. However, with more sophisticated experiments they found that the two responses differed in spatial extent of the response, dependence on spatial extent of the stimulus, and amplitude wave length relationships. They also found that a mechanical cut in the retina differentially affected the two responses. Their final conclusion was that one of these responses was mediated by a neural network and the other was spread by some other mechanism.

We have also seen evidence for these two responses, namely we can invert the ERG by changing the spatial relationship between stimulus and electrode. However, we have not seen any marked difference in dynamical properties as a function of the polarity of the response.

3.2.3.3 Response Dynamics

Our work on ERG has centered on response dynamics. Our approach has been to stimulate local regions of the retina with sinusoidally modulated light. The light to ERG transduction can then be characterized as a low pass filter with a cutoff equivalent to several (4-6) simple R-C type filter stages (Fig. 1). This is not incompatible with a lateral interacting system but it does not demand such a model.

Optic nerve responses, on the other hand, often show an initial high-frequency transient response. This is equivalent to a high pass characteristic in terms of R-C type filters. One mechanism for obtaining this type of characteristic is to connect units with lateral inhibition which itself has a low pass character. When a number of units are simultaneously stimulated they then show a high frequency on transient and high pass characteristic.[1] Therefore, such behavior seen in squid optic nerve units considered in the light of anatomy is at least suggestive of lateral inhibition.

---

[1] In order to see how a high pass characteristic is produced by a lateral inhibiting network we must first imagine a Limulus eye where each element is inhibiting its neighbors via a synapse. In general, as each element speeds up it cuts down the activity in its neighbors. This can be described in terms of the classical Hartline-Ratliff equations (Hartline and Ratliff, 1957)

$$r_i = e_i - \sum_j K_{ij}(r_j - r_j^o) \tag{1}$$

where r's are responses and e's are stimuli while r_j^o are inhibitory thresholds. Let us assume that the r^o are negligible, that all the K's (inhibitory constants) are equivalent and that we stimulate the receptors equally. Then the equation becomes:

$$r = e - nKr \tag{2}$$

where n is the number of equivalent receptors. If we further assume that lateral inhibition is a system with low pass dynamics then in the Laplace transform domain

$$R(s) = E(s) - \frac{nA}{s+a} R(s) \tag{3a}$$

or

$$R(s) = E(s) \frac{s+a}{s+a + nA} \tag{3b}$$

where s is the Laplace complex frequency and a is the corner frequency of the low pass lateral inhibition dynamics. The transfer function is $\frac{s+a}{s+a + nA}$. If we think of sinusoidal inputs to this system, then at high frequencies (large imaginary s) the transfer function is 1. At very low frequencies (s=0) the transfer function is $\frac{a}{a + nA}$. Therefore if all the quantities are positive this system favors high frequencies over low frequencies.

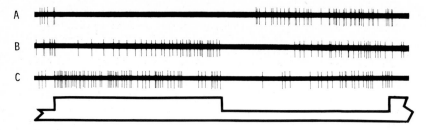

Fig. 2. Responses of a single optic nerve fiber in Dosidicus gigas. Bottom line: symbolizes light intensity (on-off period 10 s). Trace A is under dark-adapted condition. Note off type behavior. Trace B during light adaptation which is transitional toward trace C where when fully light-adapted the unit displays an on type behavior

As we stated in the introduction, the usual explanation for on, off, and on-off behavior in vertebrate ganglion cells is a combination of spatial and dynamical properties in the interaction of an excitatory process and an inhibitory process. Ratliff and Mueller (1957) showed that these response patterns can be synthesized in the Limulus eye by carefully balacing excitation and inhibition using the trick of properly arranging the positions and intensities of lights on the cornea. What their work showed was that all that is needed is the right spread function coupled with the right lateral inhibitory network in order to get the desired behavior. We have found that when we stimulate the eyes of Loligo opalescens (Hartline and Lange, 1974) or Dosidicus gigas (Fig. 2) with bright lights and record from single optic nerve units we also obtain on, on-off, and off responses. We find that these responses are not fixed, however, and the particular form obtained at any given time depends strongly on the recent history of illumination. In particular, a unit can go from a classical off response to a classical on response as it light adapts. It is far easier to conceive an interactive network where the relationships among the elements change as a function of adaptation to explain such data, than it would be to conceive a single element with these capabilities.

There are other phenomena which are compatible with our hypothesis of lateral interactions in the retina. The first of these is facilitation which we reported earlier (Lange and Hartline, 1974). Perhaps even more relevant is the related phenomenon of background enhancement. These two phenomena are manifest by an increase in the flash-induced ERG amplitude either after a conditioning flash (facilitation) or on a background (enhancement) over that which would have followed the flash alone. Since it is synapses which classically show facilitation, a natural explanation of these results would be to assume that there is lateral excitation and that it is these excitatory synapses that facilitate. In such a model, enhancement would be a steady-state case of facilitation. When experiments like this are done with short (10-100 ms) times between flashes, then a decrement rather than an increment is seen in the second response. Therefore, the entire model response is a short time constant lateral inhibition followed by a long time constant facilitation. It should be noted that this is exactly opposite to the situation in Limulus (Knight et al., 1970).

Another phenomenon is sustained oscillations. We find that when the eye of squid or octopus is dark adapted and then lit by a bright light, then the ERG will sometimes break into sustained oscillations at about 30 hz. The oscillations stop when the light is turned off. The establishment of the full amplitude of oscillation is quicker when the light is re-lit. The frequency of these oscillations is quite constant. Manipulation of light intensities and states of adaptation can modify the amplitude or the rate at which the amplitude changes, but only the temperature seems to affect the frequency and then only very slightly.

Again, a lateral interaction provides a good model for the behavior. Lateral excitation or lateral inhibition can have the properties of a feedback if they are recurrent, that is if the output of a given receptor affects the inputs of its neighbors. This is the case of Limulus, for instance. If the feedback is strong enough, then the feedback system becomes unstable and oscillates with an ever-growing amplitude. Since the retina surely shows a saturation non-linearity for large enough amplitude the oscillations do not grow indefinitely.

3.2.4 References

Cohen, A.I.: An ultrastructural analysis of the photoreceptors of the squid and their synaptic connections II. Intraretinal synapses and plexus. J. Comp. Neurol. 147, 379-398 (1973)

Hartline, H.K., Ratliff, F.: Inhibitory interaction of receptor units in the eye of Limulus. J. Gen. Physiol. 40, 357-376 (1957)

Hartline, P.H., Lange, G.D.: Optic nerve responses to visual stimuli in squid. J. Comp. Physiol. 93, 37-54 (1974)

Knight, B.W., Toyoda, J., Dodge, Jr., F.A.: A quantitative description of the dynamics of excitation and inhibition in the eye of Limulus. J. Gen. Physiol. 56, 421-437 (1970)

Lange, G.D., Hartline, P.H.: Retinal responses in squid and octopus. J. Comp. Physiol. 93, 19-36 (1974)

Ratliff, F., Mueller, C.G.: Synthesis of "on-off" and "off" responses in a visual-neural system. Science 126, 840-841 (1957)

Tasaki, Kyoji, Norton, A.C., Fukada, Y., Motokawa, K.: Further studies on the dual nature of the octopus ERG. Tohoku J. Exp. Med. 80, 75-88 (1963).

3.3 Hyperpolarizing Photoreceptors in Invertebrates

J. S. McReynolds

3.3.1 Introduction

In sensory physiology, the concept that a stimulus causes a depolarizing receptor potential was once a widespread generalization, and many textbooks still refer to the depolarizing nature of the stimulus, although it has been more than ten years since the discovery that vertebrate photoreceptors respond to light with a hyperpolarizing receptor potential - a finding that has now been confirmed in many different vertebrate species.

A revised generalization often heard now is that invertebrate photoreceptors depolarize, while those of vertebrates hyperpolarize. This article will point out that there are several kinds of invertebrate photoreceptors which respond to light with a hyperpolarizing receptor potential, and that these may have only appeared to be nonexistent or rare because anatomical considerations such as small cell size have made them unfavorable for electrophysiological investigation.

One of the most noticeable morphological differences between the photoreceptors of vertebrates and arthropods (the invertebrate phylum whose receptors have been most studied electrophysiologically) is that the outer segments of vertebrate rods and cones are formed from modified cilia, whereas the analogous membrane specializations of the depolarizing invertebrate photoreceptor cells are in the form of microvilli. Eakin (1963, 1965, 1972) has compared the ultrastructure of photoreceptor cells from different phyla and shown that there are a large number of invertebrate species with ciliary-type photoreceptor specializations. These studies revealed that the ciliary and microvillous (or rhabdomeric) types of photoreceptors generally occur in animals of different major evolutionary lines, although there are exceptions.

One of the motivations for studying this class of invertebrate receptors was that they might also have some functional properties similar to vertebrate photoreceptors and their investigation might provide much additional information about photoreceptor physiology.

This chapter will describe some of the different kinds of hyperpolarizing photoreceptors that have been found in invertebrates, and some visual systems in which their existence is strongly suspected but not yet directly confirmed. Some of the major differences between the functional properties of invertebrate and vertebrate hyperpolarizing photoreceptors will be discussed. In addition, what appear to be basic differences between these photoreceptors and the more well known depolarizing variety will be discussed, especially in Pecten and Lima, where both types occur in the same eye and can be studied together.

Fig. 1. Diagram of Pecten eye. One axon from a distal photoreceptor and two from proximal photoreceptors are shown entering the respective branches of the optic nerve

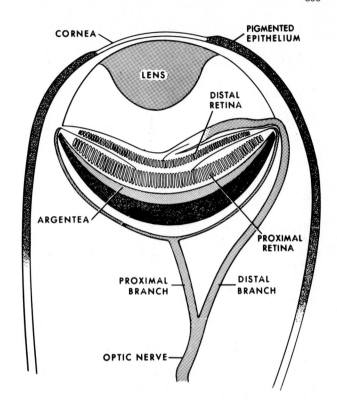

3.3.2 Pecten

The scallop has several dozen eyes, the largest of which are about a millimeter in diameter, located on the outer edge of the mantle around the edge of each shell. Many different species of scallops have similar eyes, but the photoreceptor physiology has been most extensively studied in Pecten irradians and Pecten maximus. A schematic diagram of this camera type eye is shown in Fig. 1. Besides the reflecting tapetum or argentea, which lies behind the entire retina, the most striking feature of this eye is the presence of a double retina. There are two separate layers of photoreceptor cells, each layer giving rise to a different branch of the optic nerve. Electron microscopy has revealed that the photoreceptors in the two layers are morphologically different (Miller, 1958; Barber et al., 1967). The cells of the proximal layer bear microvilli, like most invertebrate photoreceptors, but those of the distal layer have membrane specializations which are derived from modified cilia. The axons making up the two branches of the optic nerve not only arise from morphologically distinct types of photoreceptors, they also mediate different kinds of responses to light. Hartline (1938) showed that fibers of the proximal branch of the optic nerve fired action potentials in response to illumination of the eye, whereas those of the distal branch fired only in response to turning off or dimming of a light stimulus.

Intracellular recordings (McReynolds and Gorman, 1970a) have revealed that the proximal cells respond to light with a typical depolarizing receptor potential whereas the response of the distal cells is a hyperpolarization which is graded with the intensity of the stimulating

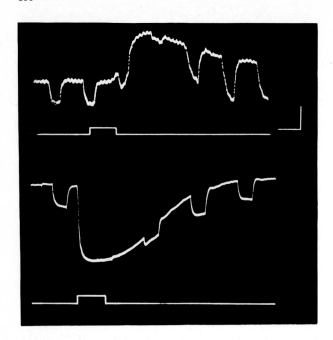

Fig. 2. Membrane conductance changes during the response to light in proximal (upper) and distal (lower) photoreceptors. The repeated downward deflections are voltage drops caused by brief current pulses passed through a bridge circuit (for details of method see McReynolds and Gorman, 1970b). Their amplitude is inversely proportional to the membrane conductance. Calibration 10 mV, 100 ms

light and can be up to 70 mV in amplitude. The resting potential of both kinds of cells is generally in the range of -20 to -40 mV.

The depolarizing receptor potential of the proximal cell is caused by an increase in membrane conductance, and has a reversal potential more positive than zero. The hyperpolarizing response of the distal photoreceptor is also produced by an increase in membrane conductance (Fig. 2) and has a reversal potential of about -80 mV (McReynolds and Gorman, 1970b). Therefore, it must be generated by a different mechanism than the hyperpolarizing response of vertebrate photoreceptors, which is associated with a decrease in membrane conductance (Toyoda et al., 1969; Baylor and Fuortes, 1970), mainly to sodium ions (Sillman et al., 1969; Korenbrot and Cone, 1972; Cervetto, 1973; Brown and Pinto, 1974). By altering the ionic concentration of the extracellular fluid it was shown that the hyperpolarizing receptor potential in Pecten is due to a selective increase in membrane permeability to potassium ions (McReynolds and Gorman, 1974). Chloride ions do not appear to be involved in maintaining the resting potential or generating the receptor potential. Removal of sodium ions from the extracellular fluid causes the membrane potential to become more negative in darkness and thereby reduces the response to light by about the same amount.

Thus the resting permeability of the membrane to sodium ions keeps the membrane depolarized in the dark. This does not imply that there is a dark current as in vertebrate photoreceptors (Penn and Hagins, 1969) but only that the ratio of sodium to potassium permeability, which determines membrane potential at any time, is higher under resting conditions than that found in most neurons. In light this ratio is decreased by an increase in the permeability to potassium, and the membrane hyperpolarizes. The absolute permeabilities are not known.

It has recently been postulated (Yoshikami and Hagins, 1973) that a light-induced increase in the intracellular free calcium ion concentration causes the decrease in sodium permeability that generates the

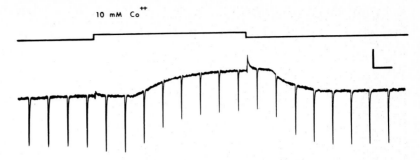

Fig. 3. Effect of cobalt ions on hyperpolarizing receptor potential. Downward deflections are receptor potential responses to brief light flashes of constant intensity. The interval between each flash is 10 s. During the time indicated by the horizontal line above, the artificial sea water perfusing the retina was changed to a sea water solution containing 10 mM cobalt chloride. The small upward deflections in the record are artifacts from switching the perfusion fluid. Calibration 10 mV, 10 s

vertebrate photoreceptor potential. Since injection of Ca^{2+} into Aplysia neurons has been shown to cause an increase in K^+ conductance (Meech, 1972; Brown and Brown, 1973) it is reasonable to investigate whether an increase in intracellular Ca^{2+} might underlie the production of the hyperpolarizing receptor potential in Pecten. Attempts to inject Ca^{2+} into these small photoreceptors has not been successful, and alterations of extracellular calcium concentration in either direction have so far resulted in irreversible deterioration and loss of the unit being recorded.

Metabolic inhibitors cause an increase in intracellular calcium concentration in many cells by blocking the energy supply for the active transport mechanisms which normally sequester Ca^{2+} in mitochondria or pump it outward across the cell membrane. Such an indirect approach has been used in an attempt to increase the intracellular calcium concentration in Pecten photoreceptors (Gorman and McReynolds, 1974). Both 2, 4-dinitrophenol and cyanide cause a rapid and reversible hyperpolarization of the photoreceptor membrane by increasing the membrane permeability to potassium ions. Although in this case it is not certain that the result is due to an increase in intracellular Ca^{2+}, it does indicate that there is an active metabolic process which keeps potassium permeability low in the dark.

Addition of cobalt or manganese ions to the extracellular fluid has been effective in blocking transmission at chemical synapses by preventing the entry of calcium ions, which is necessary for the release of transmitter. This approach has been used to try to determine whether calcium entry may be a necessary step in the light-induced increase in potassium permeability of Pecten distal photoreceptors. Application of 10 mM Co^{2+} to the extracellular fluid causes a considerable decrease in the size of the receptor potential, and also some depolarization of the cell membrane in darkness (Fig. 3). Similar results are obtained with 10 mM Mn^{2+}. These results are at least consistent with the hypothesis that light may cause an increase in Ca^{2+} permeability of the cell membrane. The reduction in resting potential might then mean that there is some Ca^{2+} entry even in darkness. As yet, however, there is no direct evidence that Ca^{2+} plays any role in the excitation of these photoreceptors.

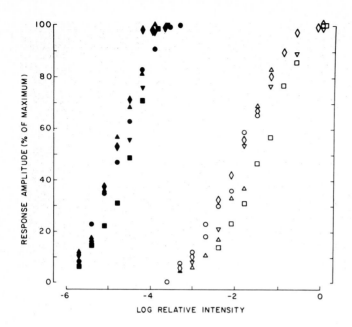

Fig. 4. Response amplitude of depolarizing (<u>filled symbols</u>) and hyperpolarizing (<u>open symbols</u>) receptor potentials as a function of light intensity. Responses are from 5 cells of each type and are plotted as per cent of the maximum response for each cell. Intensity of 100 ms duration light stimulus shown relative to brightest flash (10^{16} photons/cm^2/s). All depolarizing responses are from eyes dark-adapted at least 30 min. Hyperpolarizing responses are from both light and dark adapted preparations. (From McReynolds and Gorman, 1970a)

The depolarizing and hyperpolarizing photoreceptors have the same spectral sensitivity (McReynolds and Gorman, 1970b) and therefore presumably the same visual pigment. It is interesting to consider what differences in the chain of events following absorption of light by the visual pigment could lead to the generation of responses of opposite polarity. Clearly the ionic channels which are opened as a result of light stimulation are different in the two cells. This could be due to the production of a common substance which has different actions on the two kinds of membranes, much like the opposite effects of acetylcholine at different postsynaptic membranes (Kandel, 1967). Alternatively, the difference may be related to events that occur at an earlier stage and perhaps produce different endproducts in the two cells. The responses of these two receptor types differ not only in the polarity of their receptor potential but in several other respects which may reflect differences in the transduction mechanisms:

1. The absolute sensitivity of the depolarizing response is two to three orders of magnitude greater than that of the hyperpolarizing response. Fig. 4 shows a plot of the amplitude of the receptor potential as a function of light intensity for several photoreceptors of each type. Note that the threshold for the hyperpolarizing response is at light intensity levels which nearly saturate the depolarizing receptors. Part of the difference in sensitivity could be due to simply a difference in visual pigment concentration, but the density would have to be 100 to 1000 times greater in the proximal cells to account for all of the observed difference in sensitivity. Unfortunately, no information is available about the density of pigment in the two cell types.

2. The depolarizing receptors adapt rapidly to repeated light stimuli whose intensities are in the middle or upper part of the response range of hyperpolarizing cells. Recovery from this light adaptation may take several tens of minutes, so that with repeated bright stimuli only the hyperpolarizing receptors respond. Thus, under the appropriate conditions of dark or light adaptation and stimulus intensity it is possible to activate selectively either the depolarizing or hyperpolarizing receptors.

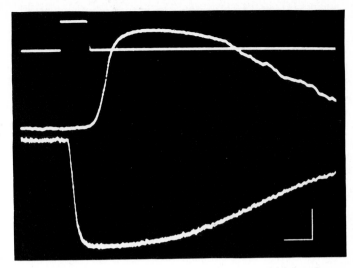

Fig. 5. Time course of depolarizing and hyperpolarizing receptor potentials. Responses of equal amplitude were chosen for comparison. Uppermost trace: zero membrane potential and signal from photocell monitoring light flash, which was of same duration in both recordings. Calibration 10 mV, 100 ms

3. The time courses of the two response types are quite different. The hyperpolarizing response has a much shorter latency of onset and much faster time to peak than does a depolarizing response of the same amplitude (Fig. 5). Although in each cell type both latency and time to peak decrease with increasing stimulus intensity, the responses of the hyperpolarizing cells are faster at all intensities.

4. Another striking, and perhaps related, difference in the two response types is that dark adapted proximal cells give an irregular, fluctuating response to very dim lights, which becomes smooth at higher intensities (Fig. 6). This is very similar to the responses of other depolarizing invertebrate photoreceptors such as those of Limulus, where the fluctuations are supposed to be the responses to the absorption of single photons. As the light intensity is increased the frequency of these quantal responses increases and they summate; but feedback mechanisms are believed to turn down the gain of the system so that the response per photon absorbed becomes smaller and the receptor potential becomes smoother (Fuortes and Hodgkin, 1964; Lisman and Brown, 1972). Light adaptation of Limulus photoreceptors also causes a faster time course of the receptor potential (Fuortes and Hodgkin, 1964).

The hyperpolarizing response in Pecten is faster than the depolarizing response, and is always smooth (Fig. 6). No fluctuating or "bump"-like responses are seen near threshold, even after prolonged dark-adaptation. The cell behaves as though it were already light adapted at light intensities which are just threshold for eliciting an electrical response. Since this threshold occurs at much higher light intensities than in the depolarizing cells, it may be that the visual pigment is actually reacting with light at lower levels of illumination, but the result of these reactions is not expressed as a change in membrane potential until a relatively large amount of some substance has been produced, and the cell has already become light adapted.

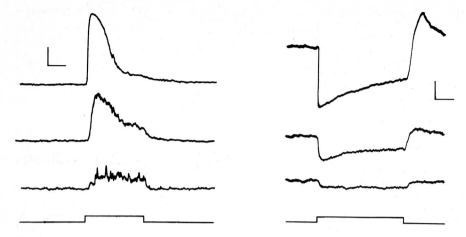

Fig. 6. Responses of dark-adapted receptors to different intensities of light. Lowest traces: response to light intensities just above threshold for each receptor type. Middle and upper traces: responses to light stimuli 16 and 256 times brighter. Left column: proximal cell; right column: distal cell. Calibration 10 mV, 2 s

These results suggest that the difference in the two types of receptor potentials may be due to changes occurring at some earlier stage in the intervening chain of events, and should certainly be investigated more thoroughly.

Although action potentials were not often observed in intracellular recordings from either type of cell, they normally occur in extracellular recordings (Hartline, 1938; Land, 1966). Even when present in intracellular recordings, action potentials usually disappear within a minute or so after penetration, although membrane potential and receptor potential sensitivity to light remain stable, suggesting that the spike-generating mechanism is easily damaged by microelectrode impalement. When action potentials are recorded intracellularly there may or may not be spontaneous firing. The frequency of impulse firing is increased by the depolarizing receptor potential and inhibited by the hyperpolarizing receptor potential. The distal cells usually give a burst of increased impulse firing when the light stimulus is turned off. This sometimes occurs superimposed on a transient depolarizing wave which follows the hyperpolarizing receptor potential, but is often observed in the absence of this slow depolarization, or even while the membrane is still hyperpolarized and is returning to the dark level (Fig. 7).

Kennedy (1960) observed "off" discharges in response to light stimulation of nerve fibers in the surf clam Spisula. He was able to show that although white light produced inhibition followed by an "off" discharge, adaptation to strong blue light caused an "on" response to be produced by red light. He interpreted this as indicating the presence of two opposing mechanisms: an inhibitory process more sensitive to short wavelength light and an excitatory process more sensitive to long wavelength light (Fig. 8). The inhibitory process has the greater overall sensitivity, since light of all wavelengths normally produces inhibition during the period of illumination, but the excitatory process, which has a much longer time course, is revealed when the light is turned off. Although no receptor potentials were recorded in his

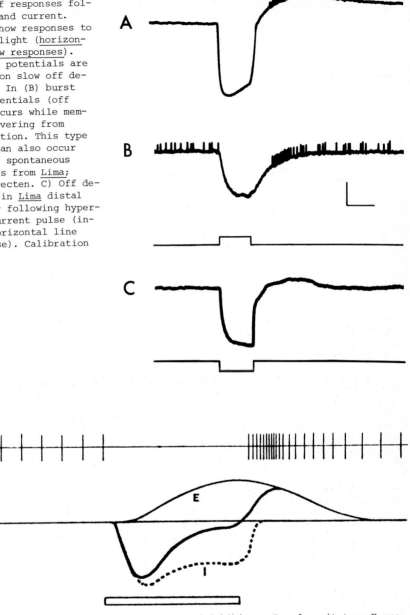

Fig. 7A-C. Off responses following light and current. (A) and (B) show responses to 1 s flash of light (horizontal line below responses). In (A) action potentials are superimposed on slow off depolarization. In (B) burst of action potentials (off discharge) occurs while membrane is recovering from hyperpolarization. This type of response can also occur in absence of spontaneous firing. (A) is from Lima; (B) is from Pecten. C) Off depolarization in Lima distal photoreceptor following hyperpolarizing current pulse (indicated by horizontal line below response). Calibration 10 mV, 1 s

Fig. 8. Theoretical scheme for contributions of inhibitory I and excitatory E processes to the generation of inhibition and off response. See text for discussion. (From Kennedy, 1960)

experiments, one could postulate a hyperpolarizing response to the short wavelength pigment and a depolarizing response to the long-wavelength pigment. It has recently been shown that the single photopigment in the distal receptors of Pecten can be driven between two states, with electrical responses of opposite polarity, by long and short wavelength illumination (Cornwall and Gorman, 1976).

It is also possible that the off response in Pecten distal cells is at
least partly generated by electrical properties of the membrane. Since
the cells are relatively depolarized in darkness, there may be considerable inactivation of voltage-sensitive sodium channels. Hyperpolarization of the membrane by light could result in the gradual removal
of this inactivation, so that when the membrane potential returns toward the dark level it is able to generate action potentials at a
lower threshold than prior to the hyperpolarization. This would be
analogous to an anode break response. The scheme of Fig. 8 could still
represent the events in this kind of mechanism if, instead of being
related to different pigments, the inhibitory process is viewed as
the membrane hyperpolarization and the excitatory process as indicating the time course of the removal of inactivation. The sum of these
would then be a measure of the excitability of the impulse generating
mechanism. The duration of the "off" discharge and the fact that it
increases with increasing duration of the preceeding illumination over
many tens of seconds (Land, 1966) would require a type of inactivation
mechanism with a very long time constant.

If such a mechanism were responsible for the generation of the off response in Pecten it should be possible to elicit off responses of different frequency and duration in the absence of light stimulation, by
appropriately hyperpolarizing the cell with current. Unfortunately,
it was not possible to do such experiments in Pecten because action
potentials were usually recorded for only a short period, if at all,
after penetration with microelectrodes.

3.3.3 Lima

The visual system of the file clam Lima scabra was studied by Mpitsos
(1973). In many ways it is similar to that of Pecten. The animal has
numerous eyes arranged around the edge of the mantle, although they
lie under a transparent portion of the mantle epithelium rather than
being on stalks as in Pecten. The eye consists of a cup of microvil-
lous-type photoreceptors, surrounded by supporting cells containing a
dense red screening pigment. The axons of these proximal cells give
an "on" response when illuminated. In front of this cup is a round
transparent structure which had previously been described as a lens.
The "lens", however, gives rise to nerve fibers which generate an off
response, and is really a distal retina analogous to that of Pecten.
Electron microscopy revealed that the cells in the distal retina bear
numerous modified cilia arranged in whorls, which could be the photoreceptive membrane specializations (Bell and Mpitsos, 1968). Since the
distal retina of Lima is physically separated from the proximal retina
it can be excised and recorded from in isolation.

Recordings from nerve fibers coming from the isolated distal retina
showed off responses to white light, but it was possible to convert
this to an on response by adapting with strong blue light. Thus, there
is evidence for two opposing photochemical effects similar to that described in Spisula. Unfortunately, this phenomenon was not studied
with intracellular recording, so it is not known whether the long and
short wavelength mechanisms operate by producing respectively depolarizing and hyperpolarizing receptor potentials. An intracellular record
from a cell in the distal retina did show a small hyperpolarization in
response to a flash of white light, however, and one from a cell in the
proximal retina showed a depolarizing response (Mpitsos, 1973).

Fig. 9. Off depolarization increases
with duration of preceding illumination. Equal intensity stimuli of
0.01, 0.1, 1, 5 and 21 s duration.
Calibration 10 mV, 2 s

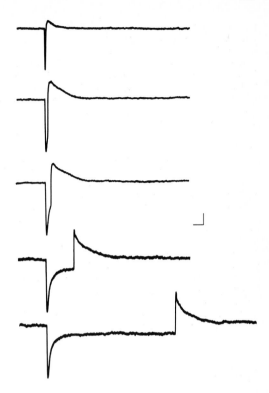

Intracellular recording in our laboratory has confirmed that the photoreceptor cells of the proximal retina are depolarizing and those of
the distal retina are hyperpolarizing. Some additional observations
are described briefly below.

As in Pecten, the proximal cells are much more sensitive to light, do
not respond to repeated bright stimuli which give responses in the
distal cells, and show "bumpy" responses to dim illumination when dark
adapted. One interesting feature not seen in Pecten is that the proximal cells often given a large transient depolarization to the onset
of a light stimulus and another large depolarization or series of oscillations when the light is turned off. This suggests synaptic interaction between photoreceptors.

The hyperpolarizing receptors of the distal retina give high-threshold,
graded responses to light, with no evidence of quantal responses near
threshold even after prolonged dark adaptation. When the light is
turned off there is often a depolarizing transient which becomes
greater with increasing duration of the stimulus (Fig. 9).

The hyperpolarizing response of Lima photoreceptors is, like that of
Pecten, caused by an increase in membrane conductance, and has a reversal potential near -80 mV. An off discharge of action potentials
sometimes occurs superimposed on the off depolarization (Fig. 7), and
both a depolarization and/or spike discharge can often be elicited
following passive hyperpolarization in the absence of light (Fig. 7).
However, if a light stimulus is given while the membrane potential is
passively hyperpolarized, the hyperpolarizing receptor potential becomes smaller but the depolarizing off response to the same light
flash becomes larger and is sometimes quite delayed (Fig. 10).

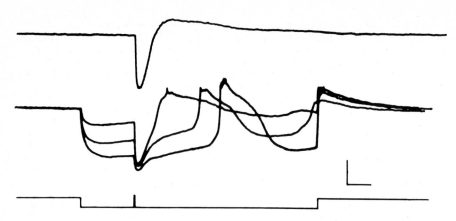

Fig. 10. Effect of passive hyperpolarization on hyperpolarizing receptor potential and off depolarization. Receptor potential and off depolarization are in response to 100 ms light flash indicated by vertical mark below responses. Upper trace: response at resting potential (-22 mV). Lower three traces: superimposed records of response to same light flash when membrane is hyperpolarized to -33, -47 and -60 mV by current step whose duration is indicated below. Note that off responses at end of current steps are much less than those following light flashes. Calibration 20 mV, 1 s

In these preliminary experiments it has not been determined to what extent a purely electrogenic membrane mechanism could account for the off responses produced by light. The experiments of Mpitsos (1973) indicate that the wavelength of the stimulating light also has an important effect on the relative strengths of inhibition and excitation, and one might expect these effects to be somewhat reflected in the membrane potential of the cell. Further experiments with chromatic adaptation and intracellular recording will be necessary to adequately explain the mechanism of the off response.

Other recent experiments indicate that the hyperpolarizing response of Lima is, like that of Pecten, due to a light-induced increase in potassium conductance (A. Gorman and C. Cornwall, pers. comm.).

3.3.4 Primitive Chordates

The photoreceptor of the larval "tadpole" of the Ascidian tunicate Amaroucium constellatum has a membrane specialization derived from a single cilium (Barnes, 1971). No extracellular recordings have been made in this eye, but a few intracellular recordings have shown that the response is, as predicted by the ciliary structure of the photoreceptors, a graded hyperpolarization (Gorman et al., 1971). This response is associated with a decrease in membrane conductance and, therefore, resembles that of the more closely related vertebrate photoreceptors rather than Pecten or Lima. However, in this case it was not possible to measure the current-voltage characteristics of the membrane, so that the possibility remains that the observed conductance decrease may be a result of the hyperpolarization rather than a direct effect of light. Recording from this preparation is extremely difficult and no additional data is available.

The Thalacian tunicate Salpa has a single large eye with numerous photoreceptors whose axons run to the neuropile of the immediately adjacent and only neural ganglion of the animal. The photoreceptor cells have a membrane specialization consisting of a mass of microvilli (Gorman et al., 1971). Intracellular recordings from the photoreceptor cells reveal some rather remarkable properties (McReynolds and Gorman, 1975). The cells have extremely low resting potentials, but give enormous hyperpolarizing responses to light. For example, a cell with a resting potential of only -4 mV, may give a response nearly 80 mV in amplitude to a bright flash. The response is graded with light intensity and is due to an increase in membrane conductance with a reversal potential of -80 mV. The increase in conductance could be to potassium and/or chloride ions; experiments to determine the ionic basis of the response have not been done.

The time course of the salp receptor potential is very slow even with bright stimuli. A response to dim illumination can only be elicited with stimuli of long duration, and may not reach peak amplitude until many seconds after the onset of illumination. Very small depolarizing deflections are occasionally seen but it is not known whether the photoreceptors normally have action potentials or whether the receptor potential itself spreads to the synaptic terminals in the ganglion to modulate transmitter release. Inhibitory post-synaptic potentials recorded from cells in the ganglion in response to light stimulation of the eye are accompanied by a reduction in amplitude of the large action potentials, suggesting that during this response the conductance of the post-synaptic membrane is increased.

A most curious phenomenon is that dark-adapted salp photoreceptors were sometimes observed to give progressively larger responses to successive brief flashes, even though membrane potential returns to the dark level between flashes. In such cases the responses could eventually reach an amplitude more than ten times as great as the response to the initial flash. Further investigation of this "faciliatory" effect of light has not been possible due to the unavailability of animals.

Some of the many interesting properties observed in salp photoreceptors, such as the extreme slowness of the response, facilitation by previous exposure to light, and possible release of transmitter by slow potentials, may yield valuable information about the mechanisms of phototransduction and transmitter release. The phenomena reported above are only very preliminary results obtained during a brief period when these animals were available.

3.3.5 Species in which there is Indirect Evidence for Hyperpolarizing Receptor Potentials

Barber and Land (1967) have shown that the photoreceptors of the cockle Cardium edule have ciliary membrane specializations and that the axons of these cells respond to light with an off discharge. It therefore seems likely that the receptor potential in these cells would be a membrane hyperpolarization.

Kennedy (1960) has shown inhibition and off responses in nerve fibers of the surf clam Spisula solidissima, as described in more detail earlier in this article. No specific photoreceptor organs or morphological membrane specializations were seen, however, and it appears that a restricted portion of the axon itself is light sensitive. The nature of the impulse discharge suggests that the initial electrical event would be a membrane hyperpolarization.

In the quahog clam <u>Mercenaria mercernaria</u>, an off discharge has been studied which resembles in many ways that of the distal nerve fibers of <u>Pecten</u>, including the ability to integrate the light stimulus over a long period of time (Wiederhold et al., 1973). As in <u>Spisula</u> no definite photoreceptor cell bodies were found, but electron microscopy has revealed invaginated whorls of membrane in the nerve fibers in the area to which light sensitivity is restricted. Although no receptor potentials were recorded, it seems likely that these cells would also show a hyperpolarizing receptor potential.

3.3.6 Discussion

A number of invertebrate species have now been shown to have photoreceptor neurons that respond to light with a hyperpolarizing receptor potential. The microvillous membrane of the salp photoreceptor shows that the presence of a ciliary membrane specialization is not necessary for a hyperpolarizing receptor potential. This is not surprising in view of the fact that this kind of response can be generated by quite different membrane mechanisms. There does appear to be an evolutionary association between the two, however, in that all ciliary photoreceptors whose physiology has been investigated so far have shown either hyperpolarizing receptor potentials or action potential firing patterns indicative of such. On this basis, it is likely that many other photoreceptors with ciliary structure will turn out to have this type of response.

A hyperpolarizing receptor potential need not be regarded as inhibitory. In the absence of action potentials, as in vertebrate rods and cones, the receptor potential itself spreads electrotonically to the synaptic terminals where it modulates transmitter release. The available evidence suggests that here, as at other synapses, transmitter is released by depolarization of the presynaptic terminals (see 3.7 of this vol.). Depending on the response of the post-synaptic cell membrane to the transmitter substance, the resulting membrane potential change can be either depolarizing or hyperpolarizing. In most of the invertebrate species described in this article the photoreceptor neuron itself generates action potentials which propagate to the distant output region of the cell. The direct inhibition of action potential firing in the receptor neuron by a hyperpolarizing receptor potential can be considered an example of primary inhibition by a stimulus. In many other vertebrate visual systems the receptors may depolarize in response to light but generate an off response in the second-order neurons by synaptic inhibition (Gwilliam, 1965; Zettler and Järvilehto, 1971; Shaw, 1972; Chappell and Dowling, 1972).

The generation of off responses following the hyperpolarizing receptor potential may be partly due to electrical properties of the excitable membrane, by a process similar to that causing anode break responses but with very long time constants. The peculiar increase in size and delay of the depolarizing transient sometimes observed in <u>Lima</u> photoreceptors when the membrane is passively hyperpolarized suggests that the complete mechanism is more complicated. Furthermore, other data clearly indicate that opposing photochemical mechanisms with different spectral sensitivities play an important role in generation of their off responses. It should also be noted that if such a system involved two opposing mechanisms with the same or spectrally very similar pigments, this would not be revealed by chromatic adaptation.

In most receptor neurons the receptor or generator potential has been thought to be produced mainly by the stimulus in some way causing a change in membrane permeability to sodium ions. Although in some cases there is suspicion that other ions may also be involved, their identification and exact role have not been clear. The ionic studies on the light response of the distal photoreceptors of Pecten demonstrate that receptor potentials can be generated by selective changes in potassium permeability, and it is likely that a similar process produces the hyperpolarizing receptor potentials of many other invertebrate visual cells. Since it has been shown that light can cause a change in potassium permeability, it is quite possible that such changes may be a component of other kinds of receptor potentials as well. If a stimulus causes a single permeability increase which affects both sodium and potassium ions the response might be depolarizing because of the larger driving force on sodium ions at normal resting potentials, but the potassium component would cause the response to have a more negative reversal potential than if it were due solely to a change in sodium permeability. This situation would be analogous to the action of acetylcholine at the neuromuscular junction.

On the other hand, if the stimulus initiates separate processes which lead to independent changes in sodium and potassium permeabilities, perhaps with differences in time course and other parameters, then the resulting potential change could be much more complicated. Two examples indicate that this may actually be the case in some receptors. In the proximal (depolarizing) photoreceptors of Pecten light causes an initial increase in membrane conductance, but with bright stimuli this is followed by a decrease in membrane conductance to a value much lower than in the dark, even though the membrane is still depolarized (Fig. 2). This suggests a second conductance change mechanism with a different time course and could, for example, be due to a decrease in potassium permeability. The fact that it does not appear to occur except with bright stimuli and is not elicited by passive depolarization suggests that it results from a separate light-induced process. Further study of the proximal cell would be useful, since this type of multi-component receptor potential may be more common than previously suspected. Detwiler (1976) has recently found that the depolarizing photoreceptor of the nudibranch mollusc Hermissenda crassicornis actually has three separate receptor potential mechanisms. In response to light there is an increase in sodium conductance, an increase in potassium conductance, and a decrease in potassium conductance, with different time courses. The three light-activated conductance changes have different sensitivities to light and different kinetics of light adaptation, so that the response in light adapted cells can be a pure hyperpolarization, due to an increase in potassium conductance, while under certain conditions dim flashes can produce a slow depolarization which is caused by a decrease in potassium conductance.

A role for calcium ions in generating the hyperpolarizing receptor potential has not been established, although indirect evidence suggests that this possibility should be investigated further. If it were indeed shown that light controls the ionic permeabilities which generate the receptor potential by first increasing the calcium permeability of the membrane, then one would still have to explain how light modulates the permeability to calcium.

3.3.7 References

Barber, V.C., Evans, E.M., Land, M.F.: The fine structure of the eye of the mollusc Pecten Maximus. Z. Zellforsch. 76, 295-312 (1967)

Barber, V.C., Land, M.F.: Eye of the cockle, Cardium edule: anatomical and physiological observations. Experientia 23, 677 (1967)

Barnes, S.N.: Fine structure of the photoreceptor and cerebral ganglion of the tadpole larva of Amaroucium constellatum (Verrill). Z. Zellforsch. 117, 1-16 (1971)

Baylor, D.A., Fuortes, M.G.F.: Electrical responses of single cones in the retina of the turtle. J. Physiol. 207, 77-92 (1970)

Bell, A.L., Mpitsos, G.J.: Morphology of the eye of the flame fringe clam, Lima scabra. Biol. Bull. 135, 414 (1968)

Brown, A.M., Brown, H.M.: Light response of a giant Aplysia neuron. J. Gen. Physiol. 62, 239-254 (1973)

Brown, J.E., Pinto, L.H.: Ionic mechanism for the photoreceptor potential of the retina of Bufo marinus. J. Physiol. 236, 575-591 (1974)

Byzov, A.L., Trifonov, Y.A.: The response to electrical stimulation of horizontal cells in the carp retina. Vis. Res. 8, 817-822 (1968)

Cervetto, L.: Influence of sodium, potassium and chloride ions on the intracellular responses of turtle photoreceptors. Nature 241, 401-403 (1973)

Cervetto, L., Piccolino, M.: Synaptic transmission between photoreceptors and horizontal cells in the turtle retina. Science 183, 417-419 (1974)

Chappell, R.L., Dowling, J.E.: Neural organization of the median ocellus of the dragonfly. I. Intracellular activity. J. Gen. Physiol. 60, 121-147 (1972)

Cornwall, M.C., Gorman, A.L.F.: Color dependent potential changes of opposite polarity in single visual receptors. Biophys. J. 16, 146a (1976)

Detwiler, P.: Multiple light-evoked conductance changes in the photoreceptors of Hermissenda crassicornis. J. Physiol. (1976, in press)

Dowling, J.E., Ripps, H.: Effect of magnesium on horizontal cell activity in the skate retina. Nature 242, 101-103 (1973)

Eakin, R.M.: Lines of evolution of photoreceptors. In: General Physiology of Cell Specialization. Mazia, D., Tyler, A. (eds.). San Francisco: McGraw Hill 1963, pp. 393-425

Eakin, R.M.: Evolution of photoreceptors. Cold Spring Harb. Symp. Quant. Biol. 30, 363-370 (1965)

Eakin, R.M.: Structure of invertebrate photoreceptors. In: Handbook of Sensory Physiology. Dartnall, H.J.A. (ed.). Berlin-Heidelberg-New York: Springer 1972, Vol. VII, pp. 625-684

Fuortes, M.G.F., Hodgkin, A.L.: Changes in time scale and sensitivity in the ommatidia of Limulus. J. Physiol. 172, 239-263 (1964)

Gorman, A.L.F., McReynolds, J.S.: Control of membrane K^+ permeability in a hyperpolarizing photoreceptor: similar effects of light and metabolic inhibitors. Science 185, 620-621 (1974)

Gorman, A.L.F., McReynolds, J.S., Barnes, S.N.: Photoreceptors in primitive chordates; fine structure, hyperpolarizing receptor potentials, and evolution. Science 172, 1052-1054 (1971)

Gwilliam, G.F.: The mechanism of the shadow reflex in Cirripedia. II. Photoreceptor cell response, second order response, and motor cell output. Biol. Bull. 129, 244-256 (1965)

Hartline, H.K.: The discharge of impulses in the optic nerve of Pecten in response to illumination of the eye. J. Cell. Comp. Physiol. 11, 465-478 (1938)

Kandel, E.R.: Dale's Principle and the functional specificity of neurons. In: Electrophysiological Studies in Neuropharmacology. Kolle, W. (ed.). Springfield, Ill.: C.C. Thomas 1967, pp. 385-398

Kennedy, D.: Neural photoreception in a lamellibranch mollusc. J. Gen. Physiol. 44, 277-299 (1960)

Korenbrot, J., Cone, R.A.: Dark ionic flux and the effects of light in isolated rod outer segments. J. Gen. Physiol. 60, 20-45 (1972)

Land, M.F.: Activity in the optic nerves of Pecten maximus in response to changes in light intensity, and to pattern and movement in the optical environment. J. Exp. Biol. 45, 83-99 (1966)

Land, M.F.: Functional aspects of the optical and retinal organization of the mollusc eye. Symp. Zool. Soc. London 23, 75-96 (1968)

Lisman, J.E., Brown, J.E.: The effects of intracellular Ca^{2+} on the light response and on light adaptation in Limulus ventral photoreceptors. In: The Visual System: Neurophysiology, Biophysics, and Their Clinical Applications. Arden, G.B. (ed.). New York: Plenum 1972, pp. 23-33

McReynolds, J.S., Gorman, A.L.F.: Photoreceptor potentials of opposite polarity in the eye of the scallop, Pecten irradians. J. Gen. Physiol. 56, 376-391 (1970a)

McReynolds, J.S., Gorman, A.L.F.: Membrane conductances and spectral sensitivities of Pecten photoreceptors. J. Gen. Physiol. 56, 392-406 (1970b)

McReynolds, J.S., Gorman, A.L.F.: Ionic basis of hyperpolarizing receptor potential in scallop eye: increase in permeability to potassium ions. Science 183, 658-659 (1974)

McReynolds, J.S., Gorman, A.L.F.: Hyperpolarizing photoreceptors in the eye of a primitive chordate, Salpa democratica. Vis. Res. 15, 1181-1186 (1975)

Meech, R.M.: Intracellular calcium injection causes increased potassium conductance in Aplysia nerve cells. Comp. Biochem. Physiol. 42A, 493-499 (1972)

Miller, W.H.: Derivatives of cilia in the distal sense cells of the retina of Pecten. J. Biophys. Biochem. Cytol. 4, 227-228 (1958)

Mpitsos, G.J.: Physiology of vision in the mollusk Lima scabra. J. Neurophysiol. 36, 371-383 (1973)

Penn, R.D., Hagins, W.A.: Signal transmission along retinal rods and the origin of the electroretinographic a-wave. Nature 223, 201-205 (1969)

Shaw, S.R.: Decremental conduction of the visual signal in barnacle lateral eye. J. Physiol. 220, 145-175 (1972)

Sillman, A.J., Ito, H., Tomita, T.: Studies on the mass receptor potential of isolated frog retina. II. On the basis of the ionic mechanism. Vis. Res. 9, 1443-1451 (1969)

Toyoda, J., Nosaki, H., Tomita, T.: Light-induced resistance changes in single photoreceptors of Necturus and Gekko. Vis. Res. 9, 453-463 (1969)

Wiederhold, M.L., MacNichol, E.F., Jr., Bell, A.L.: Photoreceptor spike responses in the hardshell clam, Mercenaria mercenaria. J. Gen. Physiol. 61, 24-55 (1973)

Yoshikami, S., Hagins, W.A.: Control of the dark current in vertebrate rods and cones. In: Biochemistry and Physiology of Visual Pigments. Langer, H. (ed.). Berlin: Springer 1973, pp. 245-255

Zettler, G., Järvilehto, M.: Decrement-free conduction of graded potentials along the axon of a monopolar neuron. Z. Vergl. Physiol. 75, 402-421 (1971).

3.4 The Economy of Photoreceptor Function in a Primitive Nervous System

D. L. ALKON

3.4.1 Introduction

The number and complexity of synaptic contacts made by photoreceptors in a relatively primitive nervous system such as that of the nudibranch mollusc Hermissenda crassicornis (Alkon, 1973a, 1974a) are not unique. Photoreceptors in vertebrate retinas also participate in elaborate synaptic organization (Dowling and Boycott, 1965; Lasansky, 1971). While the Hermissenda photoreceptors interact with each other, second-order optic ganglion cells and hair cells, cones can interact with each other as well as with bipolar and horizontal cells (Baylor and Fuortes, 1970).

Nor is the specificity of photoreceptor types in Hermissenda unique; a parallel to cones and rods can be drawn. Dark-adapted Type A cells (corresponding to cones) are less sensitive to light than the dark-adapted Type B photoreceptors (corresponding to rods). Light-adapted Type A cells, however, are more sensitive to light than light-adapted Type B photoreceptors (Alkon and Fuortes, 1972).

A particularly distinguishing feature of the Hermissenda photoreceptors is their number: five in each eye. These photoreceptors, then, cannot be as available as retinal cones and rods for specialized and refined information processing. Thus, whereas both Type A and B photoreceptors, for the visible wavelengths, have one maximum in their spectral sensitivity at 510 mμ (Alkon and Fuortes, unpublished observations), vertebrate cones have one of three distinct maxima (red, green, and blue; cf. Marks, 1965). Similarly, whereas the eye of Hermissenda resolves the position and direction of movement of a slowly moving shadow or edge, the spatially distributed matrix of photoreceptors in the vertebrate retina can discriminate visual patterns.

In more primitive eyes, therefore, specialization of photoreceptor function is limited by a relatively small number of photoreceptors. Such a lack of specialization would also be expected for the rest of the nervous system in more primitive species. In fact, recent observations indicate that the Hermissenda photoreceptors assume several diverse information processing functions. These functions in a more evolved nervous system, with a much greater number of neural elements, are performed by specialized higher order visual and vestibular cells.

In effect, then, multiplicity of neural integration functions substitutes in Hermissenda photoreceptors for specialization of integration function in photoreceptors of more complex eyes. This multiplicity constitutes the unique economy of photoreceptor function in this nudibranch mullusc.

Any analysis of what the Hermissenda photoreceptors do and how they perform their diverse functions must concern anatomic and electrical features of the cells themselves as well as their synaptic relationships.

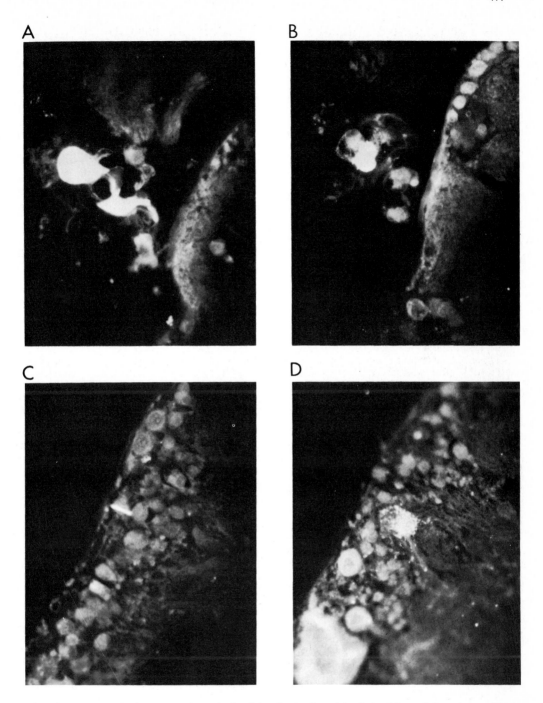

Fig. 1A-D. Type B photoreceptor stained by iontophoretic injection of Procion yellow. Dark field photomicrographs demonstrate the fluorescent dye within the cell and its processes. A) cross section of cell soma and axon. B-D) details of axon and terminal branches. x 480 (Masukawa, L., unpubl. observ.)

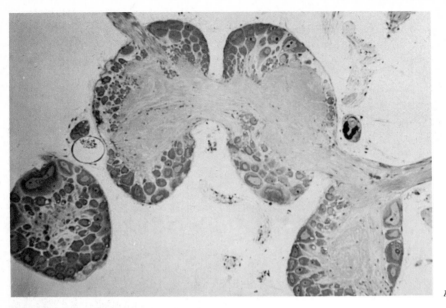

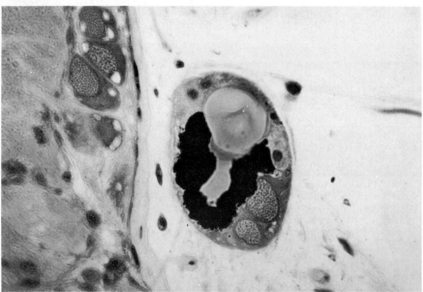

Fig. 2 A and B

3.4.2 Cell Geometry

The structure of the Hermissenda photoreceptors bears little resemblance to vertebrate cones or rods. Procion yellow intracellular staining (Fig. 1) reveals a cell geometry identical to that arrived at by
serial toluidine blue sections (Fig. 2A,B; and Stensaas et al., 1968).
The somata of the two Type A cells in each eye are approximately 25 μ
in diameter and are located in the anteroventral portion of the eye.

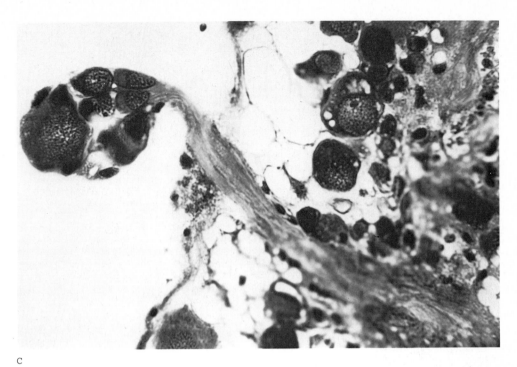

C

Fig. 2A-C. Circumesophageal nervous system stained with toluidine blue.
A) 10 μm Section through entire circumesophageal system (80 x). B) 10 μm Section
of Hermissenda eye: external epithelium, lens, pigment cup, and two photoreceptors
(480 x). C) 25 μm thick section stained with toluidine blue showing the optic ganglion (40 μm across) and "optic tract". (From Alkon, 1973a)

The somata of the three Type B cells in each eye are approximately
40 μ in the longest diameter and are located in the posterodorsal
portion of each eye. Procion stained sections indicate a larger volume of terminal branches for Type B than for Type A photoreceptors
(cf. Fig. 1, Table 1). The five photoreceptor axons as they leave the
eye form the optic nerve which remains ensheathed as it courses through
the optic ganglion and then joins the axons of the optic ganglion
cells to form the optic tract. The axon extends from the soma of each
photoreceptor for approximately 80 μ to a region of terminal branches
within the optic tract (Fig. 2C).

3.4.3 Response to Light

The primary role of a photoreceptor in any nervous system is transduction of light energy into an electrical signal and the subsequent
transmission of information contained within that signal to other
cells. Simultaneous recordings from the soma and distal axon of
Hermissenda photoreceptors (Fig. 3A,C,D) in addition to lesion experiments (Alkon, 1974a) indicate that the photoreceptor generator potential (in response to light) arises in the soma and spreads passively toward the end of the photoreceptor axon to an excitable region
responsible for generating impulses (Fig. 4,5,6). The generator poten-

Table 1. Photoreceptor Branching Volumes

	Cell	Volume (μ^3)
Type A	1	52,060
	2	58,240
	3	50,810
Type B	1	129,120
	2	94,540
	3	71,160

The extent of branching of the photoreceptor cells in the central ganglion is expressed in terms of the total volume which the arborizations encompassed. This volume was calculated from color slides of 10 μ thick histological sections of Procion-yellow marked cells. The area was determined by assuming a rectangular shape generalization for all branched areas. The maximum length (X) and mid length width (Y) of the arborization area were measured from projections of color slides of the histological sections of marked branches. The mean total volume of neuropile which contained Procion yellow marked branches were calculated from 3 Type A and 3 Type B cells.

It was found from these measurements that Type B cells branched on the average, in twice the volume of neuropile than Type A cells. However, for both the A and B cells measured, the extent of penetration into the neuropile was similar indicating an overlapping of branches of the two types of cells within the ganglion (L.M. Masukawa, unpubl. observ.)

tial, largely a depolarizing wave, has at least two distinct depolarizing components, one of which arises from an increase in conductance to an ion with a positive equilibrium potential. These depolarizing components can be distinguished by application of steady negative currents (Fig. 7).

In addition to the depolarizing waves, the generator potential also contains a hyperpolarizing component. This hyperpolarizing component is probably present as a notch or interruption in the depolarizing response of dark-adapted photoreceptors (Fig. 4) and as an isolated hyperpolarizing wave in light-adapted photoreceptors (Fig. 8). The hyperpolarizing wave has been shown to arise from an increased conductance of an ion with a negative equilibrium potential (Fig. 8; Alkon, 1972, unpublished observations). Thus, the photoreceptor generator response, a part of the primary transduction process, arises from at least two and possibly three separate conductance changes.

3.4.4 Excitable Focus

Photoreceptor impulses originating in the distal end of the axon have quite specific characteristics. Type B impulses have an average amplitude of approximately 15 mV and a threshold usually closer to the resting potential than the threshold of Type A impulses with an average amplitude of approximately 45 mV. A train of Type A impulses, elicited by a depolarizing current pulse or the depolarizing response to light, is followed by a long-lasting after-hyperpolarization. Single Type A photoreceptor impulses elicited following an impulse train have a reduced after-potential indicating extracellular potassium accumulation (Alkon, 1976).

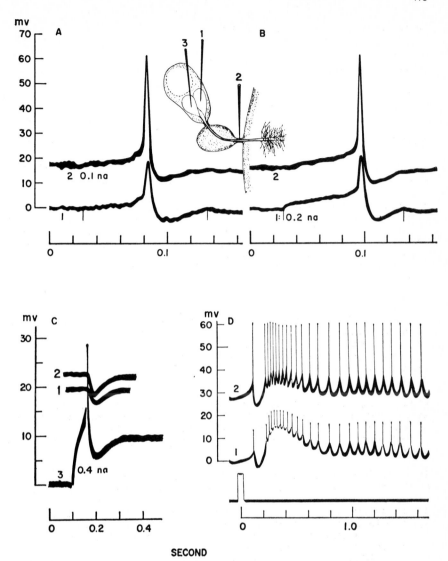

Fig. 3A-D. Responses recorded from soma and axon of same cell. The distance between the two electrodes was 80-100 μ. A) Responses to a step of depolarizing current through axonal electrode. B) Response to a step of current through the soma electrode. In either case the spike is larger and earlier in the axon. C) Two electrodes were inserted, respectively, in the soma and axon of one receptor and a third electrode was inserted in the soma of a second receptor. A step of depolarizing current through this third electrode evokes a spike in one receptor and a hyperpolarizing synaptic potential in the other. The synaptic potential is larger and has shorter time to peak in the axon. D) Responses to a flash of light. The generator potential is larger at the soma (lower trace) while the spikes are larger at the axon. Down-going bars in (A) and (B) indicate timing of current steps. (From Alkon and Fuortes, 1972)

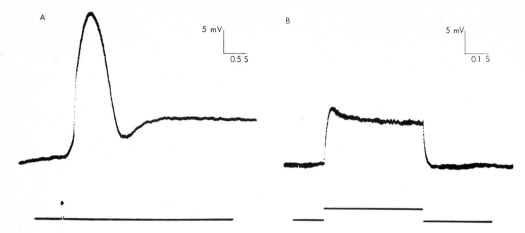

Fig. 4A and B. Responses of photoreceptor with cut axon. A) Response to 50 ms flash (2 x 10⁵ ergs/cm²/s). B) Response to positive current pulse (1.5 nA). Bottom traces: duration of light and current stimuli. (From Alkon, unpubl. observations)

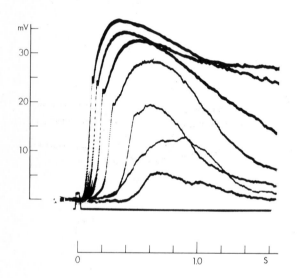

Fig. 5. Responses of photoreceptor with cut axon. Responses are to flashes (indicated by bottom trace) of quartz-iodide light source (2 x 10⁵ ergs cm⁻²/s) attenuated by neutral density filters as follows (increasing intensity): 4.5 O.D., 4.2, 3.6, 3.0, 2.4, 1.0, 1.2

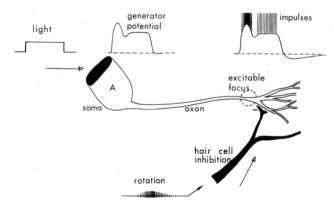

Fig. 6. Diagram of Type A photoreceptor. Cell body is located in the antero-ventral portion of the Hermissenda eye. (From Alkon, unpubl. observations)

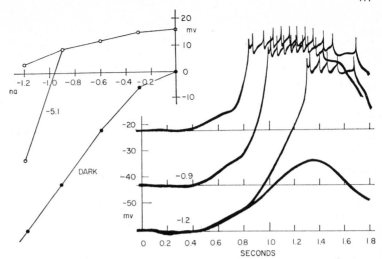

Fig. 7. Action of currents on responses evoked after section of the optic nerve. The records show responses to dim flashes (intensity 8×10^{-6}) applied at time 0, in the presence of hyperpolarizing currents of the intensities indicated. The early part of the generator potential does not become larger with hyperpolarizing currents. The later wave, however, is first increased and then critically blocked by the current. The plot at left measures the potential drop evoked by the current in darkness and at the peak of the responses. (From Alkon and Fuortes, 1972)

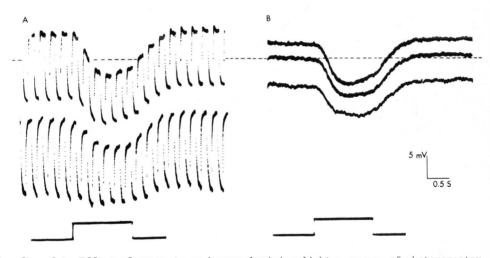

Fig. 8A and B. Effect of currents on hyperpolarizing light response of photoreceptor with cut axon. Receptor was first exposed to one min of bright light (2×10^5 ergs/ cm^2/s). Flashes (duration indicated by bottom traces) were then delivered at a frequency of 12/min. A) Negative current pulses (0.5 nA) cause smaller photoreceptor potential changes during the hyperpolarizing response to light (indicating a conductance increase). This was true with steady depolarization (using D.C. plus 0.5 nA) and steady hyperpolarization (using D.C. minus 0.5 nA) of the photoreceptor. Thus, the conductance change measured is not due to rectification of the photoreceptor membrane. B) Polarization of the photoreceptor above the resting level (using plus 0.25 nA) causes the hyperpolarizing response to increase slightly. Polarization of the photoreceptor below the resting level (using minus 0.25 nA) causes a small reduction of the response. Dashed line: level of resting membrane potential. (From Alkon, unpubl. observations)

Photoreceptor impulses could contribute to the cellular integrating and transmitting capabilities in a number of ways. Although the photoreceptor axons are short, the number of terminal branches is large (Fig. 1) and the diameters of these branches are often quite small. Decrement of passively spreading potentials within the photoreceptor axon was found to be fairly small (approximately 30% for 70 μ; cf. Alkon and Fuortes, 1972). As an inverse function of membrane resistance, the membrane space constant in the terminal branches might be quite small because of high resistances in these endings. Passively spreading potentials, such as the depolarizing responses of the photoreceptors (or the hyperpolarizing responses of cones) might decay so as to be insignificant at the presynaptic terminal endings. Impulses generated at the excitable focus could provide essential amplification of the generator signals. If the impulses are critical for transmission of photoreceptor information, the proximity of the excitable focus near the site of synaptic input (see below) would determine the efficacy of such input.

3.4.5 Synaptic Contacts

Simultaneous recordings from the soma and distal axon of photoreceptors (Fig. 3C) in addition to lesion experiments (Alkon, 1974a) also indicate that synaptic input to the photoreceptors occurs on the terminal axonal branches. Furthermore, sections of the optic tract as it begins within the cerebropleural ganglion do not reveal axons other than those of photoreceptors (5) and optic ganglion cells (14) (Sensaas et al., 1968; Alkon, 1973a). Procion injections of hair cells, which have synaptic interactions with photoreceptors (Alkon, 1973b), show no hair cell collaterals extending external to the cerebropleural ganglion, i.e. outside of the optic tract (Fig. 9). Finally, serial sections of toluidine blue stained Hermissenda nervous systems demonstrate apposition of hair cell and visual cell terminal branchings within the optic tract approximately thirty microns medial to the lateral border of the cerebropleural ganglion (Alkon, unpubl. observ.).

On the basis of simultaneous intracellular recordings and current passage in more than five hundred pairs of cells, the photoreceptors of Hermissenda have been found to make specific synaptic connections with cells within the same eye, with ipsilateral optic ganglion cells and with hair cells of both the ipsilateral and contralateral statocysts (Alkon, 1973a,b, 1974a, 1975b). In brief, the three Type B photoreceptors in each eye are mutually inhibitory and inhibit ipsilateral optic ganglion cells (Fig. 10). Type B photoreceptors inhibit some ipsilateral and contralateral hair cells as well as receive inhibition from

Fig. 9A-E. Hair cell stained by iontophoretic injection of Procion yellow. Dark-field photomicrographs demonstrate the fluorescent dye within the cell and its processes. A) Partial cross section of cell soma. B) Hair cell axon within the static nerve as it enters the pleural ganglion. C) Upper arrow: continuation of hair cell axon within the pleural ganglion. Lower arrow: optic nerve made visible by auto-fluorescence. D) The axon gives off a spray of fine branches in the ipsilateral pleural ganglion. E) It then crosses to the contralateral pleural neuropile via the connective joining the two pleural ganglia. x 250. (From Detwiler and Alkon, 1973)

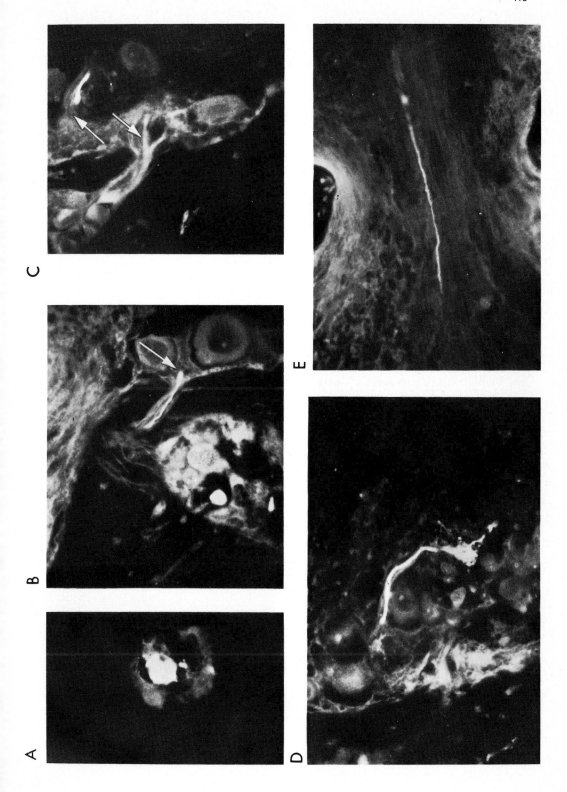

420

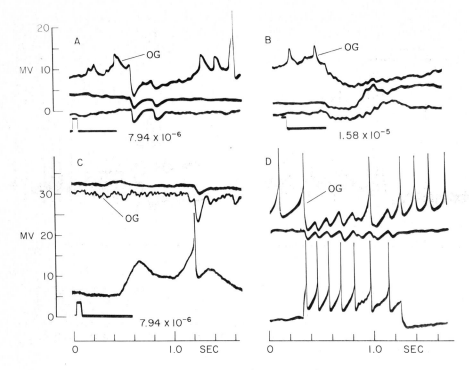

Fig. 1OA-D. A,B) Responses to dim flashes of Type A photoreceptor (middle record)
Type B photoreceptor (lowest record) and optic ganglion cell (OG) penetrated simultaneously. IPSPs occur simultaneously in all three cells. C) Responses to dim
flash. Simultaneous IPSP in Type A photoreceptor (upper record) and optic ganglion
cell (middle record) associated with spike in Type B photoreceptor (lowest record)
penetrated simultaneously. D) Depolarizing pulse (O.4 nA) to Type B photoreceptor.
Simultaneous IPSPs in Type A photoreceptor (middle record) and optic ganglion cell
(upper record) are associated with spikes in Type B photoreceptor (lowest record)
penetrated simultaneously. Optic ganglion cell IPSPs are larger and have a faster
rise to peak amplitude than IPSPs in Type A photoreceptor. (From Alkon, 1973a)

other hair cells in both statocysts. The two Type A photoreceptors in
each eye have little, and usually no, synaptic effect on each other.
Type A photoreceptors inhibit and are inhibited by Type B photoreceptors in the same eye. Type A photoreceptors also are inhibited by
ipsilateral and contralateral hair cells. Type A photoreceptors excite other hair cells in the ipsilateral statocyst. These features
of the synaptic relations of the photoreceptors are invariant from
preparation to preparation.

Thus, with only five cells in each eye, specific visual information
is transmitted through two distinct channels: second-order optic ganglion cells within the visual pathway, and hair cells within the statocysts. Optic ganglion cells, which are spontaneously active in darkness, hyperpolarize and cease firing in response to illumination.
Hair cells hyperpolarize in response to illumination of the contralateral eye. Hair cells respond to illumination of the ipsilateral
eye, with a hyperpolarization which may be followed by a prolonged
depolarization and increase of impulse activity (Fig. 11). Economy
of photoreceptor function is further demonstrated by physiologic
stimulation of the primitive vestibular or statocyst pathways.

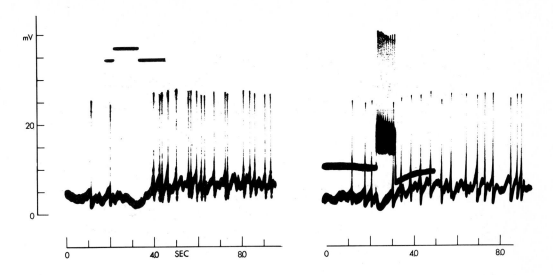

Fig. 11. Response of hair cell to illumination of the ipsilateral eye (record on left) and a train of impulses in an ipsilateral Type A photoreceptor (records on right). Bar on left indicates duration of flash (intensity: 6.0×10^3 ergs/cm^2). A 0.6 nA depolarizing pulse was given to a Type A photoreceptor penetrated simultaneously with the hair cell. Base-like activity of hair cell: 67 impulses/min. (From Alkon, 1975b)

It has recently become possible to physiologically stimulate statocyst hair cells in the isolated nervous system of Hermissenda while making intracellular recordings (Alkon, 1975c). Hair cells located in the statocyst so as to be in front of a centrifugal force vector (up to 0.9 g) depolarize in response to rotation. Hair cells in back of the vector hyperpolarize in response to rotation. These hair cell responses are independent of the statocyst's fixed orientation with respect to the center of rotation. Photoreceptor responses to rotation, synaptic in origin (Alkon, 1976), are dependent on the statocyst's fixed orientation with respect to the center of rotation. Thus, when the cephalad end of a line bisecting the circumesophageal nervous system points toward the center of rotation, Type A photoreceptors hyperpolarize in response to rotation. Furthermore, impulses of Type A photoreceptors during their steady state response to light are eliminated by rotation (Fig. 12). Some Type B photoreceptors show a small hyperpolarizing response to rotation, but this response is never sufficient to eliminate or substantially decrease impulse activity during the steady state light response.

When the caudal end of a line exactly bisecting the circumesophageal nervous system points toward the center of rotation, Type A and Type B photoreceptors depolarize in response to rotation. Impulse frequency in Type A as well as Type B photoreceptors is increased (Fig. 13).

The hyperpolarizing responses of photoreceptors in the cephalad orientation can be explained by a direct inhibition from caudal hair cells (which are in front of the centrifugal force vector for the cephalad orientation). Such direct inhibition has previously been demonstrated (Alkon, 1973b). The depolarizing response of photoreceptors in the caudal orientation can be explained, assuming that the cephalad hair cells do not inhibit photoreceptors, as a disinhibition from caudal

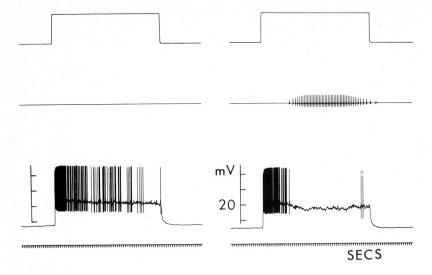

Fig. 12. Effect of rotation on Type A photoreceptor (cephalad orientation) during its response to light. Record on left: response to light alone. Record on right: response to light and rotation. Top trace: duration of light stimulus. The amplitude of monitor signals (middle trace) as well as the inverse of the intervals between monitor signals are proportional to the angular velocity of the turntable. (From Alkon, in press)

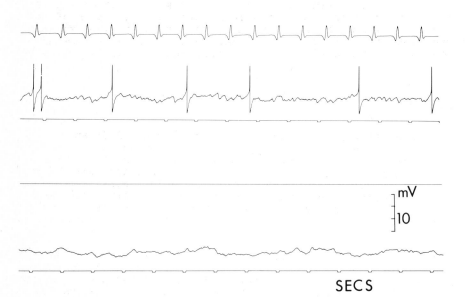

Fig. 13. Responses of Type A photoreceptor (caudal orientation) to light step. Lower record: response to light step alone (1/3 maximum intensity). Upper record: response to light step paired with rotation (indicated in top trace by artifact). Tops of impulses in upper record are not included. (From Alkon, in press)

hair cells. Thus, in the caudal orientation, caudal hair cells will hyperpolarize in response to rotation and decrease the inhibition of photoreceptors normally resulting from the spontaneous activity of these hair cells. This disinhibition effect would be facilitated by the mutual inhibition which has been demonstrated for hair cells located on a statocyst circumference at 180° with respect to each other (Detwiler and Alkon, 1973).

As discussed above, transmission of specific visual information through two neural channels is provided for by the synaptic organization of the Hermissenda eye and its input to the optic ganglion and statocysts. In a similar manner, transmission of specific vestibular information is provided for by the synaptic organization of the statocyst (Detwiler and Alkon, 1973) and its input to the eyes (Alkon, 1976).

3.4.6 Persistent Modification of Photoreceptor Signals

Another type of neural integration performed by the eye of Hermissenda and its interaction with the statocysts involves a form of primitive conditioning (Alkon, in press). Exposure of the isolated nervous system (cephalad orientation) to repeated pairing of an intermittent light stimulus with an intermittent rotation stimulus, results in a marked reduction of conducted impulses of the Type A photoreceptor during its subsequent responses to light. This change in the Type A photoreceptor response to light can persist for hours depending on the parameters for rotation and light stimulus pairing. This change does not result from an unpaired paradigm nor does it occur with either the rotation or light stimulus alone. Responses to light of Type B photoreceptors are unaffected by stimulus pairing. Reduction of conducted impulses during Type A photoreceptor responses also explains the neural correlates reported in responses of hair cells from animals previously exposed to associative training (Alkon, 1974b, 1975a).

3.4.7 Discussion

Previous comparative studies (Bullock and Horridge, 1965) suggested that, with evolution, neural integration is accomplished by progressively more centralized nervous tissue and the peripheral nervous system becomes less autonomous in its integrative function. With centralization of higher integration, we could expect increasing specialization of nerve cells and decreasing multiplicity of nerve cell functions. This evolutionary perspective on neural systems is quite consistent with the aforementioned observations of Hermissenda photoreceptors.

Photoreceptors of the Hermissenda eye fulfill multiple functions. They code light intensity and duration and can discriminate the direction of slowly moving shadows. The organization of photoreceptor inhibition of ipsilateral and contralateral optic ganglion cells enhances discrimination of differences in the intensity of illumination received by the two eyes (Alkon, 1973a). Such discrimination affords the information necessary for the animal's movement toward light sources (Alkon, 1974b).

Photoreceptors, in addition, transmit visual information to the vestibular pathway and, in turn, receive precise and comprehensive input from the statocyst hair cells. In fact, the response of the photoreceptors (particularly Type A) to light is clearly dependent on the orientation of the nervous system (and thus the intact animal) with respect to the vector of the earth's gravitational force. Finally, photoreceptor responses (as well as the animal's responses) to light are modified by repeated pairing of visual and vestibular stimuli.

The Hermissenda photoreceptors, although primary cells of the visual pathway, perform functions reserved for higher order cells in more evolved visual pathways. Thus, the photoreceptors accomplish neural integration, albeit simplified by comparison, analogous to some of the information processing of retinal horizontal and ganglion cells, lateral geniculate cells and, perhaps, even higher order elements in and between vertebrate visual and vestibular pathways.

This multiplicity of photoreceptor function is economical for accomplishing specific integration tasks. Such economy, however, is at the expense, for the same number of neural elements, of other more refined information processing by the photoreceptors. For example, instead of inhibitory interactions, electrical coupling of photoreceptors would increase the eye's sensitivity to very dim lights. As another example, synaptic input from both Type A and B photoreceptors (instead of only Type B) to second-order optic ganglion cells would increase the amount of visual information transmitted by the eye to the remainder of the visual pathway. A further refinement of visual information processing would remove the dependence of photoreceptor signals on the orientation of the animal with respect to the earth's gravitational force. Finally, if instead of discontinuous signals such as impulses, the generator potential itself were largely responsible for release of transmitter at photoreceptor presynaptic terminals (as is believed the case for vertebrate cones), a more precise transmission of light information to postsynaptic cells would be afforded. Thus, inhibitory interactions and a dependence on impulses for signal transmission would no longer distort the intensity-response relationship for the photoreceptors (Fig. 14).

The specific integrative tasks for which the photoreceptors and their synaptic organization are particularly suited are, in part, determined by the behavioral requirements of the biological niche for Hermissenda. The interdependence of the visual and vestibular responses appear to be responsible for generating behavioral sequences important to the existence of Hermissenda in an intertidal zone (Alkon, 1976). The generation of such sequences, then, takes precedence over the achievement of maximal photoreceptor sensitivity to light signals independent of gravitational orientation.

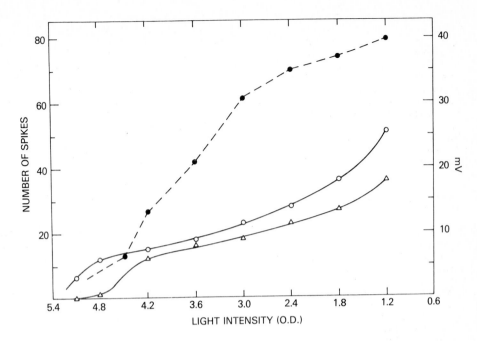

Fig. 14. Plots of photoreceptor responses as function of $\log_{10}$ of light intensity. Responses in lower two plots (<u>solid line</u>) are measured by the number of Type B photoreceptor spikes in 1.0 s interval after onset of depolarizing response to light. Responses in the upper plot (<u>dashed line</u>) are measured by the maximum generator potential of a photoreceptor with cut axon (from Fig. 5) for each intensity

3.4.8 References

Alkon, D.L.: Neural organization of a molluscan visual system.
J. Gen. Physiol. <u>61</u>, 444-461 (1973a)
Alkon, D.L.: Intersensory interactions in <u>Hermissenda</u>. J. Gen. Physiol. <u>62</u>, 185-202
(1973b)
Alkon, D.L.: Sensory interactions in the nudibranch mollusk <u>Hermissenda crassicornis</u>.
Federation Proc. <u>33</u>, 1083 (1974a)
Alkon, D.L.: Associative training of <u>Hermissenda</u>. J. Gen. Physiol. <u>64</u>, 70-84 (1974b)
Alkon, D.L.: Neural correlates of associative training in <u>Hermissenda</u>.
J. Gen. Physiol. <u>65</u>, 46-56 (1975a)
Alkon, D.L.: A dual synaptic effect on hair cells in <u>Hermissenda</u>.
J. Gen. Physiol. <u>65</u>, 385-397 (1975b)
Alkon, D.L.: Responses of hair cells to statocyst rotation. J. Gen. Physiol. <u>66</u>,
507-530 (1975c)
Alkon, D.L.: Signal transformation with pairing of sensory stimuli.
Gen. Physiol. <u>67</u>, 197-211 (1976)
Alkon, D.L.: Neural modification by paired sensory stimuli. J. Gen. Physiol.
(1976, in press)
Alkon, D.L., Fuortes, M.G.F.: Responses of photoreceptors in <u>Hermissenda</u>.
J. Gen. Physiol. <u>60</u>, 631-649 (1972)
Baylor, D.A., Fuortes, M.G.F.: Electrical responses of single cones in the retina
of the turtle. J. Physiol. (London) <u>207</u>, 77-92 (1970)

Bullock, T.H., Horridge, G.A.: Structure and Function in the Nervous System of Invertebrates. San Francisco: W.H. Freeman 1965, Vol. I, p. 1719

Detwiler, P.B., Alkon, D.L.: Hair cell interactions in the statocyst of Hermissenda. J. Gen. Physiol. $\underline{62}$, 618-642 (1973)

Dowling, J.E., Boycott, B.B.: Neural connections of the retina: fine structure of the inner plexiform layer. Cold Spr. Harb. Symp. Quant. Biol. $\underline{30}$, 393-402 (1965)

Lasansky, A.: Synaptic organization of cone cells in the turtle retina. Phil. Trans. R. Soc. London $\underline{262}$B, 365-381 (1971)

Marks, W.B.: Visual pigments of single goldfish cones. J. Physiol. (London) $\underline{178}$, 14-32 (1965)

Steenas, L.J., Stenaas, S.S., Trujillo-Cenoz, O.: Some morphological aspects of the visual system of Hermissenda crassicornis (Mollusca: Nudibranchia). J. Ultrastruct. Res. $\underline{27}$, 510 (1968).

SUBJECT INDEX

Photoreceptor Optics

Editors: A.W.Snyder,R.Menzel
259 figures.X,523 pages.1975

Contents: Photoreceptor Waveguide Optics.- Membrane and Dichroism.- Photopigment, Membrane and Dichroism.- Polarisation and Sensitivity and Dichroism.- Photomechanical Responses of Photoreceptors.- Electrophysiology of Photoreceptors.

Biochemistry and Physiology of Visual Pigments

Symposium Held at Insitut für Tierphysiologie, Ruhr-Universität Bochum/W.Germany, August 27-30,1972.
For the Organizing Committee,Editor: H.Langer
202 figures.XIII,366 pages.1973

Contents: Pigment Structure and Chemical Properties.- Photolysis and Intermediates of the Pigments.- Regeneration of the Pigments.- Excitation and Adaptation of Photoreceptor Cells.- Ionic Aspects of Excitation and Regeneration.- Enzymology and Molecular Architecture of the Light Sensitive Membrane.

Information Processing in the Visual Systems of Arthropods

Symposium Held at the Department of Zoology, University of Zurich,March 6-9,1972
Editor: R.Wehner
268 figures.XI,334 pages.1972

Contents: Anatomy of the Visual System.- Optics of the Compound Eye.- Biochemistry of Visual Pigments.- Intensity - Dependent Reactions.- Wavelength - Dependent Reactions.- Pattern Recognition.- Visual Control of Orientation Patterns.- Storage of Visual Information.- Methods of Quantifying Behavioral Data.

N.J.Strausfeld · Atlas of an Insect Brain

81 figures,partly coloured,71 plates.XII,214 pages.1976

Contents.: Introduction.- A Historical Commentary.- The Structure of Neuropil.- The Primary Compartments of the Brain.- The Coordinate System.- Some Quantitative Aspects of the Fly's Brain.- The Atlas: Sections through the Brain.- The Forms and Dispositions of Neurons in the Brain.- Appendix 1: Histological Methods.- Appendix 2: Dictionary of Terms.- References.- Subject Index.

Frog Neurobiology · A Handbook

Editors: R.Llinás,W.Precht
With contributions by numerous experts
711 figures.Approx.1110 pages.1976

This handbook is designed as an up-to-date review of all aspects of frog neurobiology. Its major emphasis is on the morphology and electro-physiology of the nervous system covering biophysics of nerve conduction and neuromuscular junction, morphology and physiology of spinal cord, brain stem, cerebellum and forebrain, autonomic nervous system, endocrinology and skin. It is the aim of this handbook to compile available information on this extremely useful preparation. It has become obvious that the simple vertebrate, especially the frog, has in the last century been a central subject of study in neurobiology. Furthermore, the frog, as a biological preparation, will become even more important in the future in view of its hardiness and the simplicity of its care. Finally, anuran cover on an enormous number of species, representing a variety adaptations which offer the neurobiologist an almost inexhaustible source of knowledge.

Springer-Verlag   Berlin   Heidelberg   New York

Springer Journals

a basic information
source

Subscription Information:

Vol. 105-112 (3 issues
each) will appear in 1976.
The publisher reserves the
right to issue additional
volumes during the calendar
year. Information about
obtaining back volumes and
microform editions avail-
able upon request.

North America:

Subscription rate: $ 524.80,
including postage and handling. Subscriptions are
entered with prepayment
only. Orders should be
addressed to: Springer-
Verlag New York Inc.,
175 Fifth Avenue, New York,
N.Y. 10010.

All other countries:

Subscription rate:
DM 1246,-, plus postage and
handling. Orders can
either be placed with your
bookdealer or sent directly
to: Springer-Verlag,
Heidelberger Platz 3,
D-1000 Berlin 33.

Sample copies available
upon request.

Journal of Comparative Physiology · A

Sensory, Neural, and Behavioral Physiology

Physiological Basis of Behavior
Sensory Physiology
Neural Physiology
Orientation, Communication,
Locomotion
Hormonal Control of Behavior

Editorial Board

Journal of Comparative Physiology · B

Metabolic and Transport Functions

Comparative Aspects of Metabolism
and Enzymology
Metabolic Regulation
Respiration and Gas Transport
Physiology of Body Fluids
Circulation
Temperature Relations
Muscular Physiology

Editorial Board

Springer-Verlag Berlin Heidelberg New York